ISNM
International Series of Numerical Mathematics
Vol. 144

Managing Editors:
K.-H. Hoffmann, München
D. Mittelmann, Tempe

Associate Editors:
R. E. Bank, La Jolla
H. Kawarada, Chiba
R. J. LeVeque, Seattle
C. Verdi, Milano

Honorary Editor:
J. Todd, Pasadena

Multivariate Polynomial Approximation

Manfred Reimer

Springer Basel AG

Author:

Manfred Reimer
Fachbereich Mathematik
Universität Dortmund
44221 Dortmund
Germany
e-mail: reimer@math.uni-dortmund.de

2000 Mathematics Subject Classification 41-02; 41A10, 41A05, 41A36, 41A55, 41A63, 43A90, 42C05, 42C15, 44
45Q05, 92C55, 41A05, 41A36, 41A55, 41A63, 43A90, 42C05, 42C15, 44A12, 45Q05, 92C55

A CIP catalogue record for this book is available from the Library of Congress, Washington D.C., USA

Bibliographic information published by Die Deutsche Bibliothek
Die Deutsche Bibliothek lists this publication in the Deutsche Nationalbibliografie; detailed bibliographic data is av
in the Internet at <http://dnb.ddb.de>.

ISBN 978-3-0348-9436-4   ISBN 978-3-0348-8095-4 (eBook)
DOI 10.1007/978-3-0348-8095-4

© 2003 Springer Basel AG

Originally published by Birkhäuser Verlag, Basel, Switzerland in 2003

Member of the BertelsmannSpringer Publishing Group

Cover graphic: Spherical polynomial approximant of degree 160, in area preserving parametrisation (see Figure 6.3

Printed on acid-free paper produced of chlorine-free pulp. TCF ∞
Cover design: Heinz Hiltbrunner, Basel

Contents

IV Applications 283

Meiner lieben Renate

Preface

Multivariate polynomials are a basic tool in the approximation of functions. When my monograph on Constructive Approximation of Multivariate Functions was published, now more than ten years ago, I could not know that a period of extremely active investigation on this field was just ahead. An important new key-word was given by Ian H. Sloan, hyperinterpolation, which is a discrete projection method, realizing the growth order of the minimal uniform projection norm on the sphere — whereas it is an open question, whether interpolatory projections do the same. In the general case, hyperinterpolation is still missing convergence. But by summation, generalized versions arise, which are convergent even at the best possible order. Anyways, the new issue caused new interest in positive quadratures of high exactness, in particular in the geometry of their nodes, and in their weight distribution. Using the new knowledge we estimate the approximation error even of certain discrete positive polynomial operators by means of the modulus of continuity. They are gained from the Newman–Shapiro operators, which attend also our new interest.

Naturally, there exists a severe complexity problem in higher dimensional spaces. We master it by a controlled truncation of the discrete operators, without destroying the approximation order.

Most of the results, gained for the sphere, define also an approximation method for the balls of lower dimensions. In the description of this process, the old orthogonal polynomials of Appell and Kampé de Feriét are very helpful. Another reason for their necessary revival is their behaviour under certain rotations, by which they become the welcome tool in the evaluation of some projections, defined by the average over a rotation group – with an important application to the Radon–, or more generally, to the k–plane transform. We finish by mentioning a high accuracy positive approximation method for the unknown density function. It constructs the approximants from the Radon–images as the data, thus solving the inversion problem of tomography by an approximation method of polynomial stability order.

In view of this development, it will not be too surprising, that I decided to write the monograph quite anew, saving what has to be saved, but rejecting what is not fitting to the new concept. A large variety of problems is attached to help the reader to become familiar with the multivariate theory. All problems are solved in a separate section.

Several parts of this book have been subject to advanced lectures which I hold at the University of Dortmund, or have been subject to discussion or cooperation with former members of my research group. I mention in particular Burkhard Sündermann, Michael Rosier and Ulrike Maier. I am aware of the motivation I owe them, and acknowlege this gratefully. I am also indepted to H. Michael Möller for valuable hints and remarks.

Finally I acknowledge the great motivation which I received at the Bommerholz conferences on multivariate approximation, which I were lucky to organize together with my friends Werner Haußmann and Kurt Jetter, the last conference also together with Joachim Stöckler. My hope is that these conferences will stimulate and promote multivariate approximation also in future.

Manfred Reimer

February 2003

Part I

Introduction

Chapter 1

Basic Principles and Facts

1.1 Preliminaries

We investigate polynomial approximations to functions

$$F : D \to \mathbb{R} \tag{1.1}$$

where D is a nonempty compact subset of $\mathbb{R}^r$, $r \in \mathbb{N}$, preferably in the uniform norm, but occasionally also in the quadratic average norm. The function is called *multivariate*, if $r \geq 2$. $C(D)$ denotes the space of all continuous functions (1.1) which is provided with the norm

$$\|F\|_\infty := \|F\|_D := \max\{|F(x)| : x \in D\}.$$

Our particular interest is directed to the case where D is one of the following sets:

$$
\begin{aligned}
B^r &:= \{x \in \mathbb{R}^r \mid |x| \leq 1\}, \\
S^{r-1} &:= \{x \in \mathbb{R}^r \mid |x| = 1\}, \\
E^r &:= \{x \in \mathbb{R}^r \mid x \geq 0,\ x_1 + \ldots + x_r \leq 1\}, \\
\Sigma^{r-1} &:= \{x \in \mathbb{R}^r \mid x \geq 0,\ x_1 + \ldots + x_r = 1\}.
\end{aligned}
$$

B^r is called the *ball*, S^{r-1} the *sphere* and E^r the *simplex* in $\mathbb{R}^r$. The sets are related by the mapping

$$x \mapsto \phi(x) = (x_1^2, x_2^2, \ldots, x_r^2)', \tag{1.2}$$

which maps B^r onto E^r, while the mapping

$$x \mapsto \psi(x) = (x_1, \ldots, x_r, \sqrt{1 - x_1^2 - \ldots - x_r^2}\,)' \tag{1.3}$$

maps B^r onto the *hemisphere*

$$S_+^r := \{x \in S^r \mid x_{r+1} \geq 0\}.$$

So most of the results which can be obtained with respect to one of these sets may be interpreted as a result with regard to each other one. Hence it suffices, in principle, to investigate the space $C(S^{r-1})$, with the advantage that we can make use of the group of *(proper) rotations* on S^{r-1}, which is a strong tool. As we are going to be *'constructive'*, we represent the elements of this group by the elements of the matrix group

$$\mathbf{A}^r := \{A \in \mathbb{R}^{r \times r} \,|\, A'A = I, \det A = +1\}.$$

Note that every basis transform in $\mathbb{R}^r$ furnishes a transform $A \mapsto AB$ in the representation of the elements of the group. A function (1.1) is called *univariate* if $D \subset \mathbb{R}$. Except for special functions we characterize univariate functions by small letters. For complexity reasons it is desirable to trace back multivariate functions to univariate ones. One way to do this is via product methods. But in rotation-invariant subspaces of $C(S^{r-1})$ *reproducing kernel functions*, which are *bizonal* and hence in principle univariate, are the more appropriate tool. The reason is that they are invariant under the subgroup of rotations which keep the other argument fixed. This gives us a first idea of how important rotations are in the theory of spherical functions.

The euclidean inner product of $x, y \in \mathbb{R}^r$ is written in each of the forms $(x, y) = x'y = xy$, depending on the context. To avoid inconsistency in the notation we define $F(x') := F(x)$ for $x \in D$. We suggest that the reader is familiar with the Γ- and with the B-*function*. Recall that

$$B(x, y) = \frac{\Gamma(x)\Gamma(y)}{\Gamma(x+y)} = \int_0^1 \xi^{x-1}(1-\xi)^{y-1}d\xi \qquad (1.4)$$

is valid for $\Re x, \Re y > 0$.

1.2   Existence of a Reproducing Kernel Function

Let $\mathbf{V}$ be a linear space of functions $D \to \mathbb{R}$, which is provided with the inner product $\langle \cdot, \cdot \rangle$.

Definition 1.1 (Reproducing Kernel). *A function $G : D \times D \to \mathbb{R}$ is called a reproducing kernel of* $\mathbf{V}$ *if the following holds:*

$$\begin{aligned}
&(i) &&G(x, \cdot) \in \mathbf{V} &&\text{for all} \quad x \in D,\\
&(ii) &&G(x, y) = G(y, x) &&\text{for all} \quad (x, y) \in D^2,\\
&(iii) &&\langle G(x, \cdot), F \rangle = F(x) &&\text{for all} \quad F \in \mathbf{V}, x \in D.
\end{aligned}$$

Theorem 1.1 (Reproducing Kernel). *If $\mathbf{V}$ is finite-dimensional, then a uniquely determined reproducing kernel exists.*

Proof. Let $S_1, ..., S_N$ be an arbitrary orthonormal basis in **V**. Define $G : D \times D \to \mathbb{R}$ by

$$G(x, y) := \sum_{j=1}^{N} S_j(x) S_j(y) \quad \text{for} \quad x, y \in D. \tag{1.5}$$

Obviously, G satisfies (i) and (ii). Next let $F \in \mathbf{V}$ and $x \in D$. Then we get

$$\langle G(x, \cdot), F \rangle = \sum_{j=1}^{N} \langle S_j, F \rangle S_j(x) = F(x),$$

and (iii) is also valid. So a reproducing kernel exists.

Next assume that H is an arbitrary repoducing kernel of **V**. Then $H(x, \cdot) \in \mathbf{V}$, $x \in D$, is reproduced at $y \in D$ by $G(y, \cdot)$, and we obtain

$$H(x, y) = \langle G(y, \cdot), H(x, \cdot) \rangle = \langle H(x, \cdot), G(y, \cdot) \rangle.$$

But $H(x, \cdot)$ reproduces $G(y, \cdot) \in \mathbf{V}$ at x, and we get

$$H(x, y) = G(y, x) = G(x, y).$$

As x and y were arbitrary this implies $H = G$. $\qquad\square$

1.3 Rotation-Invariant Spaces

Let $r \in \mathbb{N} \setminus \{1\}$. Several important function spaces are covered by the following definition.

Definition 1.2 (Rotation-Invariant Space). *D is called* rotation-invariant *if $D = AD := \{Ax \,|\, x \in D\}$ holds for all $A \in \mathbf{A}^r$. A subspace **V** of $C(D)$ is called* rotation-invariant *if D is rotation-invariant and if $F_A = F$ holds for all $F \in \mathbf{V}$ and all $A \in \mathbf{A}^r$, where $F_A(x) := F(Ax)$ for $x \in D$.*

Obviously, $C(B^r)$ and $C(S^{r-1})$ are rotation-invariant.

In what follows we provide $C(S^{r-1})$, $r \in \mathbb{N} \setminus \{1\}$, with the inner product

$$\langle E, F \rangle := \langle E, F \rangle_{S^{r-1}} := \int_{S^{r-1}} E(x) F(x) \, d\omega(x) \tag{1.6}$$

for $E, F \in C(S^{r-1})$, where the integral is the surface integral of S^{r-1} with respect to the standard measure $\omega(x)$. We use the abbreviation dx for $d\omega(x)$, if the context allows this. For later use we define

$$\omega_{r-1} := meas(S^{r-1}) = \int_{S^{r-1}} d\omega(x), \quad \Omega_r := meas(B^r) = \int_{B^r} dx \tag{1.7}$$

for $r \in \mathbb{N}$, where the measure of $S^0 = \{+1, -1\}$ is to be put to $\omega_0 = 2$. Note that

$$r \cdot \Omega_r = \omega_{r-1} = \frac{2\pi^{\frac{r}{2}}}{\Gamma(\frac{r}{2})} \tag{1.8}$$

holds, see Problem 1.1.

In a natural way the inner product can be extended to the space $L^2(S^{r-1})$ of measurable and square-integrable functions $S^{r-1} \to \mathbb{R}$.

The definition (1.6) is such that

$$\langle E_A, F_A \rangle = \langle E, F \rangle$$

holds for E, $F \in C(S^{r-1})$ and arbitrary rotations $A \in \mathbf{A}^r$, as $\det A = 1$. Exactly here we use that improper rotations are excluded by the definition of the group. The situation is not singular, and the question arises how this property of the inner product influences the structure of the reproducing kernel.

Before we answer this question we remark that a rotation-invariant set D (which is nonempty and compact in $\mathbb{R}^r$ by our general assumption) can be written in the form

$$D = \{x \in \mathbb{R}^r | \, |x| \in I_D\}$$

where I_D is a compact set in $\mathbb{R}$. If we assume that D does not consist of a single point, then we may also assume without restriction of generality that

$$S^{r-1} \subset D \subset B^r$$

holds, which in return implies

$$\{-1, +1\} \subset I_D \subset [-1, +1].$$

With these notations the following is valid.

Theorem 1.2 (Reproducing Kernel in a Rotation-Invariant Space). *Let $r \in \mathbb{N} \backslash \{1\}$. Assume $\mathbf{V}$ is a rotation-invariant subspace of $C(D)$ with reproducing kernel $G(x, y)$ with respect to the inner product $\langle \cdot, \cdot \rangle$, which satisfies*

$$\langle E_A, F_A \rangle = \langle E, F \rangle$$

for E, $F \in \mathbf{V}$ and arbitrary $A \in \mathbf{A}^r$. Then

$$G(Ax, Ay) = G(x, y)$$

holds for $(x, y) \in D \times D$, $A \in \mathbf{A}^r$, and a function $h \in C(H_D)$ exists, where $H_D := (I_D \backslash \{0\}) \times (I_D \backslash \{0\}) \times [-1, +1]$, such that

$$G(x, y) = h(|x|, |y|, xy)$$

is valid for $x, y \in D \backslash \{0\}$.

Proof. For $x, y \in D$, $A \in \mathbf{A}^r$ we get $Ax, Ay \in D$, and the following holds because of the properties $(i) - (iii)$ of the reproducing kernel:

$$
\begin{aligned}
G(Ax, Ay) &= \langle G(x, \cdot), G(A \cdot, Ay) \rangle && \text{(by properties (i), (ii), (iii)),}\\
&= \langle G(x, A^{-1} \cdot), G(\cdot, Ay) \rangle && \text{(property of the inner product),}\\
&= \langle G(\cdot, Ay), G(x, A^{-1} \cdot) \rangle && \text{(symmetry of the inner product),}\\
&= \langle G(Ay, \cdot), G(x, A^{-1} \cdot) \rangle && \text{(by property (ii)),}\\
&= G(x, A^{-1} Ay) = G(x, y) && \text{(by property (iii)).}
\end{aligned}
$$

Next let $x, y \in D \setminus \{0\}$. Moreover we assume, in the beginning, $y \notin span\{x\}$, such that

$$
\Big| |y|x \pm |x|y \Big| = \sqrt{2|x||y|(|x||y| \pm xy)} > 0
$$

is valid. We define

$$
u := u(x, y) := \frac{|y|x + |x|y}{\Big| |y|x + |x|y \Big|}, \quad v := v(x, y) := \frac{|y|x - |x|y}{\Big| |y|x - |x|y \Big|},
$$

which implies $u, v \in S^{r-1}$ and $uv = 0$. There is a rotation $A \in \mathbf{A}^r$ and an $\epsilon \in \{+1, -1\}$ such that

$$
Au = e_1, \quad Av = \epsilon e_2.
$$

Note that $\epsilon = -1$ is needed in the case $r = 2$, only. It follows that

$$
\begin{aligned}
|y|Ax + |x|Ay &= \Big| |y|x + |x|y \Big| e_1,\\
|y|Ax - |x|Ay &= \Big| |y|x - |x|y \Big| \epsilon e_2,
\end{aligned}
$$

and hence

$$
\begin{aligned}
Ax &= \tfrac{1}{|y|} \left\{ \sqrt{\tfrac{1}{2}|x||y|(|x||y| + xy)} \, e_1 + \sqrt{\tfrac{1}{2}|x||y|(|x||y| - xy)} \, \epsilon e_2 \right\},\\
Ay &= \tfrac{1}{|x|} \left\{ \sqrt{\tfrac{1}{2}|x||y|(|x||y| + xy)} \, e_1 - \sqrt{\tfrac{1}{2}|x||y|(|x||y| - xy)} \, \epsilon e_2 \right\}.
\end{aligned}
$$

So we obtain

$$
G(x, y) = \tfrac{1}{2}\Big[G(Ax, Ay) + G(Ay, Ax) \Big] = h(|x|, |y|, xy),
$$

where the function h is defined, with the abbreviation $\tau = \xi\eta$, by

$$
\begin{aligned}
h(\xi, \eta, \zeta) &:= \\
&\tfrac{1}{2} G\Big(\tfrac{1}{\eta}\big[\sqrt{\tfrac{1}{2}\tau(\tau + \zeta)}\, e_1 + \sqrt{\tfrac{1}{2}\tau(\tau - \zeta)}\, \epsilon e_2 \big], \tfrac{1}{\xi}\big[\sqrt{\tfrac{1}{2}\tau(\tau + \zeta)}\, e_1 - \sqrt{\tfrac{1}{2}\tau(\tau - \zeta)}\, \epsilon e_2 \big] \Big)\\
&+ \tfrac{1}{2} G\Big(\tfrac{1}{\xi}\big[\sqrt{\tfrac{1}{2}\tau(\tau + \zeta)}\, e_1 - \sqrt{\tfrac{1}{2}\tau(\tau - \zeta)}\, \epsilon e_2 \big], \tfrac{1}{\eta}\big[\sqrt{\tfrac{1}{2}\tau(\tau + \zeta)}\, e_1 + \sqrt{\tfrac{1}{2}\tau(\tau - \zeta)}\, \epsilon e_2 \big] \Big)
\end{aligned}
$$

for $(\xi, \eta, \zeta) \in H_D$.

As G is continuous, the additional assumption $y \notin span\{x\}$ may be dropped, and it is obvious that h is continuous in every point of the form $(|x|, |y|, xy)$, $x, y \in D \setminus \{0\}$. By the assumption $S^{r-1} \subset D$, the values of xy cover the interval $[-1, +1]$ while x and y vary in D. Therefore h is defined and continuous in H_D. Note that there is no claim concerning the values of $G(0, y)$ and $G(x, 0)$, respectively. $\square$

Later we will we able to specify some reproducing kernels by the help of Theorem 1.2 more detailed.

1.4   Rotation Principles, T-Kernels

We are interested in functions F of a rotation-invariant space which satisfy $F_A = F$ for some rotations A, say for all A of a subgroup of $\mathbf{A}^r$. This is true, for instance, if F has the form $F(x) = f(tx)$ for $x \in B^r$, where $f \in C[-1, 1]$, $t \in S^{r-1}$, and the subgroup consists of the rotations which keep t fixed. We generalize this idea as follows.

Let $t_1, ..., t_k \in S^{r-1}$, $1 \leq k \leq r - 1$, be orthogonal, and define $T := (t_1, ..., t_k)$, which is an $r \times k$ matrix with columns $t_1, ..., t_k$. Then

$$\mathbf{A}_T^r := \{A \in \mathbf{A}^r | AT = T\}$$

is the subgroup of $\mathbf{A}^r$ whose elements let $[T] := span\{t_1, ..., t_k\}$ be fixed. Note that $\mathbf{A}_T^r = \{I\}$ holds in the case $k = r - 1$.

T can be completed by the columns of a matrix $U = (u_{k+1}, ..., u_r)$ such that

$$(T, U) \in \mathbf{A}^r. \tag{1.9}$$

In the case $1 \leq k \leq r - 2$ the construction is not unique. But if T is completed likewise by $V = (v_{k+1}, ..., v_r)$ instead of U, such that $(T, V) \in \mathbf{A}^r$ holds, then we get

$$A(T, V) = (T, U),$$

if we define $A := (T, U)(T, V)'$, which implies in particular

$$A \in \mathbf{A}_T^r, \quad Av_r = u_r. \tag{1.10}$$

Lemma 1.3. *Let $r \geq 3$. Assume $T = (t_1, ..., t_k)$, $1 \leq k \leq r - 2$, is completed by $U = (u_{k+1}, ..., u_r)$ such that $(T, U) \in \mathbf{A}_T^r$. And assume that $v \in S^{r-1}$ is orthogonal to $t_1, ..., t_k$. Then there exists an element $A \in \mathbf{A}_T^r$ such that $Av = u_r$.*

Proof. T can be completed by V by the successive choice of $v_r := v$, $v_{r-1}, ..., v_{k+1}$ where the choice of v_{k+1} decides on the sign of the determinant of (T, V) to be $+1$. The remaining follows from (1.10). $\square$

Definition 1.3 (T-Kernel). *Let* $\mathbf{V}$ *be a subspace of* $C(D)$ *where* D *is rotation-invariant, and let* $T = (t_1, ..., t_k)$, $1 \leq k \leq r - 1$, *as above. Then*

$$\mathbf{V}_T := \{F \in \mathbf{V}| \, F_A = F \text{ for all } A \in \mathbf{A}_T^r\}$$

is called the T-kernel *of* $\mathbf{V}$.

Note that $\mathbf{V}_T$ is a subspace of $\mathbf{V}$ and that $k = r - 1$ implies $\mathbf{V}_T = \mathbf{V}$.

Theorem 1.4 (T-Kernel of $C(S^{r-1})$**).** *Let* $\mathbf{V} := C(S^{r-1})$, *and let* $T = (t_1, ..., t_k)$, $1 \leq k \leq r - 1$, *be as above.*

(i) For $f \in C(B^k)$ *let* $F \in \mathbf{V}$ *be defined by*

$$F(x) := f(T'x) \text{ for } x \in S^{r-1}.$$

Then $F \in \mathbf{V}_T$ *is valid.*

(ii) Vice versa let $1 \leq k \leq r - 2$ *and assume* $F \in \mathbf{V}_T$. *Then a function* $f \in C(B^k)$ *exists such that*

$$F(x) = f(T'x) \text{ for all } x \in S^{r-1}.$$

Proof. Because of $|T'x|^2 = \sum_{\nu=1}^{k}(t_\nu x)^2 \leq |x|^2 = 1$ the definition of F is correct, and for $A \in \mathbf{A}_T^r$ we get $A' \in \mathbf{A}_T^r$ and hence $F(Ax) = f(T'Ax) = f((A'T)'x) = f(T'x) = F(x)$. So we obtain $F \in \mathbf{V}_T$.

Next let $1 \leq k \leq r - 2$, $F \in \mathbf{V}_T$. We choose U such that (1.9) holds and let U be fixed, $x \in S^{r-1}$ be arbitrary. Then x can be written in the form

$$x = \sum_{\nu=1}^{k}(x, t_\nu)t_\nu + w \quad \text{where} \quad w = \sum_{\nu=k+1}^{r}(x, u_\nu)u_\nu,$$

$$|w|^2 = \sum_{\nu=k+1}^{r}(x, u_\nu)^2 = 1 - \sum_{\nu=1}^{k}(x, t_\nu)^2.$$

In the case $w \neq 0$ we define $v := w/|w|$ and obtain

$$x = \sum_{\nu=1}^{k}(x, t_\nu)t_\nu + \sqrt{1 - \sum_{\nu=1}^{k}(x, t_\nu)^2} \cdot v \tag{1.11}$$

where $v \in S^{r-1}$, v orthogonal to $t_1, ..., t_k$. In the case $w = 0$ the factor occurring with v is zero and the representation (1.11) holds with an arbitrary v, as above. Now we make use of the assumption $1 \leq k \leq r - 2$, which implies $r \geq 3$. By Lemma 1.3 there exists $A \in \mathbf{A}_T^r$ such that $Av = u_r$. Hence we obtain

$$F(x) = F(Ax) = F\left(\sum_{\nu=1}^{k}(x, t_\nu)t_\nu + \sqrt{1 - \sum_{\nu=1}^{k}(x, t_\nu)^2} \cdot u_r\right).$$

Finally we define $f \in C(B^k)$ by

$$f(\xi_1, ..., \xi_k) := F\left(\sum_{\nu=1}^{k} \xi_\nu t_\nu + \sqrt{1 - \sum_{\nu=1}^{k} \xi_\nu^2} \cdot u_r\right)$$

for $\xi_1^2 + ... + \xi_k^2 \leq 1$. The definition of f is complete as the vectors $(xt_1, ..., xt_k)'$ cover B^k as x varies in S^{r-1}, and we obtain

$$F(x) = f(xt_1, ..., xt_k) = f(x'T) = f(T'x),$$

as claimed.   □

Remark. For $k = r - 1$ the second statement of Theorem 1.4 would be that *every* $F \in C(S^{r-1})$ takes the form mentioned. For $T = (e_1, ..., e_{r-1})$ this would imply that *every* $F \in C(S^{r-1})$ has the form $F(x) = f(x_1, ..., x_{r-1})$, which is not true. So the restriction on k is necessary. We finish by the remark that if, in Theorem 1.4, **V** would be defined by $\mathbf{V} := C(B^r)$, then f would have to be replaced by a function from $C(B^r \times [0, 1])$, while $f(T'x)$ has to be replaced by $f(T'x, |x|)$.

Theorem 1.4 furnishes immediately the following corollary.

Corollary 1.5 (Characterisation of T-Kernel Functions in $C(S^{r-1})$). *Assumptions as in Theorem 1.4, but $1 \leq k \leq r - 2$. Then the following holds. $F \in C(S^{r-1})$ is contained in the T-kernel of $C(S^{r-1})$ if and only if F has the form $F(x) = f(T'x)$, where $f \in C(B^k)$.*

Proof. See Theorem 1.4, (i) and (ii).   □

Remark. Let $k = 1$ and let $t := t_1$. Then we have $T = (t)$ and the T-kernel functions take the form $F(x) = f(tx)$.

Definition 1.4 (Zonal Functions). *A function of the form $F(x) = f(tx)$ for $x \in S^{r-1}$, where $t \in S^{r-1}$ is fixed, is called a* zonal function *with* axis t.

Corollary 1.6 (Characterisation of Zonal Functions). *Let $r \geq 3$. $F \in C(S^{r-1})$ is zonal with axis $t \in S^{r-1}$ if and only if $F_A = F$ holds for all $A \in \mathbf{A}^r$ which satisfy $At = t$.*

1.5   Averages and T-Kernel Projections

We want to define projection operators $\mathbf{V} \to \mathbf{V}_T$ for subspaces of $C(D)$ by averaging the function values $F(A'x) = F(x'A)$ over the group $\mathbf{A}_T^r$. In principle we could arrange this by means of the *Haar integral* on topological groups. But as we do not want to presuppose the reader to be familiar with this notion, we use a constructive method in what follows.

In the beginning let $r \geq 2$. A rotation $A \in \mathbf{A}^r$ is uniquely determined by its first $r - 1$ column vectors $a_1, ..., a_{r-1}$, say

$$A = (a_1, ..., a_{r-1}, *),$$
$$a_1, ..., a_{r-1} \in S^{r-1},$$
$$a_j a_k = 0 \text{ for } j, k \in \{1, ..., r - 1\}, j \neq k.$$

The $*$ indicates that the last column of A is redundant. So we may consider A to be an element of the manifold

$$\left\{ (a_1, ..., a_{r-1}) \in (S^{r-1})^{r-1} \mid a_j a_k = 0 \text{ for } j \neq k \right\},$$

which is a compact metric space with respect to the euclidean distance function of $\mathbb{R}^{r(r-1)}$.

Next we introduce the *integral*, i.e., positive linear functional $I^r : C(\mathbf{A}^r) \to R$ by defining

$$I^r \Phi := \int_{S^{r-1}} \cdots \int_{S^2(a_{r-1},...,a_3)} \int_{S^1(a_{r-1},...,a_2)} \Phi(a_1, a_2, ..., a_{r-1}, *) da_1 da_2 \cdots da_{r-1} \quad (1.12)$$

for $\Phi \in C(\mathbf{A}^r)$, where the integrals which occur are surface integrals with respect to the spheres

$$S^j(a_{r-1}, ..., a_{j+1}) := S^{r-1} \cap [a_{r-1}, ..., a_{j+1}]^{\perp}$$
$$= \{x \in S^{r-1} \mid x a_k = 0 \text{ for } k = j + 1, ..., r - 1\}$$

for $j = 1, 2, ..., r - 2$, respectively. Note that (1.12) implies

$$\|I^r\|_{\infty} = I^r(1) = \omega_1 \cdots \omega_{r-1} \quad (1.13)$$

in the uniform norm. The definition of I^r is such that the following theorem holds.

Theorem 1.7 (Left-Side Invariance of I^r). *Let $r \geq 2$, $\Phi \in C(\mathbf{A}^r)$. Then*

$$I^r \Phi(C \cdot) = I^r \Phi(\cdot) \quad (1.14)$$

holds for arbitrary rotations $C \in \mathbf{A}^r$.

Proof. In (1.12) we replace Φ by $\Phi(C \cdot)$ and substitute, step by step,

$$Ca_1 = b_1, \ Ca_2 = b_2, ..., Ca_{r-1} = b_{r-1}.$$

Then we obtain

$$I^r(\Phi(C\cdot)) = \int_{S^{r-1}} \cdots \int_{S^2(a_{r-1},...,a_3)} \int_{S^1(a_{r-1},...,a_3,a_2)} \Phi(Ca_1, Ca_2, ..., Ca_{r-1}, *) da_1 da_2 \cdots da_{r-1}$$

$$= \int_{S^{r-1}} \cdots \int_{S^2(a_{r-1},...,a_3)} \int_{S^1(Ca_{r-1},...,Ca_3,Ca_2)} \Phi(b_1, Ca_2, ..., Ca_{r-1}, *) db_1 da_2 \cdots da_{r-1}$$

$$= \quad .....$$

$$= \int_{S^{r-1}} \cdots \int_{S^2(b_{r-1},...,b_3)} \int_{S^1(b_{r-1},...,b_3,b_2)} \Phi(b_1, b_2, ..., b_{r-1}, *) db_1 db_2 \cdots db_{r-1} = I^r(\Phi).$$

Remark. We may write the integral (1.12) more significantly in the form

$$I^r\Phi = \int_{\mathbf{A}^r} \Phi(A)\, dA.$$

Then (1.14) takes the form

$$\int_{\mathbf{A}^r} \Phi(CA)\, dA = \int_{\mathbf{A}^r} \Phi(A)\, dA \quad \text{for} \quad C \in \mathbf{A}^r. \tag{1.15}$$

For this reason the integral is called *left-side invariant*. We note that if we replace in the definition of the integral the column vectors by the row vectors, then we get an integral, which is *right-side invariant*. It is different in the case $r \geq 3$, where the group of rotations is noncommutative.

Next we want to define integrals which are invariant only with respect to left-side factors $C \in \mathbf{A}^r_T$ (instead of $\mathbf{A}^r$). The subgroup should not consist of the identity matrix I, only. So we suppose $r \geq 3$ and $1 \leq k \leq r - 2$ in what follows.

We complete $T = (t_1, ..., t_k)$ again by some fixed $U = (u_{k+1}, ..., u_r)$ such that

$$V := (T, U) \in \mathbf{A}^r$$

holds. U is not uniquely determined. For $A \in \mathbf{A}^{r-k}$ we define

$$\Omega(A) := V \begin{pmatrix} I & O \\ O & A \end{pmatrix} V'. \tag{1.16}$$

It is easy to see that $\Omega(A)T = T$ holds for arbitrary $A \in \mathbf{A}^{r-k}$, such that $A \mapsto \Omega(A)$ defines a mapping

$$\Omega : \mathbf{A}^{r-k} \to \mathbf{A}^r_T, \tag{1.17}$$

which is continuous and bijective, and hence defining a parameterisation of $\mathbf{A}^r_T$ with the parameter $A \in \mathbf{A}^{r-k}$. Therefore we may define the integral

$$I^r_T \Phi := \int_{\mathbf{A}^r_T} \Phi(\Omega)d\Omega := \int_{\mathbf{A}^{r-k}} \Phi(\Omega(A))dA \tag{1.18}$$

for $\Phi \in C(\mathbf{A}^r)$. Note that

$$\Omega(B)\Omega(A) = V \begin{pmatrix} I & O \\ O & B \end{pmatrix} V'V \begin{pmatrix} I & O \\ O & A \end{pmatrix} V' = V \begin{pmatrix} I & O \\ O & BA \end{pmatrix} V' = \Omega(BA)$$

(1.19)

holds for $A, B \in \mathbf{A}^{r-k}$.

Theorem 1.8 (Left-Side Invariance of I_T^r). *Let* $1 \leq k \leq r-2$, $T = (t_1, ..., t_k)$ *where* $t_1, ..., t_k \in S^{r-1}$ *are orthogonal. Then*

$$I_T^r \Phi(C \cdot) = I_T^r \Phi(\cdot)$$

holds for arbitrary $\Phi \in C(\mathbf{A}^r)$ *and* $C \in \mathbf{A}_T^r$.

Proof. Let $C \in \mathbf{A}_T^r$, say $C = \Omega(B)$, $B \in \mathbf{A}^{r-k}$, where we use that Ω is bijective. For arbitrary $\Phi \in C(\mathbf{A}^r)$ we get

$$I_T^r \Phi(C \cdot) = \int_{\mathbf{A}^{r-k}} \Phi(C(\Omega(A))dA = \int_{\mathbf{A}^{r-k}} \Phi(\Omega(B)(\Omega(A))dA$$

$$= \int_{\mathbf{A}^{r-k}} \Phi(\Omega(BA))dA = \int_{\mathbf{A}^{r-k}} \Phi(\Omega(A))dA = I_T^r \Phi(\cdot),$$

where we used Theorem 1.7 with $r - k$ instead of r. $\qquad\qquad\square$

Remark. If we choose U_ν instead of U, $\nu = 1, 2$, and define likewise $V_\nu := (T, U_\nu)$, $\Omega_\nu(A) := V_\nu \begin{pmatrix} I & O \\ O & A \end{pmatrix} V_\nu'$ for $\nu = 1, 2$, then we get with $B := U_2' U_1 \in \mathbf{A}^{r-k}$

$$\Omega_2(A) = V_1(V_1'V_2) \begin{pmatrix} I & O \\ O & A \end{pmatrix} (V_1'V_2)'V_1' = V_1 \begin{pmatrix} I & O \\ O & B'AB \end{pmatrix} V_1' = \Omega_1(B'AB),$$

and hence

$$\int_{\mathbf{A}^{r-k}} \Phi(\Omega_2(A))dA = \int_{\mathbf{A}^{r-k}} \Phi(\Omega_1(AB))dA.$$

The factor B' could be cancelled because of the left-side invariance of the integral, not so the factor B, which corresponds to a basis transform in the space $[U] = [T]^\perp$. So the integral is uniquely determined by the group of rotations which keep $[T]$ fixed, while Ω_1 and Ω_2 are different parameter representations of this group with the transform law given above.

Averages

First we note that (1.18) implies

$$\|I_T^r\|_\infty = I_T^r 1 = \omega_1 \cdots \omega_{r-k-1}$$

(1.20)

for $1 \leq k \leq r - 2$.

Next let $\mathbf{V}$ be a rotation-invariant subspace of $C(D)$. For $F \in \mathbf{V}$ and $x \in D$ we define

$$(\Pi_T^r F)(x) := \frac{1}{\omega_1 \cdots \omega_{r-k-1}} \cdot \int_{\mathbf{A}_T^r} F(\Omega' x) d\Omega, \qquad (1.21)$$

Ω as above. Obviously, by averaging the function values of F this defines a mapping from $\mathbf{V}$ onto some subspace of $C(D)$. We complete the definition by defining

$$(\Pi_T^r F)(x) := F(x) \text{ for } k = r - 1. \qquad (1.22)$$

Theorem 1.9 (T-Kernel Projections in $C(S^{r-1})$). *Let $1 \leq k \leq r - 1$. Assume $\mathbf{V}$ is a closed rotation-invariant subspace of $C(S^{r-1})$, $T = (t_1, \ldots, t_k)$ consists of orthonormal columns, and Π_T^r is defined as above. Then the following holds for arbitrary $F \in \mathbf{V}$.*

$$(i) \quad \|\Pi_T^r\|_\infty \leq \Pi_T^r 1 = 1,$$
$$(ii) \quad (\Pi_T^r F)(C \cdot) = (\Pi_T^r F)(\cdot) \quad \text{for all} \quad C \in \mathbf{A}_T^r,$$
$$(iii) \quad (\Pi_T^r F)(t) = F(t) \quad \text{for all} \quad t \in S^{r-1} \cap [T],$$
$$(iv) \quad \Pi_T^r \text{ is a projection } \mathbf{V} \to \mathbf{V}_T,$$
$$(v) \quad \Pi_T^r(FG) = F \cdot \Pi_T^r G \quad \text{for all} \quad F \in \mathbf{V}_T, G \in \mathbf{V}.$$

Proof. For $k = r - 1$ we get $\mathbf{A}_T^r = \{I\}$ and $\Pi_T^r = id_V$, and the statements are obvious. Next we assume $1 \leq k \leq r - 2$. Then (i) is obvious. Equality holds if $1 \in \mathbf{V}$. (ii) For $C \in \mathbf{A}_T^r$ we get $C' \in \mathbf{A}_T^r$, and hence

$$
\begin{aligned}
(\Pi_T^r F)(Cx) &= \frac{1}{\omega_1 \cdots \omega_{r-k-1}} \cdot \int_{\mathbf{A}_T^r} F(\Omega' Cx) d\Omega \\
&= \frac{1}{\omega_1 \cdots \omega_{r-k-1}} \cdot \int_{\mathbf{A}_T^r} F((C'\Omega)' x) d\Omega = (\Pi_T^r F)(x),
\end{aligned}
$$

where the left-side factor C' could be cancelled. (iii) is obvious as $\Omega t = t$ implies $\Omega' t = t$.

(iv) We write (1.21) in the explicit form

$$(\Pi_T^r F)(x) = \frac{1}{\omega_1 \cdots \omega_{r-k-1}} \cdot \int_{\mathbf{A}^{r-k}} F(\Omega'(A)x) dA.$$

The integrand is a continuous function in the variable A and the parameter x. Each of them is varying in a compact set. Therefore the integrand is uniformly continuous in (A, x) and the integral is a uniform limit of functions of $\mathbf{V}$. As $\mathbf{V}$ is closed, $\Pi_T^r F \in \mathbf{V}$ is valid, and (ii) implies $\Pi_T^r F \in \mathbf{V}_T$. Finally let $F \in \mathbf{V}_T$. Then we get

$$(\Pi_T^r F)(x) = \frac{1}{\omega_1 \cdots \omega_{r-k-1}} \cdot \int_{\mathbf{A}^{r-k}} F_{\Omega'(A)}(x) dA = \frac{1}{\omega_1 \cdots \omega_{r-k-1}} \cdot \int_{\mathbf{A}^{r-k}} F(x) dA = F(x),$$

where we used (1.20). So, Π_T^r is a projection onto $\mathbf{V}_T$.

(v) Next we assume $F \in \mathbf{V}_T$, $G \in \mathbf{V}$, and get from (1.21)

$$(\Pi_T^r G)(x) = \frac{1}{\omega_1 \cdots \omega_{r-k-1}} \cdot \int_{\mathbf{A}_T^r} F_{\Omega'}(x)G(\Omega'x)d\Omega = F(x)(\Pi_T^r G)(x)$$

for arbitrary $x \in S^{r-1}$, where we used (1.17), i.e., $\Omega' \in \mathbf{A}_T^r$, and Definition 1.3, i.e., $F_{\Omega'} = F$. $\square$

Definition 1.5 (T-Kernel Projections in $C(S^{r-1})$). $\mathbf{V}$ *as in Theorem 1.9. In view of Theorem 1.9, (iv), the mapping*

$$\Pi_T^r : \mathbf{V} \to \mathbf{V}_T,$$

is called the T-kernel projection of $\mathbf{V}$.

In the case $k = 1$, where $T = (t)$ holds for some $t \in S^{r-1}$, we write $\mathbf{A}_t^r$, Π_t^r and $\mathbf{V}_t$ instead of $\mathbf{A}_T^r$, Π_T^r and $\mathbf{V}_T$, respectively. Note that in this case the elements of $\mathbf{V}_t$ have the form

$$F(x) = f(tx), \quad f \in C[-1,1],$$

i.e., they are *zonal* with axis t, see Definition 1.4.

Definition 1.6 (Axial Projection). $\Pi_t^r : \mathbf{V} \to \mathbf{V}_t$, $t \in S^{r-1}$, *is the* axial projection *of* $\mathbf{V}$ *with respect to the axis t.*

As a corollary of Theorem 1.9 we get the following theorem.

Theorem 1.10 (Axial Projection). *Let $r \geq 3$, $\mathbf{V}$ a closed rotation-invariant subspace of $C(S^{r-1})$, and assume $t \in S^{r-1}$, $\dim \mathbf{V}_t = 1$. Then there is a uniquely determined $g \in C[-1,1]$, satisfying $g(1) = 1$, such that*

$$(\Pi_t^r F)(x) = F(t)g(tx)$$

holds for all $F \in \mathbf{V}$ and $x \in S^{r-1}$.

Proof. By Theorem 1.4 the elements of $\mathbf{V}_t$ have the form $f(tx)$, $f \in C[-1,1]$, and there is a nontrivial function of this kind. In view of the dimension of $\mathbf{V}_t$, we get for arbitrary $F \in \mathbf{V}$ and with some constant $c(F)$

$$(\Pi_t^r F)(x) = c(F) \cdot f(tx).$$

By Theorem 1.9, (iii), we obtain

$$F(t) = (\Pi_t^r F)(t) = c(F)f(1).$$

Now assume $f(1) = 0$ would hold. Then all $F \in \mathbf{V}$ vanish at $x = t$. Since $\mathbf{V}$ is rotation-invariant, this yields $\mathbf{V} = [0]$, and hence $\mathbf{V}_t = [0]$, which contradicts. So we get $f(1) \neq 0$, $c(F) = F(t)/f(1)$, and hence

$$(\Pi_t^r F)(x) = F(t)g(tx),$$

where g is defined by $g := f/f(1) \in C[-1,1]$. $\square$

Remark. For $r = 2$ (and $k = 1 = r - 1$) we get $\Pi_t^r = id_V$, and the statement would be that *every* $F \in \mathbf{V}$ is zonal, which is not true for $\mathbf{V} = C(S^1)$ itself. So the assumption $r \geq 3$ was necessary.

1.6   Reproducing Kernels in $C(S^{r-1})$

Let $C(S^{r-1})$, $r \in \mathbb{N} \setminus \{1\}$, be provided with the (standard) inner product $\langle \cdot, \cdot \rangle$ induced by the surface integral, see (1.6). We assume that $\mathbf{V}$ is a finite-dimensional rotation-invariant subspace. Theorem 1.1 ensures us that $\mathbf{V}$ has a reproducing kernel G. By Theorem 1.2

$$G(Ax, Ay) = G(x, y)$$

holds for $x, y \in S^{r-1}$, $A \in \mathbf{A}^r$, since the inner product satisfies

$$\langle E_A, F_A \rangle = \langle E, F \rangle.$$

The particular situation allows us to determine the structure of G more precisely.

Theorem 1.11 (Reproducing Kernel in Subspaces of $C(S^{r-1})$). *Let $r \in \mathbb{N} \setminus \{1\}$, and let $\mathbf{V}$ be a rotation-invariant subspace of $C(S^{r-1})$ with reproducing kernel G. Then a uniquely determined univariate function $K \in C[-1, 1]$ exists such that*

$$G(x, y) = K(xy)$$

holds for $x, y \in S^{r-1}$, and the following inequality is valid,

$$|K(\xi)| \leq K(1) \quad for \ -1 \leq \xi \leq 1. \tag{1.23}$$

Proof. The first statement follows from Theorem 1.2 by the identification $D := S^{r-1}$, $I_D = \{1\}$, $H_D = \{1\} \times \{1\} \times [-1, 1]$, and

$$K(\zeta) := h(1, 1, \zeta)$$

for $\zeta \in [-1, 1]$.

For $r \geq 3$ there is another, direct and more elegant proof, which we do not like to omit. Let $y \in S^{r-1}$ be fixed. Then

$$G(Ax, y) = G(x, y)$$

holds for all $A \in \mathbf{A}_y^r$. By Theorem 1.4, (ii), a function $g_y \in C[-1, 1]$ exists such that

$$G(x, y) = g_y(xy)$$

is valid for all $x \in S^{r-1}$. Now let $z \in S^{r-1}$ be an arbitrary point, say

$$z = Ay \quad where \quad A \in \mathbf{A}^r.$$

It follows by our previous result that

$$g_y(xy) = G(x, y) = G(Ax, Ay) = g_{Ay}((Ax, Ay)) = g_z(xy)$$

holds for all $x \in S^{r-1}$. The values of xy cover $[-1, 1]$ as x varies in S^{r-1}. This yields

$$g_z = g_y = \cdots = K$$

for arbitrary $y, z \in S^{r-1}$ where $K \in C[-1, 1]$. Actually, $G(x, y) = K(xy)$ holds, as claimed.

In order to prove the second statement, we use the inequality of Schwarz and obtain

$$
\begin{aligned}
K(xy)^2 &= \langle K(x \cdot), K(y \cdot) \rangle^2 \\
&\leq \langle K(x \cdot), K(x \cdot) \rangle \langle K(y \cdot), K(y \cdot) \rangle = K(xx)K(yy) = K(1)^2.
\end{aligned}
$$

It follows that (1.23) holds, since the values of xy cover the interval $[-1, 1]$. $\square$

Remark 1. In (1.23) equality occurs if and only if $\xi = xy$ and $K(x \cdot) = \gamma K(y \cdot)$ is valid with some constant γ. Inserting y and x, respectively, we get

$$K(xy) = \gamma K(1), \quad K(1) = \gamma K(yx),$$

and hence $K(xy) = \gamma^2 K(xy)$, where $K(xy)^2 = K(1)^2 \neq 0$ holds in view of (1.23) and $K \neq 0$. This implies $\gamma = \pm 1$. Therefore, equality occurs in (1.23) if and only if $\xi = xy$ and

$$K(x \cdot) = \pm K(y \cdot)$$

is valid. Note that this equation does not imply $x \in \{+y, -y\}$, as the example $K = T_\mu$, $\mu \geq 2$, (Cebysev polynomial of the first kind) demonstrates.

Remark 2. The kernel $K(xy)$ is zonal in the variable x with axis y, and vice versa. For this reason we call it a *bizonal function*.

Some Integration Formulas

We finish this section by presenting three basic and often used integration formulas, which are offered for proof in Problem 1.2 – Problem 1.4. The proofs can be found in the section Solutions.

Let $r \in \mathbb{N} \setminus \{1\}$. First we define $\bar{x} \in \mathbb{R}^{r-1}$ for $x = (x_1, \ldots, x_r)' \in \mathbb{R}^r$ by $\bar{x} := (x_1, \ldots, x_{r-1})'$. If $F \in C(S^{r-1})$ does not depend on x_r, this means, if a function $\bar{F} \in C(B^{r-1})$ exists such that $F(x) = \bar{F}(\bar{x})$ holds for $x \in S^{r-1}$, then

$$\int_{S^{r-1}} F(x) \, d\omega(x) = 2 \int_{B^{r-1}} \bar{F}(\bar{x}) \frac{d\bar{x}}{\sqrt{1 - |\bar{x}|^2}} \tag{1.24}$$

is valid

This result can be generalized as follows. For $s \in \{0, 1, \ldots, r-2\}$ we define $\bar{x} \in \mathbb{R}^{r-s-1}$ by $\bar{x} := (x_1, \ldots, x_{r-s-1})'$. If $F \in C(S^{r-1})$ does not depend on $x_{r-s}, \ldots, x_r$, this means, if a function $\bar{F} \in C(B^{r-s-1})$ exists such that $F(x) = \bar{F}(\bar{x})$ holds for $x \in S^{r-1}$, then

$$\int\limits_{S^{r-1}} F(x)\, d\omega(x) = \omega_s \int\limits_{B^{r-s-1}} \bar{F}(\bar{x})\, (1 - |\bar{x}|^2)^{\frac{s-1}{2}}\, d\bar{x} \tag{1.25}$$

is valid, where we recall that $\omega_0 = 2$ holds by definition.

(1.25) can be used to prove the formula

$$\int\limits_{S^{r-1}} (tx)^{\mu}\, dx = 2\pi^{\frac{r-1}{2}} \frac{\Gamma(\frac{\mu+1}{2})}{\Gamma(\frac{\mu+r}{2})} \tag{1.26}$$

for $t \in S^{r-1}$ and even $\mu \in \mathbb{N}_0$. More generally, let $f \in C[-1,1]$ and $t \in S^{r-1}$. The integral

$$\int\limits_{S^{r-1}} f(tx)\, d\omega(x)$$

does not depend on the choice of t. So we may choose $t := e_1$, and get from (1.25) the equations

$$\int\limits_{S^{r-1}} f(tx)\, d\omega(x) = \omega_{r-2} \int\limits_{-1}^{1} f(\xi)\, (1 - \xi^2)^{\frac{r-3}{2}}\, d\xi = \omega_{r-2} \int\limits_{0}^{\pi} f(\cos\phi)\, (\sin\phi)^{r-2}\, d\phi.$$
$$\tag{1.27}$$

In particular, putting $f(\xi) := \xi^{\mu}$ we get (1.26).

1.7   Problems

Problem 1.1. Prove formula (1.8), i.e., calculate the values of Ω_r and of ω_{r-1}.

Problem 1.2. Prove formula (1.24).

Problem 1.3. Use the result of Problem 1.2 to prove formula (1.25).

Problem 1.4. Prove formula (1.26). Which value takes the integral if μ is odd?

Chapter 2

Gegenbauer Polynomials

In the constructive theory of spherical functions the *Gegenbauer polynomials* play an important role. Apart from constant factors they are certain *Jacobi polynomials*. For $\alpha, \beta > -1$, the *indices*, the Jacobi polynomials $P_\mu^{(\alpha,\beta)}$ of degree $\mu \in \mathbb{N}_0$ are defined as orthogonal polynomials with respect to the inner product

$$(f, g) := \int_{-1}^{1} f(x)g(x)(1 - x)^\alpha(1 + x)^\beta dx, \qquad (2.1)$$

$\alpha, \beta > -1$, which are normalized by the condition

$$P_\mu^{(\alpha,\beta)}(1) = \binom{\mu + \alpha}{\mu}. \qquad (2.2)$$

Apart from constant factors, the Gegenbauer polynomials of the *Gegenbauer index* $\lambda > -\frac{1}{2}$ are obtained by putting

$$\alpha := \beta := \lambda - \frac{1}{2}.$$

There is no need in our theory to use Jacobi polynomials in their generality. On the contrary, the Gegenbauer polynomials are basic, and we want to make this transparent in what follows. So we begin with a short introduction to Gegenbauer polynomials and a presentation of their most important properties. Throughout in what follows $\mathbb{P}_\mu^1$ denotes the space of univariate polynomials of degree $\mu \in \mathbb{N}_0$.

2.1 Generating Function

First we assume $0 \neq \lambda \in \mathbb{R}$. The *generating function* G^λ is defined by

$$G^\lambda(x, z) := \frac{1}{(1 - 2xz + z^2)^\lambda} \qquad (2.3)$$

for $x \in [-1, +1]$, $z \in \mathbb{C}$, $|z| < 1$. For fixed x the function is holomorphic in $|z| < 1$, so it can be expanded in a Taylor series

$$G^\lambda(x, z) = \sum_{\mu=0}^{\infty} C_\mu^\lambda(x) z^\mu, \tag{2.4}$$

which *generates* the coefficients $C_\mu^\lambda(x)$. For sufficiently small $|z|$ we get

$$G^\lambda(x, z) = [1+z(z-2x)]^{-\lambda} = \sum_{k=0}^{\infty} \binom{-\lambda}{k} z^k (z-2x)^k = \sum_{k=0}^{\infty} \binom{-\lambda}{k} \sum_{j=0}^{k} \binom{k}{j} z^{2k-j} (-2x)^j.$$

Substituting $\mu = 2k - j$, $\nu = k - j$ we obtain

$$G^\lambda(x, z) = \sum_{\mu=0}^{\infty} z^\mu \sum_{\nu=0}^{\lfloor \frac{\mu}{2} \rfloor} \binom{-\lambda}{\mu - \nu} \binom{\mu - \nu}{\mu - 2\nu} (-2x)^{\mu-2\nu},$$

and comparing the coefficients in this formula and in (2.4), we get

$$C_\mu^\lambda(x) = \sum_{\nu=0}^{\lfloor \frac{\mu}{2} \rfloor} (-1)^\nu \frac{(\lambda)_{\mu-\nu}}{(1)_\nu (1)_{\mu-2\nu}} \cdot (2x)^{\mu-2\nu} \tag{2.5}$$

for $\lambda \neq 0$ and $\mu \in \mathbb{N}_0$, where we use, for convenience, the *Pochhammer symbol*

$$(a)_k := \prod_{j=0}^{k-1} (a + j)$$

for $a \in \mathbb{C}$ and $k \in \mathbb{N}_0$, in the representation. Note that $(a)_0 = 1, (a)_1 = a$ $(a)_2 = a(a + 1)$, and so on, while *Stirling's formula* yields the useful formula

$$\frac{(a)_k}{(b)_k} = \frac{\Gamma(b)}{\Gamma(a)} \cdot \frac{\Gamma(k + a)}{\Gamma(k + b)} \sim \frac{\Gamma(b)}{\Gamma(a)} \cdot k^{a-b}, \tag{2.6}$$

which holds for $a > 0$, $b > 0$ as k tends to ∞. Here and in what follows the symbol $\sim$ indicates asymptotic equality.

(2.5) shows that $C_\mu^\lambda(x)$ is a polynomial of exact degree μ. It is called the *Gegenbauer polynomial* of degree μ and *index* λ.

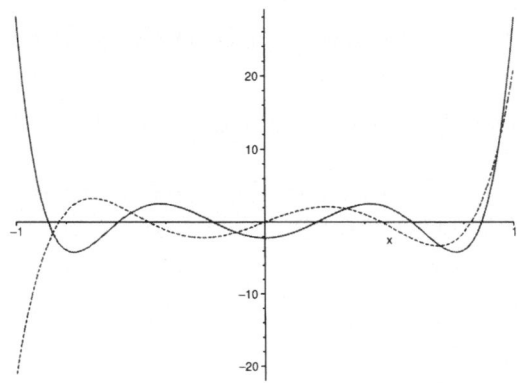

Figure 2.1. Gegenbauer Polynomials $C_5^{\frac{3}{2}}$ and $C_6^{\frac{3}{2}}$.

For $\lambda = 0$ the function (2.3) degenerates, and (2.4) is unsuitable for a proper definition. So we change the definition of the generating function and of the corresponding Gegenbauer polynomials by replacing (2.3) and (2.4) by

$$G^0(x, z) := 1 - \log(1 - 2xz + z^2) = \sum_{\mu=0}^{\infty} C_\mu^0(x) z^\mu, \qquad (2.7)$$

which is valid again for $x \in [-1, +1]$ and $|z| < 1$. By differentiating (2.7) with respect to z and some additional calculation we can derive the equation

$$\frac{1 - xz}{1 - 2xz + z^2} = 1 + \sum_{\mu=1}^{\infty} \frac{\mu}{2} \cdot C_\mu^0(x) z^\mu.$$

Now we refer to the well-known equations

$$\frac{1 - xz}{1 - 2xz + z^2} = \sum_{\mu=0}^{\infty} T_\mu(x) z^\mu,$$

$$\frac{1}{1 - 2xz + z^2} = \sum_{\mu=0}^{\infty} U_\mu(x) z^\mu,$$

which are satisfied by the *Chebyshev polynomials* T_μ and U_μ of the first and of the second kind, respectively. A comparison of the coefficients yields

$$\begin{aligned} C_0^0 &= T_0, \\ C_\mu^0 &= \tfrac{2}{\mu} \cdot T_\mu \quad \text{for } \mu \in \mathbb{N}, \end{aligned} \qquad (2.8)$$

and
$$C_\mu^1 = U_\mu$$

for arbitrary $\mu \in \mathbb{N}_0$. In particular, C_μ^0 is again a polynomial of exact degree μ, and as a summary we add that

$$C_\mu^\lambda \in \mathbb{P}_\mu^1 \setminus \mathbb{P}_{\mu-1}^1 \tag{2.9}$$

holds for arbitrary $\lambda > -1$ and $\mu \in \mathbb{N}_0$. Here we have put $\mathbb{P}_{-1}^1 := [0]$. Moreover, for later use we put also $C_{-1}^\lambda := 0$.

The Gegenbauer polynomials have numerous important properties.

Derivatives

Differentiating (2.5) we get

$$\frac{d}{dx} C_\mu^\lambda = 2\lambda C_{\mu-1}^{\lambda+1} \tag{2.10}$$

for $\lambda \neq 0$ and $\mu \in \mathbb{N}_0$. Differentiating (2.7) we obtain, together with (2.3) and (2.4),

$$\sum_{\mu=0}^\infty \frac{d}{dx} C_\mu^0(x) z^\mu = \frac{2z}{1 - 2xz + z^2} = \sum_{\mu=1}^\infty 2C_{\mu-1}^1(x) z^\mu,$$

and a comparison of the coefficients yields

$$\frac{d}{dx} C_\mu^0 = 2\,C_{\mu-1}^1 \tag{2.11}$$

for $\mu \in \mathbb{N}_0$.

Special Values

For $\lambda \neq 0$ we get from (2.4) and (2.3)

$$\sum_{\mu=0}^\infty C_\mu^\lambda(0) z^\mu = \frac{1}{(1 + z^2)^\lambda} = \sum_{\nu=0}^\infty (-1)^\nu \frac{(\lambda)_\nu}{(1)_\nu} z^{2\nu},$$

and a comparison of the coefficients yields

$$\begin{aligned} C_{2\nu}^\lambda(0) &= (-1)^\nu \frac{(\lambda)_\nu}{(1)_\nu}, \\ C_{2\nu+1}^\lambda(0) &= 0 \end{aligned} \tag{2.12}$$

for $\nu \in \mathbb{N}_0$. Similar we obtain

$$\sum_{\mu=0}^\infty C_\mu^\lambda(1) z^\mu = \frac{1}{(1 - z)^{2\lambda}} = \sum_{\mu=0}^\infty \frac{(2\lambda)_\mu}{(1)_\mu} z^\mu$$

and hence

$$C_\mu^\lambda(1) = \frac{(2\lambda)_\mu}{(1)_\mu} \tag{2.13}$$

for $\mu \in \mathbb{N}_0$ and $\lambda \neq 0$. The corresponding values of C_μ^0 are well known from (2.8). In particular we get $C_\mu^\lambda(1) > 0$ for arbitrary $\lambda > -\frac{1}{2}$, and for later use we may define the *normalized Gegenbauer polynomials*

$$\tilde{C}_\mu^\lambda := C_\mu^\lambda / C_\mu^\lambda(1) \tag{2.14}$$

for $\mu \in \mathbb{N}_0$, but now for $\lambda > -\frac{1}{2}$, only. The normalisation is such that $\tilde{C}_\mu^\lambda(1) = 1$ holds.

2.2 Differential Equation

For partially differentiable functions $F(x, z)$ we denote the partial derivatives with regard to x and z by F_x and F_z, respectively. It is possible to prove that

$$(1 - x^2)G_{xx}^\lambda + z^2 G_{zz}^\lambda + (2\lambda + 1)[zG_z^\lambda - xG_x^\lambda] = 0$$

is valid for $\lambda \neq 0$, see Problem 2.5. Because of this partial differential equation, the following theorem holds.

Theorem 2.1 (Differential Equation of C_μ^λ). *For arbitrary $\lambda \in \mathbb{R}$, $\mu \in \mathbb{N}_0$, the Gegenbauer polynomial C_μ^λ is a solution of the differential equation*

$$(1 - x^2)y'' - (2\lambda + 1)xy' + \mu(\mu + 2\lambda)y = 0. \tag{2.15}$$

Proof. For $\lambda = 0$, where, apart from constant factors, the Gegenbauer polynomials are the Chebyshev polynomials of the first kind, the result is well known. So we may assume $\lambda \neq 0$ in what follows.

We put $y_\mu := C_\mu^\lambda$ such that (2.4) takes the form $\sum_{\mu=0}^\infty y_\mu(x)z^\mu = G^\lambda(x, z)$, and define the differential operator D by $DF := zF_z$. Using $Dz^\mu = \mu z^\mu$ we get easily the equation

$$\sum_{\mu=0}^\infty [(1 - x^2)y_\mu'' - (2\lambda + 1)xy_\mu' + \mu(\mu + 2\lambda)y_\mu] z^\mu$$

$$= (1 - x^2)G_{xx}^\lambda - (2\lambda + 1)xG_x^\lambda + D(D + 2\lambda)G^\lambda.$$

Some further calculations yield

$$D(D + 2\lambda)G^\lambda = z^2 G_{zz}^\lambda + (2\lambda + 1)zG_z^\lambda,$$

and inserting this above we obtain

$$\sum_{\mu=0}^{\infty} \left[(1-x^2)y_\mu'' - (2\lambda+1)xy_\mu' + \mu(\mu+2\lambda)y_\mu\right]z^\mu$$

$$= (1-x^2)G_{xx}^\lambda + z^2 G_{zz}^\lambda + (2\lambda+1)[zG_z^\lambda - xG_x^\lambda] = 0$$

by the result from above. Since the coefficient which occurs with z^μ must vanish, we get (2.15), as claimed. $\qquad\qquad\qquad\Box$

The Gegenbauer polynomials are not characterized by the differential equation alone. However the following characterisation theorem is valid.

Theorem 2.2 (Characterisation of Gegenbauer polynomials). *Let $\lambda > -\frac{1}{2}$, $\mu \in \mathbb{N}_0$. Every solution y of the differential equation (2.15) which satisfies*

$$\begin{array}{ll}(i) & y \in \mathbb{P}_\mu^1, \\ (ii) & y(-x) = (-1)^\mu y(x),\end{array}$$

has the form $y = const \cdot C_\mu^\lambda$. The Gegenbauer polynomial C_μ^λ is the uniquely determined solution which satisfies (i), (ii), and

$$(iii) \qquad y(1) = C_\mu^\lambda(1).$$

Proof. Assume y is a solution of the differential equation and satisfies (i) and (ii). Then it has the form

$$y(x) = \sum_{\nu=0}^{\lfloor \frac{\mu}{2} \rfloor} a_\nu x^{\mu-2\nu}$$

and satisfies

$$(1-x^2)\sum_{\nu=0}^{\lfloor \frac{\mu}{2} \rfloor}(\mu-2\nu)(\mu-2\nu-1)a_\nu x^{\mu-2\nu-2}$$

$$- (2\lambda+1)\sum_{\nu=0}^{\lfloor \frac{\mu}{2} \rfloor}(\mu-2\nu)a_\nu x^{\mu-2\nu} + \mu(\mu+2\lambda)\sum_{\nu=0}^{\lfloor \frac{\mu}{2} \rfloor}a_\nu x^{\mu-2\nu} = 0.$$

This equation can be brought to the form

$$\sum_{\nu=0}^{\lfloor \frac{\mu}{2} \rfloor}\Big\{(\mu-2\nu+2)(\mu-2\nu+1)a_{\nu-1}$$

$$-[(\mu-2\nu)(\mu-2\nu-1)+(2\lambda+1)(\mu-2\nu)-\mu(\mu+2\lambda)]a_\nu\Big\}x^{\mu-2\nu} = 0,$$

with $a_{-1} := 0$, and a comparison of the coefficients yields

$$4\nu(\mu-\nu+\lambda)a_\nu = -(\mu-2\nu+2)(\mu-2\nu+1)a_{\nu-1}$$

for $\nu = 1, ..., \lfloor \frac{\mu}{2} \rfloor$. So the leading coefficient a_0 is arbitrary, while the remaining coefficients are uniquely determined multiples of a_0. This means that y has the form

$$y(x) = a_0 z(x),$$

where the leading coefficient of z equals 1. But C_μ^λ is a solution of (2.15), so we get

$$C_\mu^\lambda(x) = \alpha_0 z(x),$$

where α_0 is the leading coefficient of C_μ^λ. In view of (2.5) it is nonvanishing, and we get $z(x) = \alpha_0^{-1} C_\mu^\lambda(x)$. Inserting this above we obtain the first statement. Because of $C_\mu^\lambda \neq 0$, the second statement is an immediate consequence of it, where the constant is uniquely determined by the value of $y(1)$. $\qquad\square$

Next we assume $\lambda > 0$ and $\mu \in \mathbb{N}$. Using the abbreviation $y := C_\mu^\lambda$, we define the polynomial z by

$$z(x) := (1 - x^2){y'}^2(x) + \mu(\mu + 2\lambda)y^2(x).$$

It is even and nonnegative on $[-1, 1]$. It follows from (2.15) by some calculation that

$$z'(x) = 4\lambda x {y'}^2(x)$$

holds. So $z(x)$ is strictly monotonically increasing for $0 \leq x < \infty$, and we obtain in particular

$$0 \leq z(x) < z(1) = z(-1)$$

for $-1 < x < 1$, which implies

$$(1 - x^2){y'}^2(x) + \mu(\mu + 2\lambda)y^2(x) < \mu(\mu + 2\lambda)y^2(1)$$

and hence $y^2(x) < y^2(1)$, in this interval. In other words, we obtain

$$|C_\mu^\lambda(x)| \leq C_\mu^\lambda(1) \qquad (2.16)$$

for $-1 \leq x \leq 1$, $\mu \in \mathbb{N}$, and $\lambda > 0$, where equality implies $x \in \{-1, +1\}$.

In the Chebyshev case $\lambda = 0$ we get $z(x) = z(1) = const.$ (2.16) remains valid for $-1 \leq x \leq 1$, $\mu \in \mathbb{N}$, but equality occurs for $\mu \geq 2$ even in interior points of the interval.

2.3   Orthogonality

We assume $\lambda > \frac{1}{2}$, again, and use the abbreviation $y_\nu := C_\nu^\lambda$ for $\nu \in \mathbb{N}_0$. From the differential equation (2.15) it follows that

$$[(1 - x^2)^{\lambda + \frac{1}{2}} y_\nu ']' + \mu(\mu + 2\lambda)(1 - x^2)^{\lambda - \frac{1}{2}} y_\nu = 0$$

holds. So we obtain for $\nu, \mu \in \mathbb{N}_0$, $0 \leq \nu < \mu$, using the differential equation and integration by parts,

$$\mu(\mu + 2\lambda) \int\limits_{-1}^{1} y_\nu y_\mu (1 - x^2)^{\lambda - \frac{1}{2}} dx$$

$$= -\left[y_\nu (1 - x^2)^{\lambda + \frac{1}{2}} y_\mu' \right]_{-1}^{1} + \int\limits_{-1}^{1} y_\nu' y_\mu' (1 - x^2)^{\lambda + \frac{1}{2}} dx = \int\limits_{-1}^{1} y_\nu' y_\mu' (1 - x^2)^{\lambda + \frac{1}{2}} dx.$$

For $\nu = 0$ we get $y'_\nu = 0$, and hence

$$\int_{-1}^{1} C_\nu^\lambda C_\mu^\lambda (1 - x^2)^{\lambda - \frac{1}{2}} dx = 0 \qquad (2.17)$$

is valid for this particular value of ν. For $1 \leq \nu < \mu$ we get by the help of (2.10) or (2.11), respectively,

$$\int_{-1}^{1} C_\nu^\lambda C_\mu^\lambda (1 - x^2)^{\lambda - \frac{1}{2}} dx = \frac{(2\bar\lambda)^2}{\mu(\mu + 2\lambda)} \int_{-1}^{1} C_{\nu-1}^{\lambda+1} C_{\mu-1}^{\lambda+1} (1 - x^2)^{\lambda + \frac{1}{2}} dx,$$

where

$$\bar\lambda := \begin{cases} \lambda, & \text{if } \lambda \neq 0, \\ 1, & \text{if } \lambda = 0. \end{cases}$$

The right side vanishes by the result just proved, and (2.17) is valid for $\nu = 0$ and $\nu = 1$. For $1 < \nu < \mu$ we repeat our procedure successively, with the final result that (2.17) is valid for arbitrary $\nu \in \{0, 1, ..., \mu - 1\}$.

Definition 2.1 (Inner Product belonging to the C_μ^λ). *For* $\lambda > -\frac{1}{2}$, $f, g \in C[-1, 1]$ *we define the* inner product

$$[f, g]_\lambda := \int_{-1}^{1} f(x) g(x)(1 - x^2)^{\lambda - \frac{1}{2}} dx.$$

With this definition (2.17) takes the following form.

Theorem 2.3 (Orthogonality of the C_μ^λ)). *Let* $\lambda > -\frac{1}{2}$. *Then* $[C_\nu^\lambda, C_\mu^\lambda]_\lambda = 0$ *holds for* $\nu, \mu \in \mathbb{N}_0$ *and* $\nu \neq \mu$.

Because of Theorem 2.3, the Gegenbauer polynomials possess all the well-known properties which orthogonal polynomials have in general. In particular, they satisfy a *recurrence relation*, see Problem 2.1, and their zeros are *interlacing*, which is demonstrated by Figure 2.1. The *formula of Christoffel–Darboux* takes the following form.

Theorem 2.4 (Formula of Christoffel–Darboux). *Let* $\lambda > -\frac{1}{2}$, *and let* $\| \cdot \|$ *denote the norm induced by the inner product* $[\cdot, \cdot]_\lambda$. *Moreover, for* $\nu \in \mathbb{N}_0$ *let* k_ν *denote the leading coefficient of the polynomial* $P_\nu(x) := C_\nu^\lambda(x)/\|C_\nu^\lambda\| = k_\nu x^\nu + TLD$. *Then*

$$K_\mu(x, y) := \frac{k_\mu}{k_{\mu+1}} \cdot \frac{P_{\mu+1}(x) P_\mu(y) - P_\mu(x) P_{\mu+1}(y)}{x - y} = \sum_{\nu=0}^{\mu} P_\nu(x) P_\nu(y)$$

is valid. In other words, $K_\mu(x, y)$ *is the reproducing kernel of* $\mathbb{P}_\mu^1$ *with respect to* $[\cdot, \cdot]_\lambda$. *The leading coefficients* k_ν *are positive.*

Remark. We write TLD for *'terms of lower degree'*, here and in what follows.

Proof. For the sign of the leading coefficients we refer to (2.5), if $\lambda \neq 0$, and to (2.8), if $\lambda = 0$. So it suffices, in view of (1.5), to prove that $K_\mu(x,y)$ is the reproducing kernel of $\mathbb{P}^1_\mu$, this means that the requirements (i), (ii) and (iii) of Definition 1.1 are satisfied.

Actually, (i) and (ii) are valid. To prove (iii), let $f \in \mathbb{P}^1_\mu$ be arbitrary and $y \in \mathbb{R}$. Then we get

$$[K_\mu(\cdot,y), f(\cdot) - f(y)]_\lambda = \frac{k_\mu}{k_{\mu+1}} \cdot [P_{\mu+1}(\cdot)P_\mu(y) - P_\mu(\cdot)P_{\mu+1}(y), \phi(\cdot)]_\lambda,$$

where $\phi(x) := \frac{f(x)-f(y)}{x-y}$ is a polynomial of degree $\mu - 1$ in the variable x. Therefore the integral vanishes, and we get

$$[K_\mu(\cdot,y), f]_\lambda = h_\mu(y)f(y)$$

where

$$h_\mu(y) := [K_\mu(\cdot,y), 1]_\lambda.$$

$K_\mu(x,y)$ is a polynomial in y of degree μ for every fixed x. Therefore $h_\mu(y)f(y)$ is also a polynomial of this degree. This implies that h_μ is a constant, and $h_\mu^{-1}K_\mu(x,y)$ has all defining properties of the reproducing kernel, see Definition 1.1. So we get from (1.5)

$$h_\mu^{-1}K_\mu(x,y) = \sum_{\nu=0}^{\mu} P_\nu(x)P_\nu(y).$$

The constant can be determined by integration, this means from

$$h_\mu^{-1} \int_{-1}^{1} K_\mu(x,x)(1-x^2)^{\lambda-\frac{1}{2}}dx = \sum_{\nu=0}^{\mu} \|P_\nu\|^2 = \mu + 1.$$

Actually, from the definition of $K(x,y)$ we get for $y \to x$,

$$K_\mu(x,x) = \frac{k_\mu}{k_{\mu+1}}\left(P'_{\mu+1}(x)P_\mu(x) - P'_\mu(x)P_{\mu+1}(x)\right),$$

where

$$\frac{k_\mu}{k_{\mu+1}}P'_{\mu+1}(x) = (\mu+1)P_\mu(x) + TLD.$$

Inserting this above and using orthogonality again we obtain

$$h_\mu = \frac{1}{\mu+1} \cdot \frac{k_\mu}{k_{\mu+1}}[P'_{\mu+1}, P_\mu]_\lambda = [P_\mu, P_\mu]_\lambda = 1,$$

and the theorem is proved. $\qquad\qquad\qquad\qquad\qquad\qquad\qquad\qquad\qquad\qquad\qquad\square$

Special Systems

As mentioned above, the Gegenbauer polynomials of index $\lambda = 0$ are, apart from constant factors, the *Chebyshev polynomials* T_μ of the *first kind,* and the $C_\mu^1 = U_\mu$ are the *Chebyshev polynomials* of the *second kind.* The polynomials $C_\mu^{\frac{1}{2}}$ are known as the *Legendre polynomials.* The index $\lambda = 0$ plays an extra role only with respect to the normalisation. This follows from the equation

$$\lim_{\lambda \to 0+} \frac{1}{\lambda} C_\mu^\lambda(x) = \frac{2}{\mu} T_\mu(x) \tag{2.18}$$

which can be proved by the help of (2.5).

Zeros

For $\mu \in \mathbb{N}$ it follows from Theorem 2.3, together with (2.9), that C_μ^λ is orthogonal to the whole space $\mathbb{P}_{\mu-1}^1$. It is well known that this implies that C_μ^λ has exactly μ *simple zeros* η_ν^λ, which can be numbered such that their order is

$$-1 < \eta_\mu^\lambda < \eta_{\mu-1}^\lambda < \ldots < \eta_1^\lambda < +1.$$

We write them also in the *trigonometric form* $\eta_\nu^\lambda = \cos \psi_\nu^\lambda$, where the order is

$$0 < \psi_1^\lambda < \ldots < \psi_{\mu-1}^\lambda < \psi_\mu^\lambda < \pi.$$

2.4   Bessel Functions

In the asymptotic representation of Gegenbauer polynomials the *Bessel functions* of *index* $\alpha > -1$, $\alpha = \lambda - \frac{1}{2}$, are very helpful. They are defined by

$$J_\alpha(x) := \left(\frac{x}{2}\right)^\alpha \sum_{k=0}^\infty \frac{(-1)^k}{\Gamma(k+1)\Gamma(\alpha+k+1)} \left(\frac{x}{2}\right)^{2k}. \tag{2.19}$$

The series converges for arbitrary $x \in \mathbb{C}$. For reasons to become clear later we prefer to use the *normalized Bessel functions*

$$Z_\alpha(x) := \Gamma(\alpha+1) \left(\frac{2}{x}\right)^\alpha J_\alpha(x) \tag{2.20}$$

with the representation

$$Z_\alpha(x) = \sum_{k=0}^\infty \frac{(-1)^k \Gamma(\alpha+1)}{\Gamma(k+1)\Gamma(\alpha+k+1)} \left(\frac{x}{2}\right)^{2k}. \tag{2.21}$$

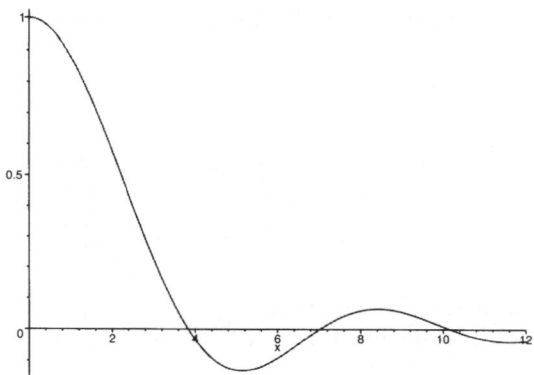

Figure 2.2. Normalized Bessel Function Z_1.

Z_α is an entire function, and it is normalized such that

$$Z_\alpha(0) = 1, \quad Z_\alpha'(0) = 0 \tag{2.22}$$

holds. See Figure 2.2. Besides we note that

$$Z_\alpha'(x) = -\frac{x}{2\alpha + 2} Z_{\alpha+1}(x) \tag{2.23}$$

is valid, and it is easy to see that $z := Z_\alpha$ solves the differential equation

$$xz'' + (2\alpha + 1)z' + xz = 0. \tag{2.24}$$

In what follows we assume $\alpha > -\frac{1}{2}$. Then (2.24) implies immediately that the function $F := z^2 + z'^2$, with the derivative

$$F'(x) = -\frac{2(2\alpha + 1)}{x} z'^2(x),$$

is nonnegative and monotonically nonincreasing in the interval $0 < x < \infty$. And as a nonzero entire function, it cannot vanish on an interval. So it must be positive everywhere, i.e.,

$$z^2(x) + z'^2(x) > 0 \tag{2.25}$$

holds for $0 \leq x < \infty$. Therefore we may introduce 'polar coordinates' $R(x) > 0$ and $\phi(x)$ by

$$\begin{aligned} z(x) &= R(x) \cos \phi(x), \\ z'(x) &= -R(x) \sin \phi(x), \end{aligned} \tag{2.26}$$

where $\phi(0) = 0$, see (2.22), and where R and ϕ are continuously differentiable. Differentiating both equalities, eliminating z'' by the help of (2.24), z and z' by the help of (2.26), and eliminating R' finally from the system obtained, we get

$$\phi'(x) = 1 - \frac{2\alpha + 1}{2x} \cdot \sin(2\phi(x)).$$

For $x \geq 2\alpha + 1$ we get $\phi'(x) \geq \frac{1}{2}$, and by integration it follows that

$$\phi(x) \geq \phi(2\alpha + 1) + \frac{1}{2}(x - [2\alpha + 1]).$$

Together this yields that ϕ is monotonically increasing in $[2\alpha + 1, \infty)$, where $\lim_{x \to \infty} \phi(x) = +\infty$ holds. Inserting this in (2.26) we see that $z(x)$ has countably many zeros in $[2\alpha + 1, \infty)$. But as an entire function it has finitely many zeros in $[0, 2\alpha + 1)$, at most. Therefore the positive zeros of $z = Z_\alpha$ — and hence of J_α — can be enumerated in the form

$$0 < j_{\alpha,1} < j_{\alpha,2} < \ldots \qquad (\to +\infty).$$

They are called *Bessel zeros*. Moreover, (2.25) shows that all zeros are simple. Because of (2.23) the same holds for the positive zeros

$$0 < j_{\alpha+1,1} < j_{\alpha+1,2} < \ldots \qquad (\to +\infty)$$

of z', $Z_{\alpha+1}$ and hence of $J_{\alpha+1}$. Note that for $\alpha > 0$, J_α has the additional zero $j_{\alpha,0} := 0$. We want to compare the location of both systems.

Lemma 2.5. *Let $\alpha > -\frac{1}{2}$, and assume that Z_α is nonvanishing in the interval $a < x < b$ where $a \geq 0$. Then Z_α' has at most one zero in this interval.*

Proof. Assume ξ and η, $a < \xi < \eta < b$, are two consecutive zeros of z'. They are both simple zeros, so z must have a relative extremum at both of these points, but of different character: one being a maximum, the other one being a minimum. Therefore z'' must agree with z in sign at one of these points, say $z(\xi)z''(\xi) \geq 0$. If we insert this together with $z'(\xi) = 0$ in (2.24) then we get a contradiction. So the assumption was false, and the lemma is proved. □

Lemma 2.5 allows us to prove the following important theorem.

Theorem 2.6 (Interlacing Property of the Bessel Zeros). *Let $\alpha > -\frac{1}{2}$. Then the following holds.*

i) The Bessel zeros have the interlacing property

$$j_{\alpha+1,0} = 0 < j_{\alpha,1} < j_{\alpha+1,1} < j_{\alpha,2} < j_{\alpha+1,2} < \ldots (\to +\infty). \qquad (2.27)$$

ii) $Z_\alpha(x)$ is positive and monotonically decreasing for $0 \leq x < j_{\alpha,1}$.

Proof. Let $z = Z_\alpha$ as above, take notice of (2.24), and assume that $j_{\alpha+1,1} \leq j_{\alpha,1}$ holds. Because of $z'(0) = 0$, $z''(0) < 0$, we get $z'(x) < 0$ for $0 < x < j_{\alpha+1,1}$, while $z'(x) > 0$ holds in a right-side neighbourhood of $j_{\alpha+1,1}$. But $z(0) = 1$ implies that $z(x)$ is positive for $0 \leq x < j_{\alpha,1}$ and must attain a positive relative minimum at $j_{\alpha+1,1}$. However, $z(x)$ vanishes at $j_{\alpha,1}$, so there must be a relative maximum between. This implies that z' has at least two zeros in an interval where z is positive, which is impossible by Lemma 2.5. So we obtain $j_{\alpha,1} \leq j_{\alpha+1,1}$, where equality is excluded by (2.25). Together this proves assertion (ii). Next let $k \in \mathbb{N}$. z' has at least one zero in the interval $(j_{\alpha,k}, j_{\alpha,k+1})$, but cannot have additional zeros, again by Lemma 2.5. Finally we get (2.27) in full by counting and comparing the location of the zeros $j_{\alpha,k}$ and $j_{\alpha+1,k}$ successively. $\qquad\square$

Corollary 2.7. $\qquad Z_{-\frac{1}{2}}(x) = \cos x, \quad Z_{\frac{1}{2}}(x) = \frac{\sin x}{x}.$

Proof. It is easy to see that the functions on the right side satisfy the corresponding differential equation (2.24) together with the initial values (2.22). $\qquad\square$

2.5 Asymptotics

For the normalized Gegenbauer polynomials, as defined in (2.14), or to be more precise, for the functions $\tilde{C}_\mu^\lambda(\cos \phi)$, $\mu \in \mathbb{N}$, two types of asymptotic formulas exist for $\mu \to \infty$, which are valid either under the restriction

$$(a) \qquad -\frac{c}{\mu} \leq \phi \leq \frac{c}{\mu},$$

or under the restriction

$$(b) \qquad \frac{c}{\mu} \leq \phi \leq \pi - \frac{c}{\mu},$$

respectively, where c is an arbitrary positive constant.

Type (a)

Let $\lambda > 0$, $c > 0$ be fixed, $\mu \in \mathbb{N}$. We replace ϕ by $\frac{\phi}{\mu}$ such that the aim is to determine the asymptotics of $\tilde{C}_\mu^\lambda(\cos \frac{\phi}{\mu})$, subject to the restriction

$$(a') \qquad -c \leq \phi \leq c.$$

From (2.14) and (2.16) we obtain $\|\tilde{C}_\mu^\lambda\|_\infty = 1$, in the maximum norm on $[-1, 1]$, and using Markov's inequality we get $\|\frac{d}{dx}\tilde{C}_\mu^\lambda\|_\infty < \mu^2$. It follows that

$$\left| \tilde{C}_\mu^\lambda \left(\cos \tfrac{\phi}{\mu} \right) - \tilde{C}_\mu^\lambda \left(1 - \tfrac{1}{2}(\tfrac{\phi}{\mu})^2 \right) \right| \leq const \cdot \mu^{-2} \qquad (2.28)$$

with some constant, which depends on c, only. So it suffices to investigate the expression $\tilde{C}_\mu^\lambda \left(1 - \frac{1}{2}(\frac{\phi}{\mu})^2 \right)$, which we expand into a Taylor series.

By a repeated application of (2.10), we get for $k \in \{0, 1, \ldots, \mu\}$, in view of (2.14) and of (2.13),

$$
\begin{aligned}
\frac{1}{k!}\left(\frac{d}{dx}\right)^k \tilde{C}_\mu^\lambda(1) &= \frac{(1)_\mu}{(2\lambda)_\mu} \cdot 2^k \frac{(\lambda)_k}{(1)_k} \cdot C_{\mu-k}^{\lambda+k}(1) \\
&= 2^k \frac{(\lambda)_k}{(1)_k} \cdot \frac{(1)_\mu}{(1)_{\mu-k}} \cdot \frac{(2\lambda + 2k)_{\mu-k}}{(2\lambda)_\mu} \\
&= 2^k \frac{(\lambda)_k}{(1)_k (2\lambda)_{2k}} \cdot (\mu - k + 1)_k (\mu + 2\lambda)_k \\
&= 2^{-k} \frac{\Gamma(\lambda + \frac{1}{2})}{\Gamma(k+1)\Gamma(\lambda + \frac{1}{2} + k)} \prod_{\kappa=0}^{k-1} \left[(\mu + \lambda)^2 - (\kappa + \lambda)^2\right].
\end{aligned}
$$

So we get, with $\alpha = \lambda - \frac{1}{2}$,

$$
\tilde{C}_\mu^\lambda\left(1 - \frac{1}{2}\left(\frac{\phi}{\mu}\right)^2\right) = \sum_{k=0}^\mu \frac{(-1)^k \Gamma(\alpha + 1)}{\Gamma(k+1)\Gamma(\alpha + k + 1)} \prod_{\kappa=0}^{k-1} \frac{(\mu + \lambda)^2 - (\kappa + \lambda)^2}{\mu^2} \cdot \left(\frac{\phi}{2}\right)^{2k}.
$$

Obviously, for fixed k the coefficient occurring with x^{2k} converges for $\mu \to \infty$ to the corresponding coefficient of Z_α, see (2.21). Actually, a comparison of the functions yields

Theorem 2.8 (Asymptotics of $\tilde{C}_\mu^\lambda(\cos \frac{\phi}{\mu})$).

$$
\lim_{\mu \to \infty} \tilde{C}_\mu^\lambda(\cos \frac{\phi}{\mu}) = Z_\alpha(\phi)
$$

holds uniformly on compact sets for $\lambda \geq 0$, where $\alpha = \lambda - \frac{1}{2}$.

Proof. In the case $\lambda = 0$ we get

$$
\tilde{C}_\mu^0(\cos \frac{\phi}{\mu}) = T_\mu(\cos \frac{\phi}{\mu}) = \cos \phi = Z_{-\frac{1}{2}}(\phi),
$$

see Corollary 2.7, and the statement is true, obviously.

Next let $\lambda > 0$, this means $\alpha > -\frac{1}{2}$, and let K be a compact set in $\mathbb{R}$, say $K \subset [-c, c]$ holds for some $c > 0$. For $\mu \in \mathbb{N}$, $\mu \geq m$, we get

$$
\left|\tilde{C}_\mu^\lambda\left(1 - \frac{1}{2}(\frac{\phi}{\mu})^2\right) - Z_\alpha(\phi)\right| \leq A + B
$$

where

$$
A = \sum_{k=0}^m \frac{\Gamma(\alpha + 1)}{\Gamma(k+1)\Gamma(\alpha + k + 1)} \left|\prod_{\kappa=0}^{k-1} \frac{(\mu + \lambda)^2 - (\kappa + \lambda)^2}{\mu^2} - 1\right| \left(\frac{\phi}{2}\right)^{2k},
$$

$$
B = \sum_{k=m+1}^\infty \frac{\Gamma(\alpha + 1)}{\Gamma(k+1)\Gamma(\alpha + k + 1)} [1 + (1 + \lambda)^{2k}] \left(\frac{\phi}{2}\right)^{2k}.
$$

Now let $\epsilon > 0$ be given. We choose m so large that

$$B \leq \frac{\epsilon}{3}$$

is valid for all $\phi \in K$. Next we choose $\mu_0 \geq m$ so large that

$$A \leq \frac{\epsilon}{3}$$

and

$$\left| \tilde{C}_\mu^\lambda \left(\cos \frac{\phi}{\mu} \right) - \tilde{C}_\mu^\lambda \left(1 - \frac{1}{2}(\frac{\phi}{\mu})^2 \right) \right| \leq \frac{\epsilon}{3}$$

hold for $\mu \geq \mu_0$, again for all $\phi \in K$. This is possible because of (2.28). Then we get

$$\left| \tilde{C}_\mu^\lambda \left(\cos \frac{\phi}{\mu} \right) - Z_\alpha \right|$$

$$\leq \left| \tilde{C}_\mu^\lambda \left(\cos \frac{\phi}{\mu} \right) - \tilde{C}_\mu^\lambda \left(1 - \frac{1}{2}(\frac{\phi}{\mu})^2 \right) \right| + \left| \tilde{C}_\mu^\lambda \left(1 - \frac{1}{2}(\frac{\phi}{\mu})^2 \right) - Z_\alpha(\phi) \right|$$

$$\leq \frac{\epsilon}{3} + A + B \leq \epsilon$$

for arbitrary $\phi \in K$ and $\mu \geq \mu_0$, and convergence takes place, as claimed. $\square$

Type (t)

The following theorem contains a more general result than we have announced.

Theorem 2.9 (Upper Bound for $|\tilde{C}_\mu^\lambda(\cos \phi)|$). *Let $\lambda \geq 1$. Then a constant $k_1(\lambda)$ exists such that*

$$[(\mu + 1) \sin \phi]^\lambda |\tilde{C}_\mu^\lambda(\cos \phi)| \leq k_1(\lambda)$$

holds for all $\mu \in \mathbb{N}_0$ and $\phi \in \mathbb{R}$.

Proof. Let $u(x) := (1 - x^2)^{\frac{\lambda}{2}} \tilde{C}_\mu^\lambda(x)$. Since u is odd or even, it suffices to investigate $u(x)$ for $0 \leq x \leq 1$. It follows from (2.15) that u satisfies the differential equation

$$(1 - x^2)u'' - xu' + f(x)u = 0$$

with

$$f(x) := (\mu + \lambda)^2 - \frac{\lambda(\lambda - 1)}{1 - x^2},$$

$$f'(x) = -2\lambda(\lambda - 1)\frac{x}{(1 - x^2)^2} \leq 0$$

for $0 \leq x < 1$.

Now let us define U by

$$U(x) := (1 - x^2)u'^2(x) + f(x)u^2(x)$$

for $0 \le x < 1$. Then we get, by the help of the differential equation of u,

$$U'(x) = f'(x)u^2(x) \le 0,$$

again for $0 \le x < 1$. Therefore $U(x) \le U(0)$ is valid in this interval. Because of $f(x)u^2(x) \le U(x)$, this implies

$$f(x)(1-x^2)^\lambda |\tilde{C}_\mu^\lambda(x)|^2 \le \left| \left(\tfrac{d}{dx}\tilde{C}_\mu^\lambda \right)(0) \right|^2 + (\mu+\lambda)^2 \left| \tilde{C}_\mu^\lambda(0) \right|^2 =: R_\mu(\lambda). \quad (2.29)$$

We evaluate $R_\mu(\lambda)$ by means of (2.10), (2.12) and (2.13), and get

$$R_\mu(\lambda) = \left[\frac{(1)_\mu}{(2\lambda)_\mu} \right]^2 \cdot \begin{cases} (\mu+\lambda)^2 \left[\frac{(\lambda)_\nu}{(1)_\nu} \right]^2, & \text{if } \mu = 2\nu \text{ is even}, \\[2mm] (2\lambda)^2 \left[\frac{(\lambda+1)_\nu}{(1)_\nu} \right]^2, & \text{if } \mu = 2\nu+1 \text{ is odd}. \end{cases}$$

Note that the inequality

$$\left[\frac{(\lambda)_\nu}{(1)_\nu} \right]^2 \le \prod_{\kappa=0}^{\nu-1} \left(\frac{\lambda+\kappa}{1+\kappa} \cdot \frac{\lambda+\tfrac{1}{2}+\kappa}{1+\tfrac{1}{2}+\kappa} \right) = \frac{(2\lambda)_{2\nu}}{(2)_{2\nu}}$$

holds because of the assumption $\lambda \ge 1$. Using this we get

$$R_\mu(\lambda) \le \frac{(1)_\mu}{(2\lambda)_\mu} \cdot \begin{cases} \frac{(\mu+\lambda)^2}{\mu+1}, & \text{if } \mu \text{ is even}, \\[2mm] \frac{2\lambda}{2\lambda+1}(\mu+2\lambda), & \text{if } \mu \text{ is odd}. \end{cases}$$

In both cases we obtain finally

$$R_\mu(\lambda) \le \frac{(\mu+\lambda)^2}{\mu+1} \cdot \frac{(1)_\mu}{(2\lambda)_\mu}. \quad (2.30)$$

Now let $x_\mu \in [0,1)$ be defined by

$$1 - x_\mu^2 = \frac{\lambda^2}{(\mu+\lambda)^2}. \quad (2.31)$$

Then

$$f(x) \ge f(x_\mu) = \frac{(\mu+\lambda)^2}{\lambda} > 0 \quad (2.32)$$

holds for $0 \le x \le x_\mu$, since f is a monotonically decreasing function. It follows from (2.29), (2.30) and (2.32) that

$$(\mu+1)^{2\lambda}(1-x^2)^\lambda |\tilde{C}_\mu^\lambda(x)|^2 \le (\mu+1)^{2\lambda-1} \cdot \lambda \cdot \frac{(1)_\mu}{(2\lambda)_\mu} \sim \frac{1}{2} \cdot \Gamma(2\lambda+1) \quad (2.33)$$

holds for $0 \le x \le x_\mu$, see (2.6).

In the interval $x_\mu \leq x \leq 1$ we obtain from (2.31), together with (2.16),

$$(\mu + 1)^{2\lambda}(1 - x^2)^\lambda |\tilde{C}_\mu^\lambda(x)|^2 \leq (\mu + 1)^{2\lambda} \left(\frac{\lambda}{\mu + \lambda}\right)^{2\lambda} \leq \lambda^{2\lambda}. \tag{2.34}$$

In view of (2.33) and (2.34) it follows that

$$(\mu + 1)^{2\lambda}(1 - x^2)^\lambda |\tilde{C}_\mu^\lambda(x)|^2 \leq k_1^2(\lambda)$$

holds for $0 \leq x \leq 1$, and hence for $-1 \leq x \leq 1$, if we choose the constant $k_1(\lambda)$ such that

$$k_1^2(\lambda) = \max\left\{\Gamma(2\lambda + 1), \lambda^{2\lambda}\right\}.$$

With $x = \cos\phi$ this finishes the proof. $\square$

Corollary 2.10 (Upper Bound for $|C_\mu^\lambda(\cos\phi)|$). *Let $\lambda \geq 1$. Then a constant $k_2(\lambda)$ exists such that*

$$|C_\mu^\lambda(\cos\phi)| \leq k_2(\lambda)(\mu + 1)^{\lambda - 1}(\sin\phi)^{-\lambda}$$

holds for $0 < \phi < \pi$ and arbitrary $\mu \in \mathbb{N}_0$.

Proof. Multiplying the inequality of Theorem 2.9 by $C_\mu^\lambda(1)$ we get the statement in view of

$$C_\mu^\lambda(1) \sim \frac{1}{\Gamma(2\lambda)} \cdot (\mu + 1)^{2\lambda - 1},$$

which holds for $\mu \to \infty$, see (2.13) and (2.6). $\square$

2.6 Asymptotics of the Gegenbauer Zeros

In this section, the positive zeros of $C_\mu^\lambda(\cos\phi)$ are given, in more detailed notation, by

$$0 < \psi_{\mu,1}^\lambda < \psi_{\mu,2}^\lambda < \cdots.$$

Theorem 2.11 (Asymptotics of the Gegenbauer Zeros). *Let $\lambda = \alpha + \frac{1}{2} \geq 0$. Then $\lim_{\mu \to \infty} \mu \psi_{\mu,k}^\lambda = j_{\alpha,k}$ holds for every fixed $k \in \mathbb{N}$.*

Proof. Let $\lambda \geq 0$. First we want to prove that

$$\lim_{\mu \to \infty} \tilde{C}_{\mu-1}^{\lambda+1}\left(\cos\frac{\phi}{\mu}\right) = Z_{\alpha+1}(\phi) \tag{2.35}$$

holds uniformly on compact sets. So assume K is a compact set in $\mathbb{R}$, say $K \subset [-c, c]$ is valid for some $c > 0$. For $\phi \in K$ and $\mu \geq 2$ we get, using $\|\tilde{C}_{\mu-1}^{\lambda+1}\|_\infty = 1$ and Markov's inequality again,

$$\left|\tilde{C}_{\mu-1}^{\lambda+1}\left(\cos\frac{\phi}{\mu}\right) - \tilde{C}_{\mu-1}^{\lambda+1}\left(\cos\frac{\phi}{\mu-1}\right)\right| \leq (\mu - 1)^2 \cdot \left|\cos\frac{\phi}{\mu} - \cos\frac{\phi}{\mu-1}\right|$$

$$= 2(\mu - 1)^2 \left|\sin\frac{\phi}{\mu(\mu-1)} \cdot \sin\frac{(2\mu - 1)\phi}{\mu(\mu-1)}\right|$$

$$< \frac{4}{\mu} \cdot c^2.$$

Actually, in view of Theorem 2.8 this implies (2.35), uniformly for all $\phi \in K$. Next let $k \in \mathbb{N}$ be fixed. For $\epsilon > 0$ and $\kappa = 1, \ldots, k$ we define the intervals

$$I_\kappa(\epsilon) := \left[j_{\alpha,\kappa} - \epsilon, j_{\alpha,\kappa} + \epsilon \right],$$

and consider their union

$$U(\epsilon) := \bigcup_{\kappa=1}^{k} I_\kappa(\epsilon).$$

From (2.27) we get

$$Z_{\alpha+1}(j_{\alpha,\kappa}) \neq 0.$$

Therefore, an $\epsilon_0 > 0$ exists such that $Z_{\alpha+1}$ does not vanish on $U(\epsilon_0)$, where we may even assume ϵ_0 to be so small that

$$0 \leq j_{\alpha,1} - \epsilon_0 < j_{\alpha,1} + \epsilon_0 < j_{\alpha,2} - \epsilon_0 < \cdots < j_{\alpha,k} + \epsilon_0 < j_{\alpha,k+1} - \epsilon_0 \qquad (2.36)$$

holds. Because of (2.23) it follows that Z_α is monotonic on each of the intervals $I_1(\epsilon), \ldots, I_k(\epsilon)$ for $0 < \epsilon \leq \epsilon_0$.

Now let ϵ, $0 < \epsilon \leq \epsilon_0$, be arbitrary, but fixed. In view of the monotonicity of Z_α on $I_\kappa(\epsilon)$,

$$Z_\alpha(j_{\alpha,\kappa} - \epsilon) \cdot Z_\alpha(j_{\alpha,\kappa} + \epsilon) < 0$$

is valid for $\kappa = 1, \ldots, k$, and because of Theorem 2.8, some $\mu_1 \in \mathbb{N}$ exists such that $C_\mu^\lambda \left(\cos \frac{\phi}{\mu} \right)$ has a zero $\phi_{\mu,\kappa}^\lambda$ in the interior of $I_\kappa(\epsilon)$ for $\kappa = 1, \ldots, k$ and $\mu \geq \mu_1$, where

$$0 < \phi_{\mu,1}^\lambda < \phi_{\mu,2}^\lambda < \cdots < \phi_{\mu,k}^\lambda < j_{\alpha,k+1} - \epsilon_0 \qquad (2.37)$$

holds in view of (2.36).

Moreover, from (2.10) or (2.11), respectively, we obtain

$$\frac{d}{d\phi} C_\mu^\lambda \left(\cos \frac{\phi}{\mu} \right) = -\frac{2\bar{\lambda}}{\mu} \cdot \sin \frac{\phi}{\mu} \cdot C_\mu^{\lambda+1} \left(\cos \frac{\phi}{\mu} \right)$$

for $\mu \in \mathbb{N}$, and from (2.35) it follows that some $\mu_2 > \mu_1$ exists such that

$$C_\mu^{\lambda+1} \left(\cos \frac{\phi}{\mu} \right) \neq 0 \quad and \quad \frac{\phi}{\mu} < \pi$$

holds for $\phi \in U(\epsilon)$ and $\mu \geq \mu_2$. This implies that

$$\frac{d}{d\phi} C_\mu^\lambda \left(\cos \frac{\phi}{\mu} \right) \neq 0$$

is valid, again for $\phi \in U(\epsilon)$ and $\mu \geq \mu_2$. Therefore, $C_\mu^\lambda \left(\cos \frac{\phi}{\mu} \right)$ is monotonic on the intervals $I_\kappa(\epsilon)$, $\kappa = 1, \ldots, k$, for $\mu \geq \mu_2$, such that $\phi_{\mu,\kappa}^\lambda$ is the unique zero of $C_\mu^\lambda \left(\cos \frac{\phi}{\mu} \right)$ in the corresponding interval.

Next we consider the set

$$V := [0, j_{\alpha,k+1} - \epsilon_0] \setminus U(\epsilon).$$

Z_α does not vanish on the closure $\bar{V}$ of V. Again by Theorem 2.8 it follows that some $\mu_\epsilon \geq \mu_2$ exists such that $C_\mu^\lambda(\cos \frac{\phi}{\mu})$ does not vanish for $\mu \geq \mu_3$ on $\bar{V}$, and hence on V. It follows that all zeros of $C_\mu^\lambda(\cos \frac{\phi}{\mu})$ in $[0, j_{\alpha,k+1} - \epsilon_0]$ are already given by (2.37). This implies

$$\mu \psi_{\mu,\kappa}^\lambda = \phi_{\mu,\kappa}^\lambda \in I_\kappa^\lambda(\epsilon),$$

and hence

$$|\mu \psi_{\mu,\kappa}^\lambda - j_{\alpha,\kappa}| \leq \epsilon$$

for $\kappa = 1, \ldots, k$ and all $\mu \geq \mu_3$. In particular,

$$\lim_{\mu \to \infty} \mu \psi_{\mu,k}^\lambda = j_{\alpha,k}$$

holds for arbitrary $k \in \mathbb{N}$. $\qquad \square$

For further information on Gegenbauer polynomials we refer to *Szegö* [73] and to *Tricomi* [75]. A comprehensive work on Bessel functions is due to *Watson* [77]. The original work of L. B. Gegenbauer (1849–1903) is scattered over the last three decades of the nineteenth century. Some references can be found in [21].

2.7 Problems

Problem 2.1. For $\mu \in \mathbb{N}$ the following recurrence relations hold:

$$(\mu + 1)C_{\mu+1}^\lambda - 2(\mu + \lambda)xC_\mu^\lambda + (\mu + 2\lambda - 1)C_{\mu-1}^\lambda = 0, \quad 0 \neq \lambda \in \mathbb{R},$$
$$(\mu + 2\lambda)\tilde{C}_{\mu+1}^\lambda - 2(\mu + \lambda)x\tilde{C}_\mu^\lambda + \mu\tilde{C}_{\mu-1}^\lambda = 0, \quad \lambda > -\tfrac{1}{2},$$

where $C_0^\lambda = 1$, $C_1^\lambda(x) = 2\lambda x$ for $\lambda \neq 0$, $\tilde{C}_0^\lambda = 1$, $\tilde{C}_1^\lambda(x) = x$ for arbitrary $\lambda > -\tfrac{1}{2}$.

Problem 2.2. For arbitrary $\lambda \in \mathbb{R}$ and $\mu \in \mathbb{N}_0$,

$$(1 - x^2)\frac{d}{dx}C_\mu^\lambda = -\mu x \, C_\mu^\lambda + (\mu + 2\lambda - 1) \, C_{\mu-1}^\lambda$$

holds, where $C_{-1}^\lambda = 0$, $C_\mu^\lambda = C_\mu^\lambda(x)$, and so on.

Problem 2.3. For $\lambda \geq 0$, $\mu \in \mathbb{N}_0$, the coefficients in the expansion

$$x^\mu = \sum_{\nu=0}^{\lfloor \frac{\mu}{2} \rfloor} a_{\mu-2\nu} C_{\mu-2\nu}^\lambda(x)$$

are positive.

Problem 2.4. (Rodrigues Formula) Let $\lambda > -\frac{1}{2}$, $\mu \in \mathbb{N}_0$. Then

$$\tilde{C}_\mu^\lambda(x) = \frac{(-1)^\mu}{2^\mu(\lambda + \frac{1}{2})_\mu}(1 - x^2)^{-\lambda+\frac{1}{2}}\left(\frac{d}{dx}\right)^\mu(1 - x^2)^{\mu+\lambda-\frac{1}{2}}.$$

Problem 2.5. Prove that G^λ satisfies for $\lambda \neq 0$ the following partial differential equation:

$$(1 - x^2)G_{xx}^\lambda + z^2 G_{zz}^\lambda + (2\lambda + 1)[zG_z^\lambda - xG_x^\lambda] = 0.$$

Problem 2.6. Let $\alpha > \frac{1}{2}$, $j_{\alpha,0} := 0$ as above. Prove

$$j_{\alpha,k+1} - 2j_{\alpha,k} + j_{\alpha,k-1} < 0, \quad k \in \mathbb{N}.$$

Hint: Let $u(x) := x^{\frac{\alpha+1}{2}} Z_\alpha(x)$. u satisfies the differential equation

$$u'' + [1 + (\frac{1}{4} - \alpha^2)\frac{1}{x^2}]u = 0.$$

Compare the location of the zeros of u and of v, $v(x) := u(x + j_{\alpha,k-1} - j_{\alpha,k})$, $k \in \mathbb{N}$, in the interval $[j_{\alpha,k}, 2j_{\alpha,k} - j_{\alpha,k-1}]$ by means of *Sturm's method*, see Tricomi [75], p.175, for instance.

Part II

Approximation Means

Chapter 3

Multivariate Polynomials

The theory of multivariate polynomial approximation is characterized by a great variety of polynomials which can be used, and also by a great richness of geometric situations which occur. This chapter presents the most important facts on multivariate polynomials.

3.1 The Zoo of Multivariate Polynomials

Let $r \in \mathbb{N}$ be a fixed space dimension. Naturally, we are interested in the case $r \geq 2$, but do not exclude the case $r = 1$ in the beginning. The elements

$$m = (m_1, \ldots, m_r) \in \mathbb{Z}^r$$

are called *multiindex (multiindices)*. We define a *semi-order* on $\mathbb{Z}^r$ by the definitions

$$m \geq 0 \quad :\Longleftrightarrow \quad m \in \mathbb{N}_0^r,$$
$$m \geq n \quad :\Longleftrightarrow \quad m - n \geq 0,$$

$n, m \in \mathbb{Z}^r$. Note that, for convenience, multiindices are written horizontally. A *monomial* is an expression

$$M_m(x) := x^m := x_1^{m_1} x_2^{m_2} \cdots x_r^{m_r} \tag{3.1}$$

where $m \in \mathbb{N}_0^r$, i.e., $m \geq 0$, and where

$$x = (x_1, \ldots, x_r)' \in \mathbb{R}^r.$$

In addition to (3.1) we define

$$M_m(x) := x^m := 0 \quad \text{for } m \not\geq 0, \tag{3.2}$$

i.e., if $m_j < 0$ holds for one j at least. A *(real) polynomial in r variables* is a finite sum

$$P(x) = \sum c_m x^m, \quad c_m \in \mathbb{R}^r. \tag{3.3}$$

By $\mathbb{P}^r$ we denote the real linear space of all polynomials (3.3).

A polynomial is well defined by its *coefficients* $c_m = c_m(P)$, $m \geq 0$, which we complete by the definition $c_m = c_m(P) := 0$ for $m \not\geq 0$. The linear functionals c_m defined by

$$\mathbb{P}^r \ni P \mapsto c_m(P) \in \mathbb{R}, \quad m \in \mathbb{Z}^r,$$

are called *coefficient functionals*.

The following   subspaces of $\mathbb{P}^r$ are of particular interest:

$$\begin{aligned}
\mathbb{P}^r_\mu &:= \big\{ P \,|\, P(x) = \sum_{|m| \leq \mu} c_m x^m \big\}, \quad \mu \in \mathbb{N}_0, \\
\overset{*}{\mathbb{P}}{}^r_\mu &:= \big\{ P \,|\, P(x) = \sum_{|m| = \mu} c_m x^m \big\}, \quad \mu \in \mathbb{N}_0, \\
\mathbb{P}^r_m &:= \big\{ P \,|\, P(x) = \sum_{n \leq m} c_n x^n \big\}, \quad m \in \mathbb{N}_0^r, \\
\mathbb{H}^r &:= \big\{ P \,|\, P \in \mathbb{P}^r, \, \Delta P = 0 \big\},
\end{aligned}$$

where Δ is the *Laplace operator*, i.e., where

$$\Delta P = \frac{\partial}{\partial x_1^2} P + \frac{\partial}{\partial x_2^2} P + \cdots + \frac{\partial}{\partial x_r^2} P.$$

We agree that a polynomial space occurring with a subindex $\not\geq 0$ is always the null space $[0]$. To make the notation more convenient, we agree also that, consistent with (3.2), every polynomial occurring with a subindex $\not\geq 0$ is defined by 0.

A polynomial $P \in \mathbb{P}^r$ is said

     – to have *total degree* μ, if $P \in \mathbb{P}^r_\mu$, $\mu \in \mathbb{N}_0$,

     – to have *degree* m, if $P \in \mathbb{P}^r_m$, $m \in \mathbb{N}_0^r$,

     – to be *homogeneous of degree* μ, if $P \in \overset{*}{\mathbb{P}}{}^r_\mu$, $\mu \in \mathbb{N}_0$,

     – to be *harmonic*, if $P \in \mathbb{H}^r$.

A polynomial P is homogeneous of degree μ if and only if

$$P(\lambda x) = \lambda^\mu P(x) \tag{3.4}$$

holds for arbitrary $x \in \mathbb{R}^r$ and $\lambda \in \mathbb{R}$. By combination of the spaces defined above we obtain the additional spaces

$$\mathbb{H}^r_\mu := \mathbb{P}^r_\mu \cap \mathbb{H}^r, \quad \overset{*}{\mathbb{H}}{}^r_\mu := \overset{*}{\mathbb{P}}{}^r_\mu \cap \mathbb{H}^r,$$

again for $\mu \in \mathbb{N}_0$. In view of (3.4), the elements P of $\mathbb{P}^r_\mu$ have a uniquely determined representation

$$P(x) = \sum_{\nu=0}^{\mu} \sum_{|n|=\nu} c_n x^n.$$

This corresponds to the space decomposition

$$\mathbb{P}_\mu^r = \overset{*}{\mathbb{P}}_0^r \oplus \overset{*}{\mathbb{P}}_1^r \oplus \cdots \oplus \overset{*}{\mathbb{P}}_\mu^r. \tag{3.5}$$

In Section 4.1 it will be proved that

$$\mathbb{H}_\mu^r = \overset{*}{\mathbb{H}}_0^r \oplus \overset{*}{\mathbb{H}}_1^r \oplus \cdots \oplus \overset{*}{\mathbb{H}}_\mu^r \tag{3.6}$$

holds likewise. These structures define relations between the dimensions of the different spaces and show that the homogeneous and the homogeneous-harmonic polynomials play a basic role. We note that by the maximum principle or by (3.4), respectively, a harmonic or homogeneous polynomial which vanishes on S^{r-1} is always the *null polynomial*. This implies that

$$\|P\|_{S^{r-1}} = \max\left\{|P(x)| : x \in S^{r-1}\right\}$$

defines a *norm* on each of the spaces $\mathbb{H}^r$, $\overset{*}{\mathbb{P}}_\mu^r$ and $\overset{*}{\mathbb{H}}_\mu^r$.

In order to determine some dimensions we define the mapping $\mathbb{P}_\mu^r \to \overset{*}{\mathbb{P}}_\mu^{r+1}$ by

$$\mathbb{P}_\mu^r \ni P_\mu \mapsto \overset{*}{P}_\mu \in \overset{*}{\mathbb{P}}_\mu^{r+1},$$

$$\overset{*}{P}_\mu(x_1, \ldots, x_r, x_{r+1}) := x_{r+1}^\mu P_\mu\left(\tfrac{x_1}{x_{r+1}}, \ldots, \tfrac{x_r}{x_{r+1}}\right).$$

The mapping is linear and because of $P_\mu(x_1, \ldots, x_r) = \overset{*}{P}_\mu(x_1, \ldots, x_r, 1)$ bijective. This means that the spaces $\mathbb{P}_\mu^r$ and $\overset{*}{\mathbb{P}}_\mu^{r+1}$ are isomorphic,

$$\mathbb{P}_\mu^r \cong \overset{*}{\mathbb{P}}_\mu^{r+1}. \tag{3.7}$$

We denote their common dimension by

$$M_\mu^r := \dim \mathbb{P}_\mu^r = \dim \overset{*}{\mathbb{P}}_\mu^{r+1}.$$

From (3.5) we get the *recurrence relation*

$$M_\mu^r = M_0^{r-1} + M_1^{r-1} + \cdots + M_\mu^{r-1}$$

for $r \in \mathbb{N} \setminus \{1\}$, $\mu \in \mathbb{N}_0$, with the *initial values*

$$M_\mu^1 = \mu + 1,$$

which together have a unique solution. It is easy to confirm that this is given by the binomial coefficient $\binom{\mu+r}{r}$, such that we obtain

$$\dim \mathbb{P}^r_\mu = \binom{\mu+r}{r}, \tag{3.8}$$

$$\dim \overset{*}{\mathbb{P}}{}^r_\mu = \binom{\mu+r-1}{r-1} \tag{3.9}$$

for $r \in \mathbb{N}$, $\mu \in \mathbb{N}_0$.

The dimensions of the corresponding harmonic spaces are determined later, see Corollary 3.11.

Homogenisation

Above we mapped *r-variate* polynomials onto $(r+1)$ *-variate* homogeneous polynomials. In what follows we discuss a similar map where, however, the image is again *r-variate*. It is defined for *even* and for *odd* polynomials, only, which depends on μ being *even* or *odd*, respectively. So let us define the additional spaces

$$\mathbb{Q}^r_\mu := \{Q \in \mathbb{P}^r_\mu \mid Q(-x) = (-1)^\mu Q(x), \ x \in \mathbb{R}^r\}$$

for $r \in \mathbb{N}$, $\mu \in \mathbb{N}_0$. Then we get

$$\overset{*}{\mathbb{P}}{}^r_\mu \subset \mathbb{Q}^r_\mu$$

and

$$\mathbb{P}^r_\mu = \mathbb{Q}^r_\mu \oplus \mathbb{Q}^r_{\mu-1},$$

with $\mathbb{Q}^r_{-1} := [0]$. The elements Q of $\mathbb{Q}^r_\mu$ have a representation

$$Q(x) = \sum_{\nu=0}^{\lfloor \frac{\mu}{2} \rfloor} \sum_{|n|=\mu-2\nu} c_n \, x^n$$

with real coefficients c_n. It follows that

$$|x|^\mu Q(\tfrac{x}{|x|}) = \sum_{\nu=0}^{\lfloor \frac{\mu}{2} \rfloor} \sum_{|n|=\mu-2\nu} c_n \, (x_1^2 + \cdots + x_r^2)^\nu x^n$$

is an element of $\overset{*}{\mathbb{P}}{}^r_\mu$, which permits the following definition.

Definition 3.1 (Operator of Homogenisation). *Let* $r \in \mathbb{N} \setminus \{1\}$, $\mu \in \mathbb{N}_0$. *The operator of homogenisation* $\overset{*}{\cdot}: \mathbb{Q}^r_\mu \to \overset{*}{\mathbb{P}}{}^r_\mu$ *is defined by*

$$Q \mapsto \overset{*}{Q}, \quad \overset{*}{Q}(x) := |x|^\mu Q(\tfrac{x}{|x|}).$$

Multinomial Coefficients

The polynomial $(x_1 + x_2 + \cdots + x_r)^\mu$, $\mu \in \mathbb{N}_0$, is an element of $\overset{*}{\mathbb{P}}{}^r_\mu$. The expansion

$$(x_1 + x_2 + \cdots + x_r)^\mu = \sum_m \binom{\mu}{m} x^m, \tag{3.10}$$

defines the *multinomial coefficients* $\binom{\mu}{m}$ for $r \in \mathbb{N}$, $m \in \mathbb{N}_0^r$. It is easy to see that the usual *binomial coefficients* $\binom{\mu}{\nu}$, $\mu, \nu \in \mathbb{N}_0$, $0 \le \nu \le \mu$, have to be identified with

$$\binom{\mu}{\nu} = \binom{\mu}{\nu, \mu - \nu}.$$

The multinomial coefficients satisfy

$$\binom{\mu}{m} = \frac{\mu!}{m!} \tag{3.11}$$

for $m \ge 0$, $|m| = \mu \in \mathbb{N}_0$, where $m!$ is defined by

$$m! := m_1! \cdots m_r! \tag{3.12}$$

for $m \in \mathbb{N}_0^r$. Moreover, by our agreement from above we get

$$\binom{\mu}{m} = 0 \quad \text{for } m \not\ge 0 \text{ or } |m| \ne \mu. \tag{3.13}$$

The Homogeneous Polynomials

Bounds for the Coefficients of a Homogeneous Polynomial

We agreed in defining $c_m = c_m(P)$ for $m \not\ge 0$ by *zero*. Often this allows us to omit the indication that m is in $\mathbb{N}_0^r$. For the coefficients of the homogeneous polynomials the following bounds are known.

Theorem 3.1 (Bound on $|c_m(P)|$, Kellogg). *Assume* $P \in \overset{*}{\mathbb{P}}{}^r_\mu$, $r \in \mathbb{N} \setminus \{1\}$, $\mu \in \mathbb{N}_0$, *is bounded by* $\|P\|_{S^{r-1}} \le 1$. *Then*

$$|c_m(P)| \le \binom{\mu}{m} \tag{3.14}$$

holds for $|m| = \mu$. *In the case* $r = 2$ *all bounds are lowest possible.*

Proof. First we consider the case $r = 2$. So let

$$P(\xi, \eta) = \sum_{\nu=0}^{\mu} a_\nu \, \xi^{\mu-\nu} \eta^\nu,$$

and assume in the beginning that even the stronger condition

$$\|P\|_{S^{r-1}} < 1 \tag{3.15}$$

holds. Each coefficient occurs either with

$$\tfrac{1}{2}[P(\xi,\eta) + P(\xi,-\eta)] = a_0\xi^\mu + a_2\xi^{\mu-2}\eta^2 + \cdots ,$$

or with

$$\tfrac{1}{2}[P(\xi,\eta) - P(\xi,-\eta)] = a_1\xi^{\mu-1}\eta + a_3\xi^{\mu-3}\eta^3 + \cdots ,$$

where both homogeneous polynomials satisfy the norm condition put onto P. Therefore it suffices to investigate the coefficients of homogeneous polynomials which are either even or odd with respect to the second argument.

First we assume P to be even in the second argument. We write it in the form

$$P(\xi,\eta) = b_0\xi^\mu - b_2\xi^{\mu-2}\eta^2 + b_4\xi^{\mu-4}\eta^4 \mp \cdots .$$

There are two particular polynomials of this kind, namely

$$A(\xi,\eta) = \Re(\xi + i\eta)^\mu = \binom{\mu}{0}\xi^\mu - \binom{\mu}{2}\xi^{\mu-2}\eta^2 + \binom{\mu}{4}\xi^{\mu-4}\eta^4 \mp \cdots \tag{3.16}$$

and

$$
\begin{aligned}
A(\xi,\eta) - P(\xi,\eta) &= c_0\xi^\mu - c_2\xi^{\mu-2}\eta^2 + c_4\xi^{\mu-4}\eta^4 \mp \cdots \\
&= \eta^\mu\Big\{c_0\big(\tfrac{\xi}{\eta}\big)^\mu - c_2\big(\tfrac{\xi}{\eta}\big)^{\mu-2} + c_4\big(\tfrac{\xi}{\eta}\big)^{\mu-4} \mp \cdots \Big\}.
\end{aligned}
$$

Note that $c_0 = A(1,0) - P(1,0) > 0$ holds because of (3.15), and that

$$A(\cos\phi, \sin\phi) = \cos\mu\phi$$

attains the values $+1$, -1 alternatingly $(\mu+1)$-times as ϕ runs from 0 to π. Likewise $A(\cos\phi, \sin\phi) - P(\cos\phi, \sin\phi)$ alternates $(\mu+1)$-times and hence it has μ roots in the interval $(0,\pi)$. This implies that the nonzero polynomial

$$c_0 z^\mu - c_2 z^{\mu-2} + c_4 z^{\mu-4} \mp \cdots ,$$

which is even or odd, respectively, vanishes μ-times while $z = \cot\phi$ runs from $+\infty$ to $-\infty$. So all of its roots are either zero, or occurring in pairs $(\xi, -\xi)$, $\xi \in \mathbb{R}$, and so it can be written in the form $z^\epsilon q(-z^2)$ where all roots of the polynomial q are negative. This implies that the coefficients of q do not change their signs. Because of $c_0 > 0$ this implies

$$c_0 \geq 0,\ c_2 \geq 0,\ \ldots ,$$

and hence

$$\binom{\mu}{2\nu} - b_{2\nu} \geq 0,\ \text{for } \nu = 0, 1, \ldots, \left\lfloor \frac{\mu}{2} \right\rfloor .$$

Now we replace P by $-P$, and obtain likewise

$$\binom{\mu}{2\nu} + b_{2\nu} \geq 0, \text{ for } \nu = 0, 1, \ldots, \left\lfloor \frac{\mu}{2} \right\rfloor.$$

Together with $|a_{2\nu}| = |b_{2\nu}|$ this yields

$$|a_{2\nu}| \leq \binom{\mu}{2\nu}, \text{ for } \nu = 0, 1, \ldots, \left\lfloor \frac{\mu}{2} \right\rfloor.$$

Next we assume that P is odd with respect to the second argument. Then we get the estimates

$$|a_{2\nu+1}| \leq \binom{\mu}{2\nu+1}, \text{ for } \nu = 0, 1, \ldots, \left\lfloor \frac{\mu-1}{2} \right\rfloor,$$

by a similar comparison of the polynomials

$$P(\xi, \eta) = b_1 \xi^{\mu-1} \eta - b_3 \xi^{\mu-3} \eta^3 + b_5 \xi^{\mu-5} \eta^5 \mp \cdots$$

and

$$B(\xi, \eta) = \Im(\xi + i\eta)^\mu = \binom{\mu}{1} \xi^{\mu-1} \eta - \binom{\mu}{3} \xi^{\mu-3} \eta^3 + \binom{\mu}{5} \xi^{\mu-5} \eta^5 \mp \cdots, \quad (3.17)$$

where

$$B(\cos\phi, \sin\phi) = \sin\mu\phi.$$

Together this proves (3.14) under the stricter assumption (3.15). However, by a continuity argument, (3.14) remains valid under the original assumption, and the statement of the theorem holds in the case $r = 2$.

The remaining is proved by mathematical induction with respect to r. So we assume that the statement of Theorem 3.1 is true for the dimensions $2, \ldots, r-1$. This is valid for $r = 3$. Next assume $P \in \overset{*}{\mathbb{P}}{}^r_\mu$, say

$$P(x) = \sum_{|m|=\mu} c_m x^m,$$

satisfies the assumption

$$\|P\|_{S^{r-1}} \leq 1.$$

For $x \in \mathbb{R}^r$ we define $\bar{x} := (x_1, \ldots, x_{r-1})'$. Writing it in the form

$$\bar{x} = \xi\bar{y}, \ \bar{y} \in S^{r-2},$$

we get, with $m = (\bar{m}, m_r)$,

$$P(\xi\bar{y}, \eta) = \sum_{m_r=0}^{\mu} \left(\sum_{|\bar{m}|=\mu-m_r} c_{\bar{m},m_r} \bar{y}^{\bar{m}} \right) \xi^{\mu-m_r} \eta^{m_r}.$$

For fixed $\bar{y} \in S^{r-2}$ we obtain $(\xi\bar{y})^2 + \eta^2 = \xi^2 + \eta^2$, such that $P(\xi\bar{y}, \eta)$, which is a homogeneous polynomial in the two variables ξ and η, is bounded in absolute value by unity for $(\xi, \eta) \in S^1$. So the homogeneous polynomials

$$\sum_{|\bar{m}|=\mu-m_r} c_{\bar{m},m_r}\bar{y}^{\bar{m}},$$

which occur as coefficients of $P(\xi\bar{y}, \eta)$, are bounded in absolute value by $\binom{\mu}{\mu-m_r,m_r}$, and by our assumption with respect to the dimensions $2, \ldots, (r-1)$, we get

$$|c_m| \le \binom{\mu}{\mu-m_r,m_r}\binom{\mu-m_r}{\bar{m}} = \binom{\mu}{m}$$

for $|m| = \mu$, where we used (3.11). Hence the statement holds also for the dimension r, and mathematical induction finishes the proof of the inequality (3.14). In the case $r = 2$ every bound is attained either for $P = A$, or for $P = B$, respectively. $\square$

We remark that for $r \ge 2$ every $P \in \overset{*}{\mathbb{P}}{}^r_\mu$ can be written in the form

$$P(x) = \sum_{\nu=0}^{\mu} q_{\mu-\nu}(x_2, \ldots, x_r)x_1^\nu \tag{3.18}$$

where the $q_\kappa(x_2, \ldots, x_r)$ are homogeneous polynomials in the variables $x_2, \ldots, x_r$ of degree κ. For $\mu \ge 1$ we can write (3.18) also in the form

$$P(x) = a_1 x_1^\mu + \mu(a_2 x_2 + \cdots + a_r x_r)x_1^{\mu-1} + \sum_{\nu=0}^{\mu-2} q_{\mu-\nu}(x_2, \ldots, x_r)x_1^\nu \tag{3.19}$$

with uniquely determined $a_1, \ldots, a_r \in \mathbb{R}$, which we collect in the vector

$$a := (a_1, \ldots, a_r)' \in \mathbb{R}^r. \tag{3.20}$$

Using this notation, we can supply Theorem 3.1 with the following theorem.

Theorem 3.2 (Leading Coefficients of $P \in \overset{*}{\mathbb{P}}{}^r_\mu$). $P \in \overset{*}{\mathbb{P}}{}^r_\mu, r \ge 2, \mu \in \mathbb{N}$, $\|P\|_{S^{r-1}} \le 1$ implies $|a|^2 = a_1^2 + \cdots + a_r^2 \le 1$.

Proof. (a) First let $r = 2$. Then $P(x)$ takes the form

$$P(x) = a_1 x_1^\mu + \mu a_2 x_1^{\mu-1} x_2 + \sum_{\nu=0}^{\mu-2} b_\nu x_1^\nu x_2^{\mu-\nu},$$

with real coefficients $b_0, \ldots, b_{\mu-2}$. Now we assume that $|a| > 1$ holds. Using the polynomials (3.16) and (3.17) we put $F(x) := a_1 A(x) + a_2 B(x)$, and get

$$F(x) - P(x) = \sum_{\nu=0}^{\mu-2} c_\nu x_1^\nu x_2^{\mu-\nu} = x_2^\mu \cdot q\Big(\frac{x_1}{x_2}\Big)$$

with some real coefficients $c_0, \ldots, c_{\mu-2}$, and hence with some real polynomial $q \in \mathbb{P}^1_{\mu-2}$. Next we write a in the form

$$a = |a|(\cos \epsilon, \sin \epsilon)', \quad 0 \le \epsilon < 2\pi.$$

Then we obtain

$$F(\cos \phi, \sin \phi) = |a| \cos(\mu\phi - \epsilon),$$

and so $F(\cos \phi, \sin \phi)$ attains, alternatingly, $(\mu + 1)$-times the values $\pm|a|$ while ϕ is running from $\frac{\epsilon}{\mu}$ to $\frac{\epsilon}{\mu} + \pi$. In view of the assumption on P this implies that

$$F(\cos \phi, \sin \phi) - P(\cos \phi, \sin \phi) = (\sin \phi)^\mu \cdot q(\cot \phi)$$

changes $(\mu+1)$-times its sign. So there are μ zeros in the open interval $(\frac{\epsilon}{\mu}, \frac{\epsilon}{\mu} + \pi)$, where $\sin \phi$ vanishes at most once. This implies that $q(\cot \phi)$ has at least $\mu - 1$ zeros in this interval at a finite value of $\cot \phi$. It follows that q itself has $\mu - 1$ zeros, at least, and so it vanishes. This yields $F(x) - P(x) = 0$, and hence

$$P(\cos \phi, \sin \phi) = |a| \cos(\mu\phi - \epsilon).$$

Again by the assumption on P we obtain

$$1 \ge \|P\|_{S^{r-1}} = |a| > 1,$$

but this is a contradiction. So our assumption on $|a|$ was false, and the statement of Theorem 3.2 is valid for $r = 2$.

(b) Next let $r \ge 3$, let $\bar{a} := (a_2, \ldots, a_r)'$, and choose the orthogonal matrix A such that

$$e_1 = Ae_1, \quad \begin{pmatrix} 0 \\ \bar{a} \end{pmatrix} = |\bar{a}|Ae_2$$

holds. Obviously, $x_1 = e_1'x$ is invariant under the substitution $Ax \to x$, and from (3.19) we get

$$P(Ax) = a_1 x_1^\mu + \mu|\bar{a}|x_1^{\mu-1}x_2 + \sum_{\nu=0}^{\mu-2} \bar{q}_{\mu-\nu}(x_2, \ldots, x_r)x_1^\nu,$$

where the $\bar{q}_{\mu-\nu}(x_2, \ldots, x_r)$ are homogeneous polynomials of degree $\mu - \nu$, again. Now let $F(x) := P(Ax)$. $F(x)$ is a homogeneous polynomial of degree μ and satisfies $\|F\|_{S^{r-1}} = \|P\|_{S^{r-1}} \le 1$. Moreover, a_1 and $|\bar{a}|$ occur as the particular coefficients in the bivariate homogeneous polynomial

$$F(x_1, x_2, 0, \ldots, 0) = a_1 x^\mu + \mu|\bar{a}|x_1^{\mu-1}x_2 + \sum_{\nu=0}^{\mu-2} \bar{q}_{\mu-\nu}(x_2, 0, \ldots, 0)x_1^\nu$$

of degree μ, which satisfies

$$|F(x_1, x_2, 0, \ldots, 0)| \le 1 \quad \text{for } x_1^2 + x_2^2 = 1.$$

By the result from **(a)** this yields

$$|a|^2 = a_1^2 + |\bar{a}|^2 \leq 1,$$

as claimed. ◻

Remark. For $P \neq 0$ the norm of $P/\|P\|_{S^{r-1}}$ is unity, and Theorem 3.2 yields $|a|/\|P\|_{S^{r-1}} \leq 1$, i.e.,

$$|a| \leq \|P\|_{S^{r-1}}. \tag{3.21}$$

For the original work see Kellogg [27].

The Gradient of a Homogeneous Polynomial

Theorem 3.3 (Euler's Partial Differential Equation). *Every $P \in \overset{*}{\mathbb{P}}{}^r_\mu$, $\mu \in \mathbb{N}_0$, $r \geq 2$, satisfies Euler's partial differential equation*

$$x' \operatorname{grad} P - \mu P = 0.$$

Proof. From $P(x) = \sum\limits_{|m|=\mu} c_m x^m$ we obtain

$$x'(\operatorname{grad} P)(x) = \sum_{\nu=1}^r \sum_{|m|=\mu} m_\nu c_m x^m = \sum_{|m|=\mu} \left(\sum_{\nu=1}^r m_\nu \right) c_m x^m = \mu P(x). \qquad ◻$$

Lemma 3.4. *Let $P \in \overset{*}{\mathbb{P}}{}^r_\mu$, $\mu \in \mathbb{N}_0$, $r \geq 2$, and let A be an orthogonal $(r \times r)$-matrix. Then*

$$|(\operatorname{grad} P)(Ax)| = |(\operatorname{grad} P_A)(x)|$$

holds for $x \in S^{r-1}$.

Proof. The statement follows immediately from $(\operatorname{grad} P_A)(x) = A'(\operatorname{grad} P)(Ax)$. ◻

Theorem 3.5 (Gradient of $P \in \overset{*}{\mathbb{P}}{}^r_\mu$). *Let $P \in \overset{*}{\mathbb{P}}{}^r_\mu$, $\mu \in \mathbb{N}_0$, $r \geq 2$, and define $G(x) := |(\operatorname{grad} P)(x)|^2$. Then we get*

$$\|G\|_{S^{r-1}} = \mu^2 \cdot \|P\|_{S^{r-1}}^2.$$

Proof. For $\mu = 0$ the statement is obvious. Next assume $\mu \geq 1$.

(a) First we want to prove

$$|(\operatorname{grad} P)(x)|^2 \leq \mu^2 \|P\|_{S^{r-1}}^2 \tag{3.22}$$

for arbitrary $x \in S^{r-1}$. Let us represent x in the form $x = Ae_1$, where A is a properly chosen orthogonal matrix. Because of $\|P\|_{S^{r-1}} = \|P_A\|_{S^{r-1}}$ we have to prove, equivalently,

$$|(\operatorname{grad} P_A)(e_1)|^2 \leq \mu^2 \|P_A\|_{S^{r-1}}^2.$$

P_A is again an element of $\overset{*}{\mathbb{P}}^r_\mu$, so it suffices to prove

$$|(grad\,P)(e_1)|^2 \leq \mu^2 \|P\|^2_{S^{r-1}} \qquad (3.23)$$

for arbitrary $P \in \overset{*}{\mathbb{P}}^r_\mu$. To this end we use again the representation (3.19). It follows that

$$\left(\tfrac{\partial}{\partial x_1}P\right)(e_1) \;=\; \mu\,q_0(0) \;=\; \mu a_1,$$
$$\left(\tfrac{\partial}{\partial x_\nu}P\right)(e_1) \;=\; \left(\tfrac{\partial}{\partial x_\nu}q_1\right)(0) \;=\; \mu a_\nu \text{ for } \nu = 2,\ldots,r,$$

and hence

$$(grad\,P)(e_1) = \mu a.$$

Together with (3.21) this yields

$$|(grad\,P)(e_1)|^2 = \mu^2 a^2 \leq \mu^2 \|P\|^2_{S^{r-1}},$$

which implies that (3.22) is valid for arbitrary $x \in S^{r-1}$. This is the same as

$$\|G\|_{S^{r-1}} \leq \mu^2 \|P\|^2_{S^{r-1}}. \qquad (3.24)$$

(b) Next we choose $x \in S^{r-1}$ as an extreme point, i.e., such that $|P(x)| = \|P\|^2_{S^{r-1}}$ holds. Then *Lagrange's optimality condition*

$$(grad\,P)(x) = \lambda x$$

is satisfied with some $\lambda \in \mathbb{R}$. Together with Theorem 3.3 this yields

$$\lambda = x'(grad\,P)(x) = \mu\,P(x)$$

and

$$G(x) = \mu^2 P^2(x) = \mu^2 \|P\|^2_{S^{r-1}}.$$

So we get

$$\|G\|_{S^{r-1}} \geq \mu^2 \|P\|^2_{S^{r-1}}.$$

Together with (3.24) this finishes the proof. $\qquad\qquad\square$

Remark. For $F \in C(D)$, $\emptyset \neq D \subset \mathbb{R}^r$ compact, the *set of extreme points* is defined by

$$\mathcal{E}(F) := \{x \in D|\; |F(x)| = \|F\|_D\}.$$

It is easy to see that $\mathcal{E}(P) \subset \mathcal{E}(G)$ holds for $P \in \overset{*}{\mathbb{P}}^r_\mu$ in the situation of Theorem 3.5. But even $\mathcal{E}(P) = \mathcal{E}(G)$ is valid, except for particular polynomials P. For details we refer to Hakopian [24] and to Reimer [52].

The Harmonic Homogeneous Polynomials

We investigate the conditions on the coefficients of a homogeneous polynomial which make it become harmonic. Note that $\overset{*}{\mathbb{H}}{}^r_\mu = \overset{*}{\mathbb{P}}{}^r_\mu$ holds for $\mu \in \{0,1\}$ such that only the case $\mu \geq 2$ is interesting.

Let $r \geq 2$ and $\mu \in \mathbb{N}_0$ be fixed, and define the index set

$$\mathcal{M} := \mathcal{M}(r,\mu) := \{m \in \mathbb{N}_0^r : |m| = \mu\}. \tag{3.25}$$

The homogeneous polynomials of degree μ are characterized by their *coefficient vector*

$$c := (c_m)_{m \in \mathcal{M}}.$$

Now let us introduce the partial differential operators

$$D^n = \left(\frac{\partial}{\partial x_1}\right)^{n_1} \cdots \left(\frac{\partial}{\partial x_r}\right)^{n_r}$$

for $n \in \mathbb{N}_0^r$. They are *biorthogonal* with the monomials in the following sense,

$$D^n x^m = m!\, \delta_{m,n}$$

holds for $m, n \in \mathcal{M}$, where $\delta_{m,n}$ is the *Kronecker symbol*. Restricted to homogeneous polynomials

$$P(x) = \sum_{n \in \mathcal{M}} c_n(P)x^n$$

this yields

$$c_m(P) = \frac{1}{m!} D^m P \text{ for } m \in \mathcal{M}. \tag{3.26}$$

This is a realisation of the map $P \mapsto c = c(P)$ of the elements P of $\overset{*}{\mathbb{P}}{}^r_\mu$ onto their coefficient vector $c := (c_m)_{m \in \mathcal{M}}$, and the following theorem holds.

Theorem 3.6 (Characterisation of the Null Element in $\overset{*}{\mathbb{P}}{}^r_\mu$). *Let $P \in \overset{*}{\mathbb{P}}{}^r_\mu$, $r \geq 2$, $\mu \in \mathbb{N}_0$. Then $P = 0$ is valid if and only if $D^m P = 0$ holds for all $m \in \mathcal{M}(r,\mu)$.*

Proof. The map onto the coefficients is bijective. $\quad\square$

Next let $P \in \overset{*}{\mathbb{P}}{}^r_\mu$, $\mu \geq 2$, which implies $\Delta P \in \overset{*}{\mathbb{P}}{}^r_{\mu-2}$. Because of Theorem 3.6, P is harmonic if and only if

$$D^n(\Delta P) = 0 \tag{3.27}$$

holds for all $n \in \mathbb{N}_0^r$ with $|n| = \mu - 2$. But (3.27) is equivalent to

$$\sum_{\nu=1}^r D^{n+2e_\nu} P = 0,$$

and because of (3.26) also with

$$\sum_{\nu=1}^{r}(n+2e_\nu)!\,c_{n+e_\nu} = 0, \tag{3.28}$$

holding for all $n \in \mathbb{N}_0^r$ with $|n| = \mu - 2$, where we used the abbreviation $c_m = c_m(P)$. In order to make this equation more significant, we define the *normalized coefficients*

$$\tilde{c}_m := \binom{|m|}{m}^{-1} c_m$$

for $m \in \mathbb{N}_0^r$, and $\tilde{c}_m := 0$ for $m \not\geq 0$. Actually, in this notation the following theorem is valid.

Theorem 3.7 (Difference Equation of the Coefficients of $H \in \overset{*}{\mathbb{H}}{}_\mu^r$). $P \in \overset{*}{\mathbb{P}}{}_\mu^r$, $r \geq 2$, $\mu \in \mathbb{N}_0$ *is harmonic if and only if the normalized coefficients* $\tilde{c}_m = \tilde{c}_m(H)$ *satisfy the multivariate difference equation*

$$\sum_{\nu=1}^{r}\tilde{c}_{n+2e_\nu} = 0 \tag{3.29}$$

for all $n \in \mathbb{N}_0^r$ which satisfy $|n| = \mu - 2$.

Proof. For $\mu \in \{0,1\}$ the statement is evident. Next let $\mu \geq 2$. Then (3.29) is equivalent to (3.28), (3.27), and hence with P being harmonic. $\square$

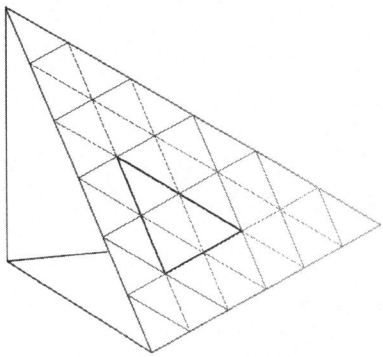

Figure 3.1. Difference Equation Lattice. Vertices of Marked Triangle contribute to $\sum \tilde{c}_m = 0$.

In view of (3.29), the restrictions of the coefficient functionals onto $\overset{*}{\mathbb{H}}{}^r_\mu$ are, in general, linearly dependent. In particular, we can write (3.29) for $m = n+2e_r \geq 2e_r$ in the explicit form

$$\tilde{c}_m = -\sum_{\nu=1}^{r-1} \tilde{c}_{m+2(e_\nu - e_r)} = -\sum_{\nu=1}^{r-1} \tilde{c}_{\bar{m}+2\bar{e}_\nu, m_r-2}, \tag{3.30}$$

with $\bar{n} := (n_1, \ldots, n_{r-1})$ for $n \in \mathbb{Z}^r$, again. For $m_r \geq 2$ this formula allows us to reduce the last index by two, step by step, by a linear combination of the coefficients, until the index m_r takes the value 0 or 1, respectively. So we get the following lemma.

Lemma 3.8 (Reduction of the Coefficients in $\overset{*}{\mathbb{H}}{}^r_\mu$). *Let $r \geq 2$, $\mu \geq 2$, $H \in \overset{*}{\mathbb{H}}{}^r_\mu$. The normalized coefficients $\tilde{c}_m = \tilde{c}_m(H)$ of H satisfy the reduction formula*

$$\tilde{c}_m = (-1)^\lambda \sum_{|\bar{n}|=\lambda} \binom{\lambda}{\bar{n}} \tilde{c}_{\bar{m}+2\bar{n}, m_r-2\lambda} \tag{3.31}$$

for $m \in \mathbb{N}_0^r$, $|m| = \mu$, $\lambda = 0, 1, \ldots, \lfloor \frac{m_r}{2} \rfloor$.

Proof. We use mathematical induction with respect to λ. For $\lambda = 0$ equation (3.31) is trivial and true. Next assume $0 \leq \lambda < \lfloor \frac{m_r}{2} \rfloor$, which implies $m_r \geq 2$. All terms on the right side of (3.31) can be reduced by the help of (3.30). This yields

$$\tilde{c}_m = (-1)^{\lambda+1} \sum_{\nu=1}^{r-1} \sum_{|\bar{n}|=\lambda} \binom{\lambda}{\bar{n}} \tilde{c}_{\bar{m}+2\bar{n}+2\bar{e}_\nu, m_r-2\lambda-2}.$$

Replacing in the inner sum $\bar{n} + \bar{e}_\nu$ by $\bar{n}$ we obtain

$$\tilde{c}_m = (-1)^{\lambda+1} \sum_{\nu=1}^{r-1} \sum_{|\bar{n}|=\lambda+1} \binom{\lambda}{\bar{n}-\bar{e}_\nu} \tilde{c}_{\bar{m}+2\bar{n}, m_r-2(\lambda+1)}.$$

For $|\bar{n}| = \lambda + 1$ we get

$$\sum_{\nu=1}^{r-1} \binom{\lambda}{\bar{n}-\bar{e}_\nu} = \sum_{\nu=1}^{r-1} \frac{\lambda! \, n_\nu}{\bar{n}!} = \frac{(\lambda+1)!}{\bar{n}!},$$

and hence

$$\sum_{\nu=1}^{r-1} \binom{\lambda}{\bar{n}-\bar{e}_\nu} = \binom{\lambda+1}{\bar{n}}. \tag{3.32}$$

This is itself an interesting formula on multinomial coefficients. Changing now the order of summation, and inserting (3.32), we see that (3.31) is valid also for $\lambda+1$, instead of λ, and mathematical induction finishes the proof. $\quad\square$

In what follows we make also use of the index set

$$\mathcal{I} := \mathcal{I}(r,\mu) := \{m \in \mathcal{M}(r,\mu) \mid m_r \in \{0,1\}\}. \tag{3.33}$$

For $\lambda := \lfloor \frac{m_r}{2} \rfloor$, Lemma 3.8 yields, in particular, that the coefficients of a harmonic homogeneous polynomial of degree μ satisfy the equations

$$\tilde{c}_m = (-1)^{\lfloor \frac{m_r}{2} \rfloor} \sum_{|\bar{n}| = \lfloor \frac{m_r}{2} \rfloor} \binom{|\bar{n}|}{\bar{n}} \tilde{c}_{\bar{m}+2\bar{n}, m_r - 2\lfloor \frac{m_r}{2} \rfloor} \tag{3.34}$$

for $m \in \mathcal{M}$. So the complete coefficient vector $c = (c_m)_{m \in \mathcal{M}}$ is determined already by the *initial vector* $(c_m)_{m \in \mathcal{I}}$. Moreover, the following theorem holds.

Theorem 3.9 (Solution of the Difference Equation). *For every arbitrarily given initial vector* $(c_m)_{m \in \mathcal{I}}$ *the difference equation (3.29) has a uniquely determined solution* $(c_m)_{m \in \mathcal{M}}$.

Proof. If $(c_m)_{m \in \mathcal{M}}$ solves (3.29), then (3.34) holds by the arguments from above, and the solution is uniquely determined by the initial vector. Next we assume the initial vector to be given. We have to prove that it determines a solution of (3.29). To this end we complete the vector $(c_m)_{m \in \mathcal{M}}$ by defining the numbers

$$\tilde{c}_m := (-1)^\lambda \sum_{|\bar{n}| = \lambda} \binom{\lambda}{\bar{n}} \tilde{c}_{\bar{m}+2\bar{n}, m_r - 2\lambda}, \tag{3.35}$$

for $m \in \mathcal{M} \setminus \mathcal{I}$, where we put $\lambda := \lfloor \frac{m_r}{2} \rfloor$, here and in the following. Next let $l \in \mathbb{N}_0^r$, where $|l| = \mu - 2$. For abbreviation we put $m := l + 2e_r$. Note that this implies $m_r \geq 2$. By replacing the multinomial coefficient in (3.35) by means of (3.32), and changing the order of summation, we get with λ as above

$$\tilde{c}_m = -\sum_{\nu=1}^{r-1} (-1)^{\lambda-1} \sum_{|\bar{n}| = \lambda} \binom{\lambda-1}{\bar{n} - \bar{e}_\nu} \tilde{c}_{\bar{m}+2\bar{n}, m_r - 2\lambda}.$$

Now we replace $\bar{n}$ by $\bar{n} + \bar{e}_\nu$ and get

$$\tilde{c}_m = -\sum_{\nu=1}^{r-1} (-1)^{\lambda-1} \sum_{|\bar{n}| = \lambda-1} \binom{\lambda-1}{\bar{n}} \tilde{c}_{\bar{m}+2\bar{e}_\nu+2\bar{n}, m_r - 2 - 2(\lambda-1)}.$$

Using again the definition (3.35), we bring this to the form

$$\tilde{c}_m = -\sum_{\nu=1}^{r-1} \tilde{c}_{\bar{m}+2\bar{e}_\nu, m_r - 2}.$$

Here we replace m by $l + 2e_r$, and obtain

$$\sum_{\nu=1}^{r} \tilde{c}_{l+2e_\nu} = 0.$$

In other words, $(c_m)_{\mu \in \mathcal{M}}$ solves the difference equation (3.29), as claimed. $\quad\square$

Applying Theorem 3.9 to our original problem, we get the following characterisation theorem.

Theorem 3.10 (Characterisation of $\overset{*}{\mathbb{H}}{}^r_\mu$). *To every initial vector $\{c_m\}_{\mu \in \mathcal{I}(r,\mu)}$ there exists a uniquely determined $H \in \overset{*}{\mathbb{H}}{}^r_\mu$ such that $c_m(H) = c_m$ holds for all $m \in \mathcal{I}(r,\mu)$.*

Proof. Obviously, by Theorem 3.7 the polynomial $P \in \overset{*}{\mathbb{P}}{}^r_\mu$ with the coefficients $c_m(P) = c_m$ is harmonic if and only if the difference equation (3.29) is satisfied. The remaining follows from Theorem 3.9. $\quad\square$

Corollary 3.11 (Dimension of $\overset{*}{\mathbb{H}}{}^r_\mu$).

$$\dim \overset{*}{\mathbb{H}}{}^r_\mu = \binom{\mu + r - 2}{r - 2} + \binom{\mu + r - 3}{r - 2}.$$

Proof. From (3.34) it follows that the dual space of $\overset{*}{\mathbb{H}}{}^r_\mu$ is generated by $|\mathcal{I}|$ coefficient functionals. This implies $|\mathcal{I}| \geq \dim \overset{*}{\mathbb{H}}{}^r_\mu$. Vice versa, from Theorem 3.10 it follows that we may define H_m for $m \in \mathcal{I}$ by

$$\overset{*}{\mathbb{H}}{}^r_\mu \ni H_m : \ c_n(H_m) = \delta_{m,n} \ \text{for} \ m, n \in \mathcal{I}. \tag{3.36}$$

These elements are linearly independent, so we get $|\mathcal{I}| \leq \dim \overset{*}{\mathbb{H}}{}^r_\mu$. Together this yields $\dim \overset{*}{\mathbb{H}}{}^r_\mu = |\mathcal{I}|$. But $|\mathcal{I}|$ is the number of $(r-1)$-variate monomials of degree μ or $\mu - 1$. We evaluate this number by means of (3.9), and the theorem is proved. $\quad\square$

It follows from (3.36) and Corollary 3.11 that the family

$$\big\{ H_m \big\}_{m \in \mathcal{I}(r,\mu)}$$

is a basis of $\overset{*}{\mathbb{H}}{}^r_\mu$, where every $H \in \overset{*}{\mathbb{H}}{}^r_\mu$ has the representation

$$H = \sum_{m \in \mathcal{I}(r,\mu)} c_m(H) H_m, \tag{3.37}$$

while $P \in \overset{*}{\mathbb{P}}{}^r_\mu$ is in $\overset{*}{\mathbb{H}}{}^r_\mu$ if and only if

$$P - \sum_{m \in \mathcal{I}(r,\mu)} c_m(P) H_m = 0. \tag{3.38}$$

Remark. The index set $\mathcal{I}(r, \mu)$ is chosen such that the restrictions of the corresponding coefficient functionals onto $\overset{*}{\mathbb{H}}{}^r_\mu$ form a basis in its dual space. There are many other choices possible to guarantee this.

Polynomials in the Kernel of Other Partial Differential Operators

The spaces of harmonic homogeneous polynomials can be written in the form

$$\overset{*}{\mathbb{H}}{}^r_\mu = \overset{*}{\mathbb{P}}{}^r_\mu \cap \ker(\Delta).$$

Though these spaces play a particular role in our theory, it is quite natural and worthwhile to generalize them as follows.

We replace Δ by a homogeneous partial differential operator of order κ which has the form

$$L = \sum_{\nu=1}^{r} a_\nu D^{\kappa \cdot e_\nu},$$

with constant real coefficients a_ν. Then $P \in \overset{*}{\mathbb{P}}{}^r_\mu$ implies

$$LP \in \overset{*}{\mathbb{P}}{}^r_{\mu-\kappa},$$

and $P \in \ker(L)$ holds if and only if

$$D^n LP = 0$$

is valid for all $n \in \mathbb{N}_0^r$ with $|n| = \mu - \kappa$. This is equivalent to

$$\sum_{\nu=1}^{r} a_\nu \tilde{c}_{n+\kappa \cdot e_\nu} = 0$$

for all $n \in \mathbb{N}_0^r$ with $|n| = \mu - \kappa$. Now assume $a_r \neq 0$. Then this difference equation allows a reduction to the *initial values* $\tilde{c}_m$, $m_r \in \{0, 1, \ldots, \kappa - 1\}$, as in the case where L is the Laplace operator.

Wave Equation

The wave equation is a particularly interesting example. Here the differential operator is *hyperbolic* and takes the form

$$L = D^{2e_r} - \sum_{\nu=1}^{r-1} D^{2e_\nu}.$$

The interesting space is now

$$\overset{*}{\mathbf{W}}{}^r_\mu := \overset{*}{\mathbb{P}}{}^r_\mu \cap \ker(L),$$

which consists of the homogeneous polynomial solutions of the wave equation. If we compare now the theory of $\overset{*}{\mathbf{W}}{}_{\mu}^{r}$ and of $\overset{*}{\mathbb{H}}{}_{\mu}^{r}$, we see that there is only a little change necessary, which concerns the signs: (3.29) has to be replaced by

$$\tilde{c}_{n+2e_r} - \sum_{\nu=1}^{r-1} \tilde{c}_{n+2e_\nu} = 0$$

for all $n \in \mathbb{N}_0^r$ with $|n| = \mu - 2$, and (3.34) has to be replaced by

$$\tilde{c}_m = \sum_{|\bar{n}|=\lfloor \frac{m_r}{2} \rfloor} \binom{|\bar{n}|}{\bar{n}} \tilde{c}_{\bar{m}+2\bar{n},m_r-2\lfloor \frac{m_r}{2} \rfloor}.$$

Theorem 3.9, Theorem 3.10, and Corollary 3.11 remain valid with $\overset{*}{\mathbb{H}}{}_{\mu}^{r}$ replaced by $\overset{*}{\mathbf{W}}{}_{\mu}^{r}$.

A General Principle of Generating Polynomial Families

Let $r \geq 2$. Assume $\{Q_\mu\}_{\mu \in \mathbb{N}_0}$ is a given sequence of even or odd *univariate* polynomials of exact degree μ, i.e., assume Q_μ has the representation

$$Q_\mu(\xi) = \sum_{\nu=0}^{\lfloor \frac{\mu}{2} \rfloor} a_{\mu-2\nu} \xi^{\mu-2\nu}, \qquad a_\mu \neq 0, \tag{3.39}$$

with real coefficients. For t and x in $\mathbb{R}^r$ we get

$$|t|^\mu Q_\mu\left(\tfrac{tx}{|t|}\right) = \sum_{\nu=0}^{\lfloor \frac{\mu}{2} \rfloor} a_{\mu-2\nu}(t_1 x_1 + \cdots + t_r x_r)^{\mu-2\nu}(t_1^2 + \cdots + t_r^2)^\nu. \tag{3.40}$$

This is a homogeneous polynomial in the variable t and can be written in the form

$$|t|^\mu Q_\mu\left(\tfrac{tx}{|t|}\right) = \sum_{|m|=\mu} A_m(x) t^m, \tag{3.41}$$

which defines $A_m(x)$ for $|m| = \mu$ and $x \in \mathbb{R}^r$.

In order to determine $A_m(x)$ we order (3.40) by powers of t. So we get

$$
\begin{aligned}
|t|^\mu Q_\mu\left(\tfrac{tx}{|t|}\right) &= \sum_{\nu=0}^{\lfloor \frac{\mu}{2} \rfloor} a_{\mu-2\nu} \sum_{|n|=\mu-2\nu} \binom{\mu-2\nu}{n} t^n x^n \sum_{|l|=\nu} \binom{\nu}{l} t^{2l} \\
&= \sum_{|m|=\mu} t^m \sum_{\nu=0}^{\lfloor \frac{\mu}{2} \rfloor} \sum_{|l|=\nu} a_{\mu-2\nu} \binom{\mu-2\nu}{m-2l} \binom{\nu}{l} x^{m-2l}. \tag{3.42}
\end{aligned}
$$

A comparison with (3.41) yields

$$A_m(x) = \sum_{|n| \leq \lfloor \frac{\mu}{2} \rfloor} a_{\mu - 2|n|} \binom{\mu - 2|n|}{m - 2n} \binom{|n|}{n} x^{m - 2n}, \tag{3.43}$$

and we see that the A_m form a family of *multivariate polynomials*

$$A_m \in \mathbb{P}_m^r, \quad m \in \mathbb{N}_0^r. \tag{3.44}$$

The unique homogeneous component of degree μ is obtained for $n = 0$, i.e., we get

$$A_m(x) = a_\mu \binom{\mu}{m} x^m + TLD(x), \tag{3.45}$$

where TLD is again an abbreviation for 'terms of lower degree', but now in the multivariate sense of $TLD \in \mathbb{P}_{\mu-1}^r$, here and in what follows.

The following theorem is concerned with a basic property of the A_m.

Theorem 3.12 (Basis Property of the A_m). *Let $r \geq 2$ and let the family of polynomials $\{A_m\}_{m \in \mathbb{N}_0^r}$ be defined by (3.43), where $a_\mu \neq 0$ holds for all $\mu \in \mathbb{N}_0$. Then the subfamily*

$$\{A_m\}_{m \in \{n : |n| \leq \mu\}}$$

is a basis of $\mathbb{P}_\mu^r$ for arbitrary $\mu \in \mathbb{N}_0$.

Proof. In view of $a_\mu \neq 0$ the statement follows immediately from (3.45). $\square$

Moreover, replacing in (3.41) the variable t_ν by $-t_\nu$ and x_ν by $-x_\nu$, we find that

$$A_m(\ldots, -x_\nu, \ldots) = (-1)^{m_\nu} A_m(\ldots, x_\nu, \ldots) \tag{3.46}$$

holds for $\nu = 1, \ldots, r$, which implies

$$A_m(-x) = (-1)^\mu A_m(x), \tag{3.47}$$

and hence $A_m \in \mathbb{Q}_\mu^r$ for $|m| = \mu$. Therefore we can homogenize the A_m, as explained in Definition 3.1, and get the additional family of polynomials $\overset{*}{A}_m \in \overset{*}{\mathbb{P}}_\mu^r$ defined by

$$\overset{*}{A}_m(x) = |x|^\mu A_m\left(\frac{x}{|x|}\right) \tag{3.48}$$

for $|m| = \mu$. If we homogenize (3.41) likewise with respect to the variable x, then we see that the $\overset{*}{A}_m$ satisfy the symmetric equations

$$\sum_{|m|=\mu} \overset{*}{A}_m(x) t^m = |t|^\mu |x|^\mu Q_\mu\left(\frac{tx}{|t||x|}\right) = \sum_{|m|=\mu} \overset{*}{A}_m(t) x^m. \tag{3.49}$$

We finish our consideration by evaluating the pivot coefficient for later applications.

Theorem 3.13 (Pivot Coefficient of $\overset{*}{A}_m$).

$$c_m(\overset{*}{A}_m) = \sum_{|n| \leq \lfloor \frac{\mu}{2} \rfloor} a_{\mu-2|n|} \binom{\mu - 2|n|}{m - 2n} \binom{|n|}{n}^2. \qquad (3.50)$$

Proof. From (3.43) and (3.48) we obtain

$$\begin{aligned}
\overset{*}{A}_m(x) &= \sum_{|n| \leq \lfloor \frac{\mu}{2} \rfloor} a_{\mu-2|n|} \binom{\mu - 2|n|}{m - 2n} \binom{|n|}{n} x^{m-2n} (x_1^2 + \cdots + x_r^2)^{|n|} \\
&= \sum_{|n| \leq \lfloor \frac{\mu}{2} \rfloor} a_{\mu-2|n|} \binom{\mu - 2|n|}{m - 2n} \binom{|n|}{n} x^{m-2n} \sum_{|l|=|n|} \binom{|n|}{l} x^{2l} \\
&= \sum_{|n| \leq \lfloor \frac{\mu}{2} \rfloor} \sum_{|l|=|n|} a_{\mu-2|n|} \binom{\mu - 2|n|}{m - 2n} \binom{|n|}{n} \binom{|n|}{l} x^{m-2n+2l}.
\end{aligned}$$

We collect the terms which occur with the monomial x^m, which are just the terms where $l = n$, and get immediately the statement of the theorem. $\qquad \square$

The Families of Appell and Kampé de Feriét

In this section we present two families of r-variate polynomials $\left\{V_m^{(s)}\right\}_{m \in \mathbb{N}_0^r}$ and $\left\{U_m^{(s)}\right\}_{m \in \mathbb{N}_0^r}$, which depend on a parameter s, which we call the *Appell index*. For $r = 1$ they coincide with some Gegenbauer polynomials. For $r = 2$ they were introduced already by Hermite. For arbitrary $r \in \mathbb{N}$, but restricted to the case $s \in \mathbb{N}$, these polynomials were studied intrinsically by Appell and Kampé de Feriét, [2]. However, several results can be generalised to more or less arbitrary indices. In particular, in our context the index $s = -1$ is of great importance.

The polynomials $V_m^{(s)}$ and $U_m^{(s)}$ depend on s, but also on r, which is the number of variables which occur at least formally. We make this apparent by writing $V_m^{r,s}$ and $U_m^{r,s}$ instead of $V_m^{(s)}$ and $U_m^{(s)}$, respectively.

We obtain the family $V_m^{r,s}$ if we identify Q_μ in (3.41) with the Gegenbauer polynomial $C_\mu^{\frac{r+s-1}{2}}$, where $s \in \mathbb{R}$ is arbitrary in the beginning. The $V_m^{r,s}$ are then the corresponding A_m, i.e., they are defined by the expansion

$$|t|^\mu C_\mu^{\frac{r+s-1}{2}} \left(\frac{tx}{|t|} \right) = \sum_{|m|=\mu} V_m^{r,s}(x) \, t^m, \qquad (3.51)$$

where t and x vary in $\mathbb{R}^r$. If we write $C_\mu^\lambda(\xi)$ for arbitrary $\lambda \in \mathbb{R}$ in the form

$$C_\mu^\lambda(\xi) = \sum_{\nu=0}^{\lfloor \frac{\mu}{2} \rfloor} a_{\mu-2\nu}^\lambda \xi^{\mu-2\nu}, \qquad (3.52)$$

then (3.44) and (3.45) take the form

$$V_m^{r,s} \in \mathbb{P}_m^r, \tag{3.53}$$

$$V_m^{r,s}(x) = a_\mu^{\frac{r+s-1}{2}} \binom{\mu}{m} x^m + TLD(x), \tag{3.54}$$

both for $m| = \mu \in \mathbb{N}_0$, $r \in \mathbb{R}$ and arbitrary $s \in \mathbb{R}$.

Now let us assume $r + s - 1 \neq 0$, first. Then we get from (2.3) and (2.4)

$$\frac{1}{(1 - 2\xi\tau + \tau^2)^{\frac{r+s-1}{2}}} = \sum_{\mu=0}^{\infty} C_\mu^{\frac{r+s-1}{2}}(\xi) \tau^\mu \tag{3.55}$$

for $1 \leq \xi \leq 1$, $0 < |\tau| < 1$. For $t, x \in B^r$, $0 < |t| < 1$, we may replace ξ and τ by

$$\xi := \frac{tx}{|t|}, \qquad \tau := |t|.$$

Inserting this in (3.55) we get, together with (3.51),

$$\frac{1}{(1 - 2tx + t^2)^{\frac{r+s-1}{2}}} = \sum_{\mu=0}^{\infty} \sum_{|m|=\mu} V_m^{r,s}(x)t^m, \tag{3.56}$$

which is an r-variate version of (3.55).

For $r + s - 1 = 0$ we have to change the generating function and get from (2.7) likewise

$$1 - \log(1 - 2tx + t^2) = \sum_{\mu=0}^{\infty} \sum_{|m|=\mu} V_m^{r,s}(x)t^m, \tag{3.57}$$

instead of (3.56).

For $r = 1$ the equations (3.56) and (3.57) reduce to the defining equations of the corresponding Gegenbauer polynomials, and we obtain $V_\mu^{1,s} = C_\mu^{\frac{s}{2}}$.

For $r \geq 2$ let $\bar{t} := (t_1, \ldots, t_{r-1})'$, $\bar{x} := (x_1, \ldots, x_{r-1})'$, $\bar{m} := (m_1, \ldots, m_{r-1})$. Then $t_r = 0$ implies

$$1 - 2tx + t^2 = 1 - 2\bar{t}\bar{x} + \bar{t}^2 \quad and \quad t^m = \bar{t}^{\bar{m}},$$

which corresponds to a reduction of the dimension r by one, and (3.56) or (3.57), respectively, yields

$$V_{\bar{m},0}^{r,s}(\bar{x}, *) = V_{\bar{m}}^{r-1,s+1}(\bar{x}).$$

Here and in what follows $*$ indicates a variable which does not appear — see (3.53).

Next we replace r by $r+1$ and s by $s-1$. Then we get for $r \in \mathbb{R}$ and $s \in \mathbb{N}_0$ by a repeated application of our reduction method

$$V_m^{r,s}(x) = V_{m,0}^{r+1,s-1}(x,*) = \cdots = V_{m,0,\cdots,0}^{r+s+1,-1}(x,*,\cdots,*). \qquad (3.58)$$

These equations say that it is possible to reduce the index s to the value of -1 by treating $V_m^{r,s}$ as an element of $\mathbb{P}_{m,0,\ldots,0}^{r+s+1}$.

The $U_m^{r,s}$ are obtained in a similar, though different way from the expansion

$$[(tx)^2 + t^2(1-x^2)]^{\frac{\mu}{2}} C_\mu^{\frac{s}{2}}\left(\frac{tx}{[(tx)^2 + t^2(1-x^2)]^{\frac{1}{2}}}\right) = \sum_{|m|=\mu} U_m^{r,s}(x)\, t^m, \qquad (3.59)$$

in the beginning for arbitrary $s \in \mathbb{R}$. Actually, for $\mu \in \mathbb{N}_0$ the left side is homogeneous in t of degree μ, such that the $U_m^{r,s}(x)$ are defined for $x \in \mathbb{R}^r$ as the coefficients occurring with t^m, where it is obvious that they are polynomials of degree μ, i.e., that

$$U_m^{r,s} \in \mathbb{P}_\mu^r \qquad (3.60)$$

holds for $|m| = \mu \in \mathbb{N}_0$. Replacing x_ν and t_ν in (3.59) simultaneously by $-x_\nu$ and $-t_\nu$, respectively, we get

$$U_m^{r,s}(\ldots, -x_\nu, \ldots) = (-1)^{m_\nu} U_m^{r,s}(\ldots, x_\nu, \ldots) \qquad (3.61)$$

for $\nu = 1, \ldots, r$. Moreover, we obtain

$$U_m^{r,s}(x) = C_\mu^{\frac{s}{2}}(1)\binom{|m|}{m} x^m \quad \text{for } x \in S^{r-1}. \qquad (3.62)$$

But in general $U_m^{r,s}$ is not in $\mathbb{P}_m^r$, nor does a chain like (3.58) exist.

But there are expansions for the $U_m^{r,s}$ corresponding to (3.56) and (3.57). To see this let

$$\xi := \frac{tx}{[(tx)^2 + t^2(1-x^2)]^{\frac{1}{2}}}, \qquad \tau := [(tx)^2 + t^2(1-x^2)]^{\frac{1}{2}}$$

for $x,t \in B^r$, $|x| < 1$, $0 < |t| < 1$. Actually, these assumptions imply $-1 \le \xi \le 1$ and $0 < \tau^2 \le t^2 < 1$. Therefore we may insert these values in the equations

$$\frac{1}{(1 - 2\xi\tau + \tau^2)^{\frac{s}{2}}} = \sum_{\mu=0}^{\infty} C_\mu^{\frac{s}{2}}(\xi)\, \tau^\mu, \text{ if } s \neq 0,$$

$$1 - \log(1 - 2\xi\tau + \tau^2) = \sum_{\mu=0}^{\infty} C_\mu^0(\xi)\, \tau^\mu,$$

and together with (3.59) we obtain

$$\frac{1}{[(1-tx)^2 + t^2(1-x^2)]^{\frac{s}{2}}} = \sum_{\mu=0}^{\infty} \sum_{|m|=\mu} U_m^{r,s}(x)t^m, \text{ for } s \neq 0, \quad (3.63)$$

$$1 - \log\left[(1-tx)^2 + t^2(1-x^2)\right] = \sum_{\mu=0}^{\infty} \sum_{|m|=\mu} U_m^{r,0}(x)t^m. \qquad (3.64)$$

The functions on the left sides of (3.56), (3.57), (3.63) and (3.64) are called the *generating functions* of the polynomials which occur on the corresponding right side.

For $s \in \mathbb{N}$ and $|m| = \mu \in \mathbb{N}_0$ it is possible to derive from (3.63) the formula

$$U_m^{r,s}(x) = \frac{(s)_\mu}{m!} \sum_{\nu=0}^{\lfloor \frac{\mu}{2} \rfloor} (-1)^\nu \frac{1}{2^{2\nu}(1)_\nu \left(\frac{s+1}{2}\right)_\nu} (1-x^2)^\nu \Delta^\nu x^m, \qquad (3.65)$$

where Δ is the Laplace operator, see [2], p.262 (11). It is easy to bring this to the form

$$U_m^{r,s}(x) = \Gamma(\mu+s) \frac{\Gamma(\frac{s+1}{2})}{\Gamma(s)} \sum_{|n| \leq \lfloor \frac{\mu}{2} \rfloor} \frac{(-1)^{|n|}}{2^{2|n|}\Gamma(|n| + \frac{s+1}{2}) n! (m-2n)!} x^{m-2n}(1-x^2)^{|n|}. \qquad (3.66)$$

For $s = 0$ and $|m| = \mu \in \mathbb{N}$ it follows likewise from (3.64) that

$$U_m^{r,0}(x) = 2\frac{(\mu-1)!}{m!} \sum_{\nu=0}^{\lfloor \frac{\mu}{2} \rfloor} (-1)^\nu \frac{1}{2^{2\nu}(1)_\nu \left(\frac{1}{2}\right)_\nu} (1-x^2)^\nu \Delta^\nu x^m \qquad (3.67)$$

and

$$U_m^{r,0}(x) = 2\sqrt{\pi}(\mu-1)! \sum_{|n| \leq \lfloor \frac{\mu}{2} \rfloor} (-1)^{|n|} \frac{1}{2^{2|n|}\left(\frac{1}{2}\right)_{|n|} n! (m-2n)!} x^{m-2n}(1-x^2)^{|n|} \qquad (3.68)$$

holds, see [19], formula 2.3.16, where $U_m^{r,0}$ has to be identified with $\frac{2}{\mu}T_m$. Note that $U_0^{r,0} = 1$ is valid. It is tedious to derive these formulae from the corresponding generating functions. So we omit this, in particular, since we obtain a corresponding result in a different way in Section 4.3.

Remark. The polynomial systems of Appell and Kampé de Feriét earn our interest by their important property of being biorthogonal with respect to certain weight functions on the ball. We derive this property in Section 4.3 from a basic biorthonormal system of *spherical harmonics*, which we bring down from the sphere to lower dimensional balls. The situation is comparable with the complex plane, where the orthogonality of the monomials on the unit circle is basic. Moreover, we restricted ourselves, finally, to the case $s \in \mathbb{N}_0$, where the index $s = 0$, which was not considered by Appell and Kampé de Feriét, plays a separate role. Note that it follows from (2.13) that $C_\mu^{\frac{s}{2}}(1)$ vanishes for $s = -1$ and $\mu \in \mathbb{N}$, such that $U_m^{r,s}$ degenerates on S^{r-1}, in this case, see (3.62). Nevertheless, just this index corresponds to the spherical harmonics and hence with our theory. This will become clearer later.

We finish this section with the remark that the $V_m^{r,s}$ and the $U_m^{r,s}$ are represented by certain *Rodrigues formulae*, and that they satisfy certain *partial differential equations.* In our context this is without interest. Moreover, there exist *recurrence formulae* for both systems. This we discuss in Section 7.6 in a more general setting.

3.2 Polynomials on Subsets

The elements P of $\mathbb{P}^r$ are defined by their coefficients with respect to the monomials, which means as a subject of algebra. For this reason we call them, more precisely, *algebraic polynomials*. On the other hand, an algebraic polynomial P defines a *polynomial function*

$$\mathbb{R}^r \ni x \mapsto P(x) \in \mathbb{R}$$

by the usual evaluation rules. The space of all polynomial functions is denoted by $\mathbb{P}^r(\mathbb{R}^r)$. It follows from the *Fundamental Theorem of Algebra* that a polynomial function vanishes only if P is the algebraic null polynomial, which means if all coefficients are zero. Therefore the map from the algebraic polynomials to the polynomial functions is an isomorphy,

$$\mathbb{P}^r(\mathbb{R}^r) \cong \mathbb{P}^r.$$

We can generalize this as follows. Let $\mathbf{V}$ be a subspace of $\mathbb{P}^r$ and let D be a nonempty subset of $\mathbb{R}^r$. Then we define the space of polynomial function *restrictions*

$$\mathbf{V}(D) := \{ P|_D : P \in \mathbf{V} \},$$

and by the arguments from above we get

Theorem 3.14 (D containing an Interior Point). *Assume* $\mathbf{V}$ *is a subspace of* $\mathbb{P}^r$ *and* $D \subset \mathbb{R}^r$ *contains an interior point. Then* $\mathbf{V}(D) \cong \mathbf{V}$ *holds, together with the statements*

$$(i) \quad \dim \mathbf{V}(D) = \dim \mathbf{V},$$
$$(ii) \quad \| \cdot \|_D \text{ is a norm on } \mathbf{V}.$$

If D does not contain an interior point, the situation can change, but need not.

For instance, let $r \geq 2$ and $\mu \geq 2$. The space $\mathbb{P}^r_\mu$ contains the nonzero algebraic prime polynomial

$$E(x) := x_1^2 + \cdots + x_r^2 - 1.$$

The corresponding polynomial function vanishes on S^{r-1}, so $\mathbb{P}^r_\mu(S^{r-1})$ cannot be isomorphic with $\mathbb{P}^r_\mu$ itself, while $\dim \mathbb{P}^r_\mu(S^{r-1}) < \mathbb{P}^r_\mu$ must hold. We could try to compensate for this lack by replacing the polynomial ring $\mathbb{P}^r$ with the ring $\mathbb{P}^r / _{x_1^2 + \cdots + x_r^2 - 1}$. Likewise we could proceed if D is an arbitrary *algebraic variety*. But to our knowledge this approach was not awarded by success, until now. Moreover, it has the disadvantage of promising nothing for the case where D is arbitrary.

In what follows we assume, again, that $\mathbf{V}$ is an arbitrary subspace of $\mathbb{P}^r$, and that $\emptyset \neq D_1 \subset D_2 \subset \mathbb{R}^r$ holds. Then the space $\mathbf{V}(D_1)$ consists of the restrictions of the elements of $\mathbf{V}(D_2)$ onto D_1. This furnishes immediately

Theorem 3.15 (Comparison of Dimensions). *If $\mathbf{V}$ is a subspace of $\mathbb{P}^r$ and if $\emptyset \neq D_1 \subset D_2 \subset \mathbb{R}^r$ holds, then*

$$\dim \mathbf{V}(D_1) \leq \dim \mathbf{V}(D_2) \leq \dim \mathbf{V}.$$

We add a theorem where a situation of particular interest occurs which is not covered by Theorem 3.14.

Theorem 3.16 (Harmonic and Homogeneous Spaces). *Let $r \geq 2$, $\mu \in \mathbb{N}_0$, and let $\mathbf{V}$ be either the space $\overset{*}{\mathbb{P}}{}^r_\mu$, or a subspace of $\mathbb{H}^r$. Then the following holds:*

$$(i) \quad \mathbf{V}(S^{r-1}) \cong \mathbf{V}(B^r) \cong \mathbf{V},$$
$$(ii) \quad \dim \mathbf{V}(S^{r-1}) = \dim \mathbf{V}(B^r) = \dim \mathbf{V},$$
$$(iii) \quad \|P\|_{S^{r-1}} = \|P\|_{B^r} \text{ for all } P \in \mathbf{V}.$$

Proof. (iii) follows from the maximum principle or from (3.4), respectively. The right side of (i) follows from Theorem 3.14, the left side from (iii), since $P|_{S^{r-1}}$ vanishes only for $\|P\|_{B^r} = 0$, and hence if $P|_{B^r}$ is the null element in $\mathbf{V}(B^r)$. (ii) is an immediate consequence of (i). $\square$

Corollary 3.17. *For $r \geq 2$ and $\mu \in \mathbb{N}_0$, $\| \cdot \|_{S^{r-1}}$ is a norm on the spaces $\mathbb{H}^r$, $\overset{*}{\mathbb{P}}{}^r_\mu$, and $\overset{*}{\mathbb{H}}{}^r_\mu$.*

Remark. To make the notation more convenient we agree on the following. If an element of $\mathbf{V}(D)$ is the restriction of a polynomial P onto D, then we write $P \in \mathbf{V}(D)$, though this may be incorrect in a strict sense. For instance we may write $1 \in \overset{*}{\mathbb{P}}{}^r_{2\mu}$ for $\mu \in \mathbb{N}$, since 1 is the restriction of $(x_1^2 + \cdots + x_r^2)^\mu$ onto S^{r-1}.

3.3 Problems

Problem 3.1.
$$\binom{\mu}{m} = \frac{\mu!}{m!}.$$

Problem 3.2.
$$\sum_{|n|=\nu} \binom{\nu}{n}\binom{\mu-\nu}{m-n} = \binom{\mu}{m}.$$

Problem 3.3.

$$\int_{S^{r-1}} x^m \, d\omega(x) = \begin{cases} 0, & \text{if } m \neq 2n, \ n \in \mathbb{N}_0^r, \\ 2\pi^{\frac{r-1}{2}} \dfrac{\Gamma(\nu+\frac{1}{2})}{\Gamma(\nu+\frac{r}{2})} \dfrac{\binom{\nu}{n}}{\binom{2\nu}{2n}}, & \text{if } m = 2n, \ n \in \mathbb{N}_0^r, \ |n| = \nu. \end{cases}$$

Problem 3.4. Prove for $r \in \mathbb{N}$ and $\mu \in \mathbb{N}_0$ by mathematical induction:

$$\sum_{|m|=\mu} x^m = [\xi^{\mu+r-1}; \, x_1, x_2, \ldots, x_r], \quad x \in \mathbb{R}^r,$$

where the expression on the right side is the divided difference of order $r - 1$ with respect to the monomial $\xi^{\mu+r-1}$ and the nodes $x_1, x_2, \ldots, x_r$.

Problem 3.5. Let D be the margin of a non-degenerating triangle in $\mathbb{R}^2$. Calculate the dimension of $\mathbb{P}_2^2(D)$.

Problem 3.6. Determine $\|M_m\| = \max\{|x^m| : \, x \in B^r\}$, $r \geq 2$, $m \in \mathbb{N}_0^r$.

Chapter 4

Polynomials on Sphere and Ball

In this chapter we investigate the space $\mathbb{P}^r(S^{r-1})$ of polynomial restrictions onto the unit sphere S^{r-1}, and its subspaces. The elements of $\mathbb{P}^r(S^{r-1})$ are called *spherical polynomials*, the elements of $\overset{*}{\mathbb{H}}{}^r_\mu$ *spherical harmonics* of degree μ.

The most important subspaces are finite-dimensional *rotation-invariant* subspaces (Definition 1.2). They are distinguished by a very rich theory and a numerical analysis at a reasonable degree of complexity — which is one of the most important problems in the multivariate case. Several basic results on spherical polynomials are transferrable to the ball.

4.1 The Rotation-Invariant Subspaces of $\mathbb{P}^r(S^{r-1})$

Let $r \in \mathbb{N} \setminus \{1\}$ be a fixed space dimension. It is obvious that the (unrestricted) polynomial spaces $\mathbb{P}^r$, $\mathbb{P}^r_m$ and $\overset{*}{\mathbb{P}}{}^r_\mu$ are rotation-invariant. The following lemma is also obvious.

Lemma 4.1 (Intersection of Rotation-Invariant Spaces). *If the subspaces $\mathbf{V}_1$ and $\mathbf{V}_2$ of $\mathbb{P}^r(D)$ are rotation-invariant, then $\mathbf{V}_1 \cap \mathbf{V}_2$ is rotation-invariant.*

Not so obvious is the following theorem.

Theorem 4.2 (Harmonic Space). *For $r \in \mathbb{N} \setminus \{1\}$ the space $\mathbb{H}^r$ is rotation-invariant.*

Proof. Assume $H \in \mathbb{H}^r$, $A \in \mathbf{A}^r$, and let $H_A(\cdot) := H(A\cdot)$, consistent with Definition 1.2. Obviously, $H_A \in \mathbb{P}^r$ holds as $\mathbb{P}^r$ is rotation-invariant. We calculate the Hesse matrix of H_A and get

$$\left(\frac{\partial^2 H_A}{\partial x_\nu \partial x_\kappa} \right)_{(x)} = A' \left(\frac{\partial^2 H}{\partial x_\nu \partial x_\kappa} \right)_{(Ax)} A.$$

By the invariance of the trace this yields

$$(\Delta H_A)(x) = (\Delta H)(Ax) = 0,$$

i.e., $H_A \in \text{IH}^r$. Therefore, IH^r is rotation-invariant.   □

Using Lemma 4.1 and Theorem 4.2 we get easily the following corollary.

Corollary 4.3. *For $r \in \text{IN} \setminus \{1\}$ the spaces $\text{IP}^r(S^{r-1})$, $\text{IP}^r_\mu(S^{r-1})$, $\overset{*}{\text{IP}}{}^r_\mu(S^{r-1})$, $\text{IH}^r(S^{r-1})$, and $\overset{*}{\text{IH}}{}^r_\mu(S^{r-1})$ are rotation-invariant.*

Dimensions

We want to determine the dimensions of the most important subspaces. $P \in \overset{*}{\text{IP}}{}^r_\mu$ vanishes on S^{r-1} if and only if P is the null polynomial, see (3.4). Therefore the spaces $\overset{*}{\text{IP}}{}^r_\mu$ and $\overset{*}{\text{IP}}{}^r_\mu(S^{r-1})$ are isomorphic, and(3.9) yields

$$\dim \overset{*}{\text{IP}}{}^r_\mu(S^{r-1}) = \binom{\mu + r - 1}{r - 1}. \tag{4.1}$$

Likewise, $H \in \overset{*}{\text{IH}}{}^r_\mu$ vanishes on S^{r-1} if and only if H is the null polynomial. This follows from the maximum principle. Therefore the spaces $\overset{*}{\text{IH}}{}^r_\mu$ and $\overset{*}{\text{IH}}{}^r_\mu(S^{r-1})$ are also isomorphic, and Corollary (3.11) yields

$$\dim \overset{*}{\text{IH}}{}^r_\mu(S^{r-1}) = \binom{\mu + r - 2}{r - 2} + \binom{\mu + r - 3}{r - 2}. \tag{4.2}$$

Next we intend to calculate the dimension of $\text{IP}^r_\mu(S^{r-1})$, which is not quite so easy. To this end let

$$F \in \text{IP}^r_\mu(S^{r-1})$$

be arbitrary, say

$$F = P|_{S^{r-1}} \text{ where } P \in \text{IP}^r_\mu.$$

We write P in the form

$$P = A + B$$

where $A \in \text{IP}^r_\mu$ and $B \in \text{IP}^r_{\mu-1}$ are defined by

$$\begin{aligned} A(x) &= \tfrac{1}{2}\big[P(x) + (-1)^\mu P(-x)\big], \\ B(x) &= \tfrac{1}{2}\big[P(x) - (-1)^\mu P(-x)\big]. \end{aligned}$$

It follows that

$$F(x) = \overset{*}{A}(x) + \overset{*}{B}(x) \text{ for } x \in S^{r-1},$$

where $\overset{=}{A} \in \overset{*}{\mathbb{P}}{}^r_\mu$, $\overset{*}{B} \in \overset{*}{\mathbb{P}}{}^r_{\mu-1}$. This is a decomposition of F by its even and its odd part, or vice versa, so the decomposition is unique, and we get the following theorem

Theorem 4.4 (Decomposition of $\mathbb{P}^r_\mu(S^{r-1})$).

$$\mathbb{P}^r_\mu(S^{r-1}) = \overset{*}{\mathbb{P}}{}^r_\mu(S^{r-1}) \oplus \overset{*}{\mathbb{P}}{}^r_{\mu-1}(S^{r-1}) \tag{4.3}$$

is valid for $r \in \mathbb{N} \setminus \{1\}$ and $\mu \in \mathbb{N}_0$.

As an immediate consequence of Theorem 4.4 we get now, for $r \in \mathbb{N} \setminus \{1\}$ and $\mu \in \mathbb{N}_0$,

$$\dim \mathbb{P}^r_\mu(S^{r-1}) = \binom{\mu+r-1}{r-1} + \binom{\mu+r-2}{r-1}, \tag{4.4}$$

which is the dimension wanted.

Reproducing Kernels

It is our aim to characterize in what follows all rotation-invariant subspaces of $\mathbb{P}^r(S^{r-1})$ of finite dimension. This will be possible by determining their reproducing kernels with respect to the inner product $\langle \cdot, \cdot \rangle$. By Theorem 1.11 they have the form

$$G(x,y) = K(xy), \quad x,y \in S^{r-1},$$

where $K \in C[-1,1]$. This means that $K(x \cdot)$ is contained in the corresponding axial kernel. In the actual case further information on K is available.

Lemma 4.5. Let $r \in \mathbb{N} \setminus \{1\}$, $\mathbf{V}$ a rotation-invariant subspace of $\mathbb{P}^r_\mu(S^{r-1})$, $N := \dim \mathbf{V} \in \mathbb{N}$. Then the following holds:

$$K \in \mathbb{P}^1_\mu([-1,1]), \tag{4.5}$$

$$K(1) = \frac{N}{\omega_{r-1}}. \tag{4.6}$$

Proof. Recall (1.5), i.e., recall that

$$K(xy) = \sum_{j=1}^N S_j(x) S_j(y)$$

holds for $x,y \in S^{r-1}$, provided $\{S_1, \ldots, S_N\}$ is an orthonormal basis of $\mathbf{V}$. Identifying $y = x$ and integrating the resulting equation over S^{r-1}, we get

$$K(1) \cdot \omega_{r-1} = \sum_{j=1}^N \langle S_j, S_j \rangle = N,$$

which proves (4.6). Next let $t, u \in S^{r-1}$ be fixed, where $tu = 0$. $K(t\cdot)$ is in $\mathbf{V}$, and hence in $\mathbb{P}_\mu^r(S^{r-1})$. Therefore $F \in \mathbb{P}_\mu^r$ exists such that

$$K(tx) = F(x) \text{ holds for } x \in S^{r-1}.$$

Now we define

$$\Phi(\xi, \eta) := \tfrac{1}{2}[F(\xi t + \eta u) + F(\xi t - \eta u)]$$

for $(\xi, \eta)' \in \mathbb{R}^2$. Then Φ is a bivariate polynomial of degree μ, which is even with respect to η. This implies that $\Phi(\xi, \sqrt{1 - \xi^2})$ is a polynomial of degree μ with respect to the variable ξ. But

$$\Phi(\xi, \sqrt{1 - \xi^2}) = K(\xi)$$

holds for $-1 \le \xi \le 1$. So K is a univariate polynomial of degree μ, as claimed. $\square$

Definition 4.1 (Particular Reproducing Kernel Functions). *Let* $r \in \mathbb{N} \setminus \{1\}$, $\mu \in \mathbb{N}_0$. *The reproducing kernel function* $K \in \mathbb{P}_\mu^1$ *of the spaces* $\overset{*}{\mathbb{H}}_\mu^r(S^{r-1})$, $\mathbb{P}_\mu^r(S^{r-1})$, *and* $\overset{*}{\mathbb{P}}_\mu^r(S^{r-1})$, *is denoted by* G_μ^r, Γ_μ^r, *and* $\overset{*}{\Gamma}_\mu^r$, *respectively.*

The Spaces $\overset{*}{\mathbb{H}}_\mu^r(S^{r-1})$

We get complete information on the reproducing kernel in the spaces of spherical harmonics since the axial kernels are outmost elementary in this case.

Theorem 4.6 (Axial Kernels in $\overset{*}{\mathbb{H}}_\mu^r(S^{r-1})$). *Let* $r \in \mathbb{N} \setminus \{1, 2\}$, $\mu \in \mathbb{N}_0$. *For* $t \in S^{r-1}$ *the axial kernel of* $\mathbf{V} := \overset{*}{\mathbb{H}}_\mu^r(S^{r-1})$ *with axis* t *is given by*

$$\mathbf{V}_t = span\{C_\mu^{\frac{r-2}{2}}(t\cdot)\}.$$

Proof. Let $\mathbf{V}_t$ be the axial kernel of $\mathbf{V}$ with axis t, and assume $H \in \mathbf{V}_t$, say $H = F|_{S^{r-1}}$ where $F \in \overset{*}{\mathbb{H}}_\mu^r$. By Corollary 1.5 a function $f \in C[-1, 1]$ exists such that $F(x) = f(tx)$ holds for $x \in S^{r-1}$.

Applying our arguments used in the proof of Lemma 4.5 with respect to K we obtain

$$f \in \mathbb{P}_\mu^1. \tag{4.7}$$

Besides we get

$$f(-tx) = F(-x) = (-1)^\mu F(x) = f(tx)$$

for arbitrary $x \in S^{r-1}$. Since the values of tx cover the interval $[-1, 1]$, this implies

$$f(-\xi) = (-1)^\mu f(\xi) \tag{4.8}$$

for $-1 \leq \xi \leq 1$, and hence for arbitrary $\xi \in \mathbb{R}$. It follows that $f(tx)$ can be homogenized, where

$$|x|^\mu f\left(\frac{tx}{|x|}\right) = F(x) \tag{4.9}$$

is valid, first for $x \in S^{r-1}$, but as homogeneous polynomials of degree μ occur on both sides, even for arbitrary $x \in \mathbb{R}^r$, $x \neq 0$. Now it follows from $\Delta F = 0$ that

$$\Delta |x|^\mu f\left(\frac{tx}{|x|}\right) = 0 \tag{4.10}$$

must hold for $x \in \mathbb{R}^r$, $x \neq 0$. Using the abbreviation

$$\xi := \frac{tx}{|x|} \text{ for } 0 \neq x \in \mathbb{R}^r$$

we get by a simple, but lengthy calculation that (4.10) is equivalent to

$$(1 - \xi^2)f''(\xi) - (r-1)\xi f'(\xi) + \mu(\mu + r - 2)f(\xi) = 0. \tag{4.11}$$

By Theorem 2.2 this implies, together with (4.7) and (4.8), that

$$f(\xi) = const \cdot C_\mu^{\frac{r-2}{2}}(\xi)$$

holds with some constant, and so we get

$$H(x) = F(x) = const \cdot C_\mu^{\frac{r-2}{2}}(tx)$$

for $x \in S^{r-1}$. This holds for arbitrary $H \in \mathbf{V}_t$, therefore we obtain

$$\mathbf{V}_t \subset \text{span}\left\{C_\mu^{\frac{r-2}{2}}(t\cdot)\right\}.$$

But $f = C_\mu^{\frac{r-2}{2}}$ itself satisfies (4.7), (4.8) and (4.11), and hence (4.10). This implies $\mathbf{V}_t \neq [0]$, and hence

$$\mathbf{V}_t = \text{span}\left\{C_\mu^{\frac{r-2}{2}}(t\cdot)\right\},$$

as claimed. □

Because of Theorem 4.6 we are able to prove the following theorem.

Theorem 4.7 (Reproducing Kernel of $\overset{*}{\mathbb{H}}{}_\mu^r(S^{r-1})$). *Let $r \in \mathbb{N} \setminus \{1\}$, $\mu \in \mathbb{N}_0$. The reproducing kernel of $\overset{*}{\mathbb{H}}{}_\mu^r(S^{r-1})$ is given by*

$$G_\mu^r(xy) = \frac{N}{\omega_{r-1}} \cdot \tilde{C}_\mu^{\frac{r-2}{2}}(xy) \tag{4.12}$$

for $x, y \in S^{r-1}$, where $N = \dim \overset{*}{\mathbb{H}}{}_\mu^r(S^{r-1})$.

Proof. $\overset{*}{\mathbb{H}}{}^r_\mu(S^{r-1})$ is rotation-invariant, see Corollary 4.3, so the reproducing kernel takes the form

$$G(x,y) = K(xy)$$

for $x, y \in S^{r-1}$, see Theorem 1.11.

For $r \geq 3$ this implies that $K(x \cdot)$ belongs to the axial kernel with axis x, and Theorem 4.6 yields

$$K(x \cdot) = const \cdot \tilde{C}_\mu^{\frac{r-2}{2}}(x \cdot).$$

The constant can be determined by means of Lemma 4.5, i.e., from

$$const = K(1) = \frac{N}{\omega_{r-1}}.$$

Together this yields (4.12).

In the exceptional case $r = 2$, the assumptions of Theorem 4.6 are not satisfied, so we have to look for a direct proof. It uses the homogenous harmonic polynomials

$$A_\mu(x_1, x_2) = \Re(x_1 + ix_2)^\mu \quad and \quad B_\mu(x_1, x_2) = \Im(x_1 + ix_2)^\mu.$$

Actually, from (4.2) we obtain $N = 1$ for $\mu = 0$, and $N = 2$ for $\mu \geq 1$, and using the parameter representation

$$x_1 = \cos\phi, \; x_2 = \sin\phi, \quad 0 \leq \phi < 2\pi,$$

we confirm easily that the systems

$$\left\{\frac{1}{\sqrt{2\pi}}A_0\right\} \quad and \quad \left\{\frac{1}{\sqrt{\pi}}A_\mu, \frac{1}{\sqrt{\pi}}B_\mu\right\}$$

are an orthonormal basis for $\mu = 0$ and $\mu \geq 1$, respectively. So we get from (1.5), with $y_1 = \cos\psi$, $y_2 = \sin\psi$, and first for $\mu \geq 1$ only,

$$G(x,y) = \frac{1}{\pi}\Big[A_\mu(x)A_\mu(y) + B_\mu(x)B_\mu(y)\Big] = \frac{1}{\pi}\Big[\cos\mu\phi \cdot \sin\mu\phi + \cos\mu\psi \cdot \sin\mu\psi\Big]$$

$$= \frac{2}{\omega_{r-1}}\cos\mu(\phi - \psi) = \frac{N}{\omega_{r-1}}\tilde{C}_\mu^0(xy).$$

For $\mu = 0$ we get similar

$$G(x,y) = \frac{1}{2\pi} = \frac{N}{\omega_{r-1}}\tilde{C}_0^0(x,y),$$

and in all cases the statement of the theorem is true.   □

Remark. The value of N is well known from (4.2). Using (2.13) and (2.8), respectively, we can bring (4.12) to the form

$$G_\mu^r = \begin{cases} \frac{2\mu+r-2}{(r-2)\omega_{r-1}}C_\mu^{\frac{r-2}{2}}, & \text{if } r \geq 3, \\ \frac{2}{\omega_{r-1}}T_\mu, & \text{if } r = 2, \; \mu \in \mathbb{N}, \\ \frac{1}{\omega_{r-1}}T_0, & \text{if } r = 2, \; \mu = 0. \end{cases} \qquad (4.13)$$

Theorem 4.6 enables us to prove the following well-known theorem.

Theorem 4.8 (Funck–Hecke). *Let* $r \in \mathbb{N} \setminus \{1, 2\}$, $\mu \in \mathbb{N}_0$, $F \in C[-1, 1]$, *and* $H \in \overset{*}{\mathbb{H}}{}^r_\mu(S^{r-1})$. *Then*

$$\mathrm{I}(F, t) := \int_{S^{r-1}} F(tx) H(x) d\omega(x) = \Lambda(F) \cdot H(t) \tag{4.14}$$

holds for $t \in S^{r-1}$, *where the constant is given by*

$$\Lambda(F) = \omega_{r-2} \int_{-1}^{1} F(\xi) \tilde{C}_\mu^{\frac{r-2}{2}}(\xi)(1 - \xi^2)^{\frac{r-3}{2}} d\xi. \tag{4.15}$$

Proof. For arbitrary rotations $A \in \mathbf{A}^r_t$ we obtain

$$\mathrm{I}(F, t) = \int_{S^{r-1}} F(tx) H(Ax) d\omega(x).$$

Averaging both sides over the group $\mathbf{A}^r_t$, we get

$$\mathrm{I}(F, t) = \int_{S^{r-1}} F(tx) \left(\Pi^r_t H\right)(x) d\omega(x).$$

Theorem 1.10 and Corollary 4.3 imply together

$$\left(\Pi^r_t H\right)(x) = H(t) \cdot g(tx)$$

with a uniquely determined function $g \in C[-1, 1]$ which satisfies $g(1) = 1$. Since $g(t \cdot)$ is located in the axial kernel of $\overset{*}{\mathbb{H}}{}^r_\mu(S^{r-1})$ with the axis t, it follows from Theorem 4.6 that

$$g(t \cdot) = \tilde{C}_\mu^{\frac{r-2}{2}}(t \cdot)$$

holds. Inserting this above in the integral we get

$$\mathrm{I}(F, t) = \Lambda(F) \cdot H(t)$$

with the constant

$$\Lambda(F) = \int_{S^{r-1}} F(tx) \tilde{C}_\mu^{\frac{r-2}{2}}(tx) d\omega(x) = \omega_{r-2} \int_{-1}^{1} F(\xi) \tilde{C}_\mu^{\frac{r-2}{2}}(\xi)(1 - \xi^2)^{\frac{r-3}{2}} d\xi,$$

where we used, finally, the reduction formula (1.25) with $s = r - 2$. $\qquad\square$

By the help of the Theorem of Funck–Hecke we can prove the following, most relevant corollary.

Corollary 4.9. *For* $\mu, \nu \in \mathbb{N}_0$, $r \in \mathbb{N} \setminus \{1\}$, $x, y \in S^{r-1}$ *the following holds:*

$$\int_{S^{r-1}} G_\mu^r(xz)\, G_\nu^r(zy)\, d\omega(z) = \begin{cases} G_\mu^r(xy) & , \text{ if } \nu = \mu, \\ 0 & , \text{ if } \nu \neq \mu. \end{cases}$$

Proof. For $\mu = \nu$ the statement follows from the reproducing property of $G_\mu^r(x \cdot)$.
Next assume $\nu \neq \mu$. In Theorem 4.8 we identify $F := G_\mu^r$, $H(\cdot) := G_\nu^r(\cdot y)$, and
obtain, for $x, y \in S^{r-1}$,

$$\int_{S^{r-1}} G_\mu^r(xt) G_\nu^r(ty)\, d\omega(t) = \Lambda_{\mu,\nu} \cdot G_\nu^r(xy) = 0,$$

where we use that

$$\Lambda_{\mu,\nu} = \omega_{r-2} \int_{-1}^{1} G_\mu^r(\xi) \tilde{C}_\nu^{\frac{r-2}{2}}(\xi)(1 - \xi^2)^{\frac{r-3}{2}}\, d\xi = 0$$

holds in view of (4.12) and of Theorem 2.3. $\square$

In what follows we use the abbreviation $G_\nu = G_\nu^r$ for $\nu \in \mathbb{N}_0$. Then we can write
the result of Corollary 4.9 in the form

$$\langle G_\mu(x \cdot), G_\nu(\cdot y) \rangle = G_\mu(xy) \cdot \delta_{\mu,\nu} \qquad (4.16)$$

for $\mu, \nu \in \mathbb{N}_0$ and arbitrary $x, y \in S^{r-1}$. In other words, for $\nu \neq \mu$ all zonal
elements of $\overset{*}{\mathbb{H}}_\mu^r(S^{r-1})$ are orthogonal to all zonal elements of $\overset{*}{\mathbb{H}}_\nu^r(S^{r-1})$. Actually,
this yields the following important theorem.

Theorem 4.10 (Orthogonality of the Spaces $\overset{*}{\mathbb{H}}_\mu^r(S^{r-1})$). *Let* $r \in \mathbb{N} \setminus \{1\}$. *For*
$\mu, \nu \in \mathbb{N}_0$, $\mu \neq \nu$, *the spaces* $\overset{*}{\mathbb{H}}_\mu^r(S^{r-1})$ *and* $\overset{*}{\mathbb{H}}_\nu^r(S^{r-1})$ *are orthogonal with*
respect to the inner product $\langle \cdot, \cdot \rangle$.

Proof. Let $A \in \overset{*}{\mathbb{H}}_\mu^r(S^{r-1})$ and $B \in \overset{*}{\mathbb{H}}_\nu^r(S^{r-1})$ be arbitrary. By the reproducing
property of G_μ and of G_ν we get, for $x \in S^{r-1}$,

$$A(x) = \langle G_\mu(x \cdot), A \rangle, \quad B(x) = \langle G_\nu(x \cdot), B \rangle.$$

Using shorthand notation, we obtain

$$\langle A, B \rangle = \int\limits_{|x|=1} A(x)B(x)\,dx$$

$$= \int\limits_{|x|=1} \left\{ \int\limits_{|y|=1} G_\mu(xy)A(y)\,dy \cdot \int\limits_{|z|=1} G_\nu(xz)B(z)dz \right\} dx$$

$$= \int\limits_{|y|=1} \int\limits_{|z|=1} A(y)B(z) \left\{ \int\limits_{|x|=1} G_\mu(yx)G_\nu(xz)\,dx \right\} dz\,dy$$

$$= 0,$$

where we used Corollary 4.9 to get the last equation.

In the following we use the symbol ϵV, $\epsilon \in \{0, 1\}$ in a direct sum in order to express that the subspace V actually occurs for $\epsilon = 1$, but is to be omitted for $\epsilon = 0$.

The Finite-Dimensional Rotation-Invariant Subspaces of $\mathbb{P}^r_\mu(S^{r-1})$

We are now able to describe all finite-dimensional and rotation-invariant subspaces of $\mathbb{P}^r_\mu(S^{r-1})$.

Theorem 4.11 (The Rotation-Invariant Subspaces in $\mathbb{P}^r(S^{r-1})$). *Let* $r \in \mathbb{N} \setminus \{1\}$, *and assume* V *is a finite-dimensional subspace of* $\mathbb{P}^r(S^{r-1})$. *V is rotation-invariant if and only if it has the form*

$$V = \bigoplus_{\nu=0}^{\mu} \epsilon_\nu \overset{*}{\mathbb{H}}{}^r_\nu(S^{r-1}), \tag{4.17}$$

where $\mu \in \mathbb{N}_0$, $\epsilon_0, \ldots, \epsilon_\mu \in \{0, 1\}$. *The decomposition is orthogonal with respect to the inner product* $\langle \cdot, \cdot \rangle$. *The reproducing kernel of* V *is given by*

$$G(x, y) = \sum_{\nu=0}^{\mu} \epsilon_\nu G^r_\nu(xy) \quad \text{for } x, y \in S^{r-1}. \tag{4.18}$$

Proof. Obviously, if V has the representation (4.17), then it is rotation-invariant. Next assume V is a finite-dimensional rotation-invariant subspace. It is contained in $\mathbb{P}^r_\mu(S^{r-1})$ for some $\mu \in \mathbb{N}_0$. By Theorem 1.11 its reproducing kernel has the form $G(x, y) = K(xy)$, where $K \in \mathbb{P}^1_\mu([-1, 1])$ holds by Lemma 4.5. From (4.12) we get

$$G_\nu \in \mathbb{P}^1_\nu \setminus \mathbb{P}^1_{\nu-1} \tag{4.19}$$

for $\nu \in \mathbb{N}_0$, where we recall once more the definition $\mathbb{P}_{-1} = [0]$. It follows that K has a uniquely determined representation

$$K(\xi) = \sum_{\nu=0}^{\mu} \epsilon_\nu G_\nu(\xi) \tag{4.20}$$

with real coefficients ϵ_ν. Since $K(xy)$ is the reproducing kernel of $\mathbf{V}$, we get

$$K(xy) = \langle K(x\cdot), K(\cdot y)\rangle$$

for $x, y \in S^{r-1}$. Here we insert (4.20) on both sides and obtain

$$K(xy) = \sum_{\nu=0}^{\mu}\sum_{\kappa=0}^{\mu} \epsilon_\nu \epsilon_\kappa \langle G_\nu(x\cdot), G_\kappa(\cdot y)\rangle.$$

In view of (4.16) it follows that

$$K(xy) = \sum_{\nu=0}^{\mu} \epsilon_\nu^2 G_\nu(xy)$$

for arbitrary $x, y \in S^{r-1}$. The values of $\xi = xy$ cover the interval $-1 \le \xi \le 1$ while x and y vary in S^{r-1}. This implies

$$K(\xi) = \sum_{\nu=0}^{\mu} \epsilon_\nu^2 G_\nu(\xi). \tag{4.21}$$

A comparison of the coefficients in (4.21) and in (4.20) yields $\epsilon_\nu^2 = \epsilon_\nu$, which is equivalent to $\epsilon_\nu \in \{0, 1\}$. So (4.18) is valid.

Finally let $F \in \mathbf{V}$ be arbitrary. Then we get for $x \in S^{r-1}$

$$F(x) = \langle K(x\cdot), F\rangle = \sum_{\nu=0}^{\mu} \epsilon_\nu \langle G_\nu(x\cdot), F\rangle.$$

The function F_ν defined by $F_\nu(x) := \langle G_\nu(x\cdot), F\rangle$ for $x \in S^{r-1}$ is contained in $\overset{*}{\mathbb{H}}{}_\nu^r(S^{r-1})$. This follows immediately by representing $G_\nu(x\cdot)$ as in (1.5) by an orthonormal basis of $\overset{*}{\mathbb{H}}{}_\nu^r(S^{r-1})$. Therefore,

$$F = \sum_{\nu=0}^{\mu} \epsilon_\nu F_\nu,$$

is an orthogonal decomposition, which corresponds to the decomposition (4.17). $\square$

The Spaces $\mathbb{P}_\mu^r(S^{r-1})$ and $\overset{*}{\mathbb{P}}{}_\mu^r(S^{r-1})$

Let $\mathbf{V}$ be defined as above but by the choice of $\epsilon_0 = \epsilon_1 = \ldots = 1$. Then

$$\dim \mathbf{V} = \sum_{\nu=0}^{\mu} \dim \overset{*}{\mathbb{H}}{}_\mu^r(S^{r-1})$$

holds. In view of (4.2) and of (4.4) this yields

$$\dim \mathbf{V} = \dim \mathbb{P}^r_\mu(S^{r-1}).$$

But $\mathbf{V}$ is a subspace of $\mathbb{P}^r_\mu(S^{r-1})$. So $\mathbf{V} = \mathbb{P}^r_\mu(S^{r-1})$ must hold, and we obtain

$$\mathbb{P}^r_\mu(S^{r-1}) = \bigoplus_{\nu=0}^{\mu} \overset{*}{\mathbb{H}}{}^r_\nu(S^{r-1}). \tag{4.22}$$

Next we recall (4.3), which is a decomposition of $\mathbb{P}^r_\mu(S^{r-1})$ by its even and its odd component, or vice versa. If we decompose (4.22) correspondingly, then we get

$$\overset{*}{\mathbb{P}}{}^r_\mu(S^{r-1}) = \bigoplus_{\nu=0}^{\lfloor \frac{\mu}{2} \rfloor} \overset{*}{\mathbb{H}}{}^r_{\mu-2\nu}(S^{r-1}), \tag{4.23}$$

$$\overset{*}{\mathbb{P}}{}^r_{\mu-1}(S^{r-1}) = \bigoplus_{\nu=0}^{\lfloor \frac{\mu-1}{2} \rfloor} \overset{*}{\mathbb{H}}{}^r_{\mu-1-2\nu}(S^{r-1}). \tag{4.24}$$

Equation (4.22) says that every spherical polynomial is the sum of spherical harmonics and hence the restriction of a harmonic polynomial of the same degree. It follows that

$$\mathbb{H}^r_\mu(S^{r-1}) \subset \mathbb{P}^r_\mu(S^{r-1}) \subset \mathbb{H}^r_\mu(S^{r-1}),$$

where the left-side inclusion is trivial. Together this yields

$$\mathbb{H}^r_\mu(S^{r-1}) = \mathbb{P}^r_\mu(S^{r-1}). \tag{4.25}$$

This result is interpreted by

Theorem 4.12 (Harmonic Extension of a Spherical Polynomial). *Let* $r \in \mathbb{N} \setminus \{1\}$. *The harmonic extension of a spherical polynomial is a harmonic polynomial of the same degree.*

We summarize the most essential results by the following space diagram,

$$\mathbb{P}^r_\mu(S^{r-1}) = \overset{*}{\mathbb{P}}{}^r_\mu(S^{r-1}) \oplus \overset{*}{\mathbb{P}}{}^r_{\mu-1}(S^{r-1})$$
$$\|$$
$$\mathbb{H}^r_\mu(S^{r-1}) = \bigoplus_{\nu=0}^{\mu} \overset{*}{\mathbb{H}}{}^r_\nu(S^{r-1}). \tag{4.26}$$

Finally we recall that the restricted harmonic spaces are isomorphic with the unrestricted ones. This proves the anticipated formula (3.6).

The reproducing kernels of the spaces $\overset{*}{\mathbb{P}}{}^r_\mu(S^{r-1})$ and $\mathbb{P}^r_\mu(S^{r-1})$ are known from Theorem 4.11, in principle. In view of Definition 4.1 they are given by

$$\overset{*}{\varGamma}{}^r_\mu = \sum_{\nu=0}^{\lfloor \frac{\mu}{2} \rfloor} G^r_{\mu-2\nu} \tag{4.27}$$

and
$$\Gamma_\mu^r = \sum_{\nu=0}^{\mu} G_\nu^r, \tag{4.28}$$

respectively. These representations can be simplified as follows.

First we assume $r \geq 3$ and let $\lambda := \frac{r-2}{2}$. For $\mu \in \mathbb{N}_0$ we get from (4.13)

$$G_\mu^r = \frac{\mu + \lambda}{\lambda \, \omega_{r-1}} C_\mu^\lambda,$$

and hence

$$
\begin{aligned}
\omega_{r-1} \sum_{\mu=0}^{\infty} G_\mu^r(\xi) \tau^\mu
&= \tfrac{1}{\lambda} \sum_{\mu=0}^{\infty} (\mu + \lambda) C_\mu^\lambda(\xi) \tau^\mu \\
&= \left(\tfrac{1}{\lambda} \cdot \tau \tfrac{\partial}{\partial \tau} + 1 \right) \frac{1}{(1 - 2\xi\tau + \tau^2)^\lambda} \\
&= \frac{1 - \tau^2}{(1 - 2\xi\tau + \tau^2)^{\lambda+1}} \\
&= \sum_{\mu=0}^{\infty} \left[C_\mu^{\lambda+1}(\xi) - C_{\mu-2}^{\lambda+1}(\xi) \right] \tau^\mu,
\end{aligned}
$$

again with $C_{-1}^{\lambda+1} = C_{-2}^{\lambda+1} = 0$. By a comparison of the coefficients we obtain

$$G_\mu^r = \tfrac{1}{\omega_{r-1}} \left[C_\mu^{\frac{r}{2}} - C_{\mu-2}^{\frac{r}{2}} \right],$$

and together with (4.27) this yields finally

$$\overset{*}{\Gamma}_\mu^r = \tfrac{1}{\omega_{r-1}} C_\mu^{\frac{r}{2}}, \tag{4.29}$$

again for $\mu \in \mathbb{N}_0$.

Next let $r = 2$. From (4.13) we obtain

$$
\begin{aligned}
G_\mu^2 &= \tfrac{2}{\omega_1} T_\mu &= \tfrac{1}{\omega_1} [U_\mu - U_{\mu-2}], \quad \text{if } \mu \in \mathbb{N}, \\
G_0^2 &= \tfrac{1}{\omega_1} T_0 &= \tfrac{1}{\omega_1} [U_0 - U_{-2}]
\end{aligned}
$$

with $U_{-1} = U_{-2} = 0$, and in view of $U_\mu = C_\mu^1$, see Section 2.1, (4.29) follows from (4.27), again. In other words, (4.29) is valid for arbitrary $r \in \mathbb{N} \setminus \{1\}$.

Finally we get from (4.27) – (4.29)

$$\Gamma_\mu^r = \frac{1}{\omega_{r-1}} \left[C_\mu^{\frac{r}{2}} + C_{\mu-1}^{\frac{r}{2}} \right]. \tag{4.30}$$

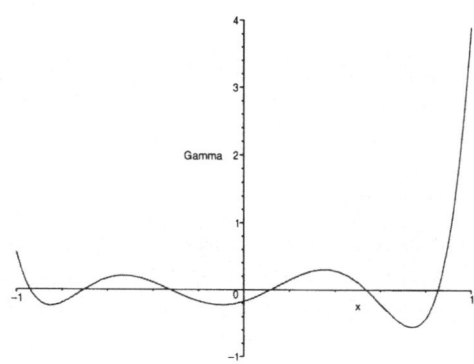

Figure 4.1. Reproducing Kernel Function Γ_6^3.

The formulae (4.29) and (4.30) are the kernel representations wanted. However, sometimes it is important to know that Γ_μ^r can also be expressed by means of a unique Jacobi polynomial in the form

$$\Gamma_\mu^r = \frac{2}{\omega_{r-1}} \cdot \frac{(r)_{\mu-1}}{(\frac{r+1}{2})_{\mu-1}} \cdot P_\mu^{(\frac{r-1}{2}, \frac{r-3}{2})} \tag{4.31}$$

for $r \in \mathbb{N} \setminus \{1\}$ and $\mu \in \mathbb{N}$. We prove this formula as follows. For arbitrary $f \in \mathbb{P}_{\mu-1}^1$ and $t = e_1 \in S^{r-1}$ we get by the reproducing property of Γ_μ^r,

$$
\begin{aligned}
0 &= \int_{S^{r-1}} \Gamma_\mu^r(tx) \big[f(tx)(1 - tx) \big] d\omega(x) \\
&= \int_{S^{r-1}} \Gamma_\mu^r(x_1) \big[f(x_1)(1 - x_1) \big] d\omega(x) \\
&= \omega_{r-2} \int_{-1}^1 \Gamma_\mu^r(\xi) f(\xi)(1 - \xi)^{\frac{r-1}{2}} (1 + \xi)^{\frac{r-3}{2}} d\xi,
\end{aligned}
$$

where we used (1.25) with $s = r - 2$ in order to get the last equation. So, Γ_μ^r is orthogonal to $\mathbb{P}_{\mu-1}^1$, and this implies

$$\Gamma_\mu^r(\xi) = const \cdot P_\mu^{(\frac{r-1}{2}, \frac{r-3}{2})}(\xi).$$

The constant can be calculated from the values for $\xi = 1$. For, $P_\mu^{(\frac{r-1}{2}, \frac{r-3}{2})}(1)$ is known from (2.2), and $\Gamma_\mu^r(1)$ from (4.6), which takes the form

$$\Gamma_\mu^r(1) = \frac{1}{\omega_{r-1}} \dim \mathbb{P}_\mu^r(S^{r-1}).$$

Note that the dimension on the right side is known from (4.4). Together this yields (4.31).

Remark. In view of (4.29) and of (4.31), the kernel functions $\overset{*}{\Gamma^r_\mu}$ and Γ^r_μ are orthogonal polynomials with respect to the interval $-1 \leq \xi \leq 1$. Therefore they have μ simple roots in the interior of the interval.

The Laplace–Beltrami Operator

The spaces of spherical harmonics are related to the Laplace–Beltrami operator on the sphere. To explain this, we assume that F is a real function which is defined in an open neighbourhood of S^{r-1} and twice continuously differentiable. In every point $x \in S^{r-1}$ we define for arbitrary $a \in S^{r-1}$ with $ax = 0$ the directional derivative

$$(D_a^{(k)} F)(x) := \left[\left(\frac{d}{d\phi} \right)^k F(x \cos \phi + a \sin \phi) \right]_{\phi=0}$$

for $k \in \{0, 1, 2\}$. Using the abbreviations

$$F^{(\nu)} = \frac{\partial}{\partial x_\nu} F \quad \text{and} \quad F^{(\nu,\kappa)} = \frac{\partial^2}{\partial x_\nu \partial x_\kappa} F$$

for $\nu, \kappa = 1, \ldots, r$, we get by some calculation the equation

$$\left(D_a^{(2)} F \right)(x) = \sum_{\nu=1}^{r} \sum_{\kappa=1}^{r} a_\nu F^{(\nu,\kappa)}(x) a_\kappa - x'(gradF)(x).$$

By means of the Hesse matrix $F''(x) := \left(F^{(\nu,\kappa)}(x) \right)_{\nu.\kappa=1,\ldots,r}$ we bring this to the form

$$(D_a^{(2)} F)(x) = a' F''(x) a - x'(gradF)(x). \tag{4.32}$$

Now we complete $u_r := x$ by $u_1, \ldots, u_{r-1} \in S^{r-1}$ such that $U = (u_1, \ldots, u_r) \in \mathbf{A}^r$. Then we get

$$(\Delta F)(x) = trace\left(F''(x) \right) = trace\left(U'F''(x)U \right), \tag{4.33}$$

independently of the choice of $u_1, \ldots, u_{r-1}$. In view of (4.32) this implies

$$(\Delta F)(x) = \sum_{\nu=1}^{r-1} \left(D_{u_\nu}^{(2)} F \right)(x) + (r-1)x'(gradF)(x) + x'F''(x) x. \tag{4.34}$$

It follows that the expression

$$(\tilde{\Delta}F)(x) := \sum_{\nu=1}^{r-1} \left(D_{u_\nu}^{(2)} F \right)(x) \tag{4.35}$$

depends on x, only, and not on the choice of the system $u_1, \ldots, u_{r-1}$. Therefore it defines a map

$$F \mapsto \tilde{\Delta}F.$$

$\tilde{\Delta}$ is called the *Laplace–Beltrami operator* on S^{r-1}. Note that the definition of $\tilde{\Delta}F$ uses function values taken from S^{r-1}, only. So $\tilde{\Delta}$ belongs to the *inner geometry* of the sphere. Actually, it allows us a better understanding of the spaces of spherical harmonics.

Theorem 4.13 (Eigenspaces of $\tilde{\Delta}$). *For $F \in \overset{*}{\mathbb{P}}{}_\mu^r$, $r \in \mathbb{N} \setminus \{1\}$, $\mu \in \mathbb{N}_0$,*

$$\Delta F = \tilde{\Delta}F + \mu(\mu + r - 2)F$$

holds. $\overset{*}{\mathbb{H}}{}_\mu^r(S^{r-1})$ *is an eigenspace of $\tilde{\Delta}$ with respect to the eigenvalue $-\mu(\mu+r-2)$.*

Proof. Assume $F \in \overset{*}{\mathbb{P}}{}_\mu^r$. Then $x'(gradF)(x) = \mu F(x)$ holds, which is Euler's partial differential equation. Since the components of $grad\, F$ are in $\overset{*}{\mathbb{P}}{}_{\mu-1}^r$, we get likewise

$$x' F''(x)\, x = \mu(\mu - 1)F(x).$$

Inserting this in (4.34) we obtain, together with (4.35), the first statement. But for $F \in \overset{*}{\mathbb{H}}{}_\mu^r$ we get $\Delta F = 0$ and hence $\tilde{\Delta}F = -\mu(\mu + r - 2)F$, as claimed. $\qquad\square$

4.2 Biorthonormal Systems on the Sphere

a) Biorthonormal Systems of Homogeneous Polynomials

Let $r \in \setminus\{1\}$, $\mu \in \mathbb{N}_0$. The polynomial $P \in \overset{*}{\mathbb{P}}{}_\mu^r$ vanishes on S^{r-1} if and only if $P = 0$. Therefore $\langle \cdot, \cdot \rangle$ is an inner product on $\overset{*}{\mathbb{P}}{}_\mu^r$, and

$$\| \cdot \|_2 := \sqrt{\langle \cdot, \cdot \rangle}$$

is a norm by which $\overset{*}{\mathbb{P}}{}_\mu^r$ and $\overset{*}{\mathbb{P}}{}_\mu^r(S^{r-1})$ are isometric.

Because of (4.29) we may identify Q_μ in (3.39) with $\overset{*}{\Gamma}{}_\mu^r$, and (3.41) takes the form

$$|t|^\mu \overset{*}{\Gamma}{}_\mu^r \left(\frac{tx}{|t|} \right) = \sum_{|m|=\mu} P_m(x)t^m. \tag{4.36}$$

This defines the polynomials $P_m \in \mathbb{P}_m^r$ for $|m| = \mu$, which correspond to the A_m. Using the notation of (3.52) we find that the coefficients of Q_μ are now given by

$$a_{\mu-2\nu} = \frac{1}{\omega_{r-1}} a_{\mu-2\nu}^{\frac{r}{2}},$$

and for $t, x \in S^{r-1}$, we get from (3.49)

$$\overset{*}{\Gamma}_\mu^r(tx) = \sum_{|m|=\mu} \overset{*}{P}_m(t) M_m(x). \tag{4.37}$$

Recall that M_m is defined by $M_m(x) = x^m$.

Next let $P \in \overset{*}{\mathbb{P}}_\mu^r$ be arbitrary. By the reproducing property of $\overset{*}{\Gamma}_\mu^r$ we get

$$P(x) = \langle \overset{*}{\Gamma}_\mu^r(x \cdot), P \rangle = \sum_{|m|=\mu} \langle \overset{*}{P}_m, P \rangle M_m(x)$$

for $x \in S^{r-1}$. But the polynomials on both sides are homogeneous. So we obtain the even stronger result

$$P = \sum_{|m|=\mu} \langle \overset{*}{P}_m, P \rangle M_m. \tag{4.38}$$

Theorem 4.14 (Biorthonormal Systems in $\overset{*}{\mathbb{P}}_\mu^r$). *Let* $r \in \mathbb{N} \setminus \{1\}$, $\mu \in \mathbb{N}_0$, $\mathcal{M}(r, \mu) := \{m \in \mathbb{N}_0^r : |m| = \mu\}$. *The system* $\{\overset{*}{P}_m\}_{m \in \mathcal{M}(r,\mu)}$ *and the monomial system* $\{M_m\}_{m \in \mathcal{M}(r,\mu)}$ *are biorthonormal, i.e.,*

$$\langle M_m, \overset{*}{P}_n \rangle = \delta_{m,n} \tag{4.39}$$

holds for $m, n \in \mathcal{M}(r, \mu)$. *Each of the systems forms a basis in* $\overset{*}{\mathbb{P}}_\mu^r$.

Proof. Obviously, the monomials form a basis. Next choose $P = M_m$ where $|m| = \mu$. Then (4.38) takes the form

$$M_m = \sum_{|n|=\mu} \langle \overset{*}{P}_n, M_m \rangle M_n, \tag{4.40}$$

and (4.39) is valid. Vice versa (4.39) implies that the $\overset{*}{P}_n$ are linearly independent, but their number equals the dimension. This finishes the proof. $\square$

From Theorem 4.14 it follows that M_m has a representation

$$M_m = \sum_{|l|=\mu} c_{m,l} \overset{*}{P}_l$$

with some real coefficients $c_{m,l}$. Together with (4.39) this yields, for $|m| = \mu = |n|$,

$$\langle M_m, M_n \rangle = \sum_{|l|=\mu} c_{m,l} \langle \overset{*}{P}_l, M_n \rangle = c_{m,n},$$

and hence

$$M_m = \sum_{|n|=\mu} \langle M_m, M_n \rangle \overset{*}{P}_n. \tag{4.41}$$

But inserting $P = \overset{*}{P}_n$ in (4.38) we get likewise

$$\overset{*}{P}_n = \sum_{|m|=\mu} \langle \overset{*}{P}_m, \overset{*}{P}_n \rangle M_m, \tag{4.42}$$

again for $|n| = \mu$. These are basis transforms, so

$$\left(\langle \overset{*}{P}_m, \overset{*}{P}_n \rangle \right)^{-1} = \left(\langle M_m, M_n \rangle \right) \tag{4.43}$$

must hold. Both matrices are Gram matrices and hence positive definite. The matrix $(\langle M_m, M_n \rangle)$ can be evaluated since

$$\langle M_m, M_n \rangle = \int_{S^{r-1}} x^{m+n} d\omega(x)$$

is known, see Problem 3.3.

Coefficient Functionals in $\overset{*}{\mathbb{P}}{}^r_\mu$, Representers and Norms

Formula (4.38) can be written in the form

$$c_m(P) = \langle \overset{*}{P}_m, P \rangle \tag{4.44}$$

for $P \in \overset{*}{\mathbb{P}}{}^r_\mu$, which is saying that $\overset{*}{P}_m$ is the *representer* of the coefficient functional

$$\overset{*}{c}{}^P_m := c_m \big|_{\overset{*}{\mathbb{P}}{}^r_\mu}$$

for $|m| = \mu$. With

$$\| \overset{*}{c}{}^P_m \|_{2,\infty} := \max\{ |c_m(P)| : P \in \overset{*}{\mathbb{P}}{}^r_\mu, \| P \|_2 \leq 1 \}$$

we get

Theorem 4.15 (Norm of $\overset{*}{c}{}_m^P$). Let $r \in \mathbb{N} \setminus \{1\}$, $m \in \mathbb{N}_0^r$, $|m| = \mu$. Then

$$|c_m(P)| \leq c_m(\overset{*}{P}_m) \tag{4.45}$$

holds for all $P \in \overset{*}{\mathbb{P}}_\mu^r$ which satisfy $\|P\|_2 \leq \|\overset{*}{P}_m\|_2$. Equality holds if and only if $P \in \{+\overset{*}{P}_m, -\overset{*}{P}_m\}$. Moreover,

$$\|\overset{*}{c}{}_m^P\|_{2,\infty} = \|\overset{*}{P}_m\|_2 \tag{4.46}$$

is valid.

Proof. Using the assumptions, we get from (4.44),

$$|c_m(P)| \leq \|\overset{*}{P}_m\|_2 \cdot \|P\|_2 \leq \|\overset{*}{P}_m\|_2^2,$$

and hence

$$|c_m(P)| \leq c_m(\overset{*}{P}_m).$$

Equality holds if and only if P has the form $P = \gamma \overset{*}{P}_m$, where

$$|\gamma| \cdot |c_m(\overset{*}{P}_m)| = c_m(\overset{*}{P}_m).$$

Because of $c_m(\overset{*}{P}_m) = \|\overset{*}{P}_m\|_2^2 > 0$, this holds if and only if $\gamma \in \{+1, -1\}$. Next assume $\|P\|_2 \leq 1$. Then we get, using the result from above,

$$\|\overset{*}{P}_m\|_2 |c_m(P)| = |c_m(\|\overset{*}{P}_m\|_2 \cdot P)| \leq c_m(\overset{*}{P}_m) = \|\overset{*}{P}_m\|_2^2,$$

and hence

$$|c_m(P)| \leq \|\overset{*}{P}_m\|_2.$$

Equality holds for $P = \overset{*}{P}_m / \|\overset{*}{P}_m\|_2$. Together this implies (4.46). $\qquad\square$

We evaluate $\|\overset{*}{P}_m\|_2^2 = c_m(\overset{*}{P}_m)$ by means of Theorem 3.13, where we identify $\overset{*}{A}_m$ and $\overset{*}{P}_m$. It follows that

$$c_m(\overset{*}{P}_m) = \frac{1}{\omega_{r-1}} \sum_{|n| \leq \lfloor \frac{\mu}{2} \rfloor} a_{\mu-2|n|}^{\frac{r}{2}} \binom{\mu - 2|n|}{m - 2n} \binom{|n|}{n}^2, \tag{4.47}$$

where we used the particular choice of Q_μ together with (4.29) and (3.52).

b) Biorthonormal Systems of Harmonic Homogeneous Polynomials

Let $r \in \setminus \{-1\}$, $\mu \in \mathbb{N}_0$, again. But now we identify Q_μ in (3.39) with G_μ^r, which is possible in view of (4.12). In this case (3.41) takes the form

$$|t|^\mu G_\mu^r\left(\frac{tx}{|t|}\right) = \sum_{|m|=\mu} R_m(x)t^m,$$ (4.48)

which defines the polynomials

$$R_m \in \mathbb{P}_m^r$$ (4.49)

for $|m| = \mu$ by identification with the corresponding A_m. Likewise (3.49) takes the form

$$\sum_{|m|=\mu} \overset{*}{R}_m(x)t^m = |t|^\mu|x|^\mu G_\mu^r\left(\frac{tx}{|t||x|}\right) = \sum_{|m|=\mu} \overset{*}{R}_m(t)x^m.$$ (4.50)

For fixed $t \in S^{r-1}$ we get

$$\Delta|x|^\mu G_\mu^r\left(\frac{tx}{|x|}\right) = 0,$$

since $f = G_\mu^r$ satisfies the differential equation (4.11), such that (4.10) is valid. It follows that

$$0 = \Delta \sum_{|m|=\mu} \overset{*}{R}_m(x)t^m = \sum_{|m|=\mu} (\Delta \overset{*}{R}_m)(x)t^m$$

for $x \in \mathbb{R}^r$ and $t \in S^{r-1}$. This yields $(\Delta \overset{*}{R}_m)(x) = 0$, and hence

$$\overset{*}{R}_m \in \overset{*}{\mathbb{H}}_\mu^r \text{ for } |m| = \mu.$$ (4.51)

But for $\mu \geq 2$ the situation is different now, since the number of the $\overset{*}{R}_m$ with $|m| = \mu$ is now greater than the dimension of $\overset{*}{\mathbb{H}}_\mu^r$, i.e. the $\overset{*}{R}_m$ are no longer linearly independent. But in spite of this the following remains valid.

Coefficient Functionals in $\overset{*}{\mathbb{H}}_\mu^r$, Representers and Norms

For $H \in \overset{*}{\mathbb{H}}_\mu^r$ and $x \in S^{r-1}$ we get, using the right side of (4.50),

$$H(x) = \langle G_\mu^r(x \cdot), H \rangle = \sum_{|m|=\mu} \langle \overset{*}{R}_m, H \rangle x^m,$$

which implies

$$H = \sum_{|m|=\mu} \langle \overset{*}{R}_m, H \rangle M_m.$$ (4.52)

This result is comparable with (4.38) and says that

$$c_m(H) = \langle \overset{*}{R}_m, H \rangle \tag{4.53}$$

holds for $H \in \overset{*}{\mathbb{H}}{}^r_\mu$ and $|m| = \mu$. So $\overset{*}{R}_m$ is the *representer* of the coefficient functional

$$\overset{*}{c}{}^H_m := c_m \big|_{\overset{*}{\mathbb{H}}{}^r_\mu} .$$

With

$$\| \overset{*}{c}{}^H_m \|_{2,\infty} := \max\{ |c_m(H)| : H \in \overset{*}{\mathbb{H}}{}^r_\mu, \ \|H\|_2 \le 1 \}$$

we get the following theorem, which corresponds to Theorem 4.15.

Theorem 4.16 (Norm of $\overset{*}{c}{}^H_m$). *Let $r \in \mathbb{N} \setminus \{1\}$, $m \in \mathbb{N}^r_0$, $|m| = \mu$. Then*

$$|c_m(H)| \le c_m(\overset{*}{R}_m) \tag{4.54}$$

holds for all $H \in \overset{*}{\mathbb{H}}{}^r_\mu$ satisfying $\|H\|_2 \le \|\overset{*}{R}_m\|_2$. Equality holds if and only if $H \in \{ +\overset{*}{R}_m, -\overset{*}{R}_m \}$. Moreover,

$$\| \overset{*}{c}{}^H_m \|_{2,\infty} = \| \overset{*}{R}_m \|_2 \tag{4.55}$$

is valid.

Proof. The proof follows the lines of the proof of Theorem 4.15.   □

We can evaluate $\| \overset{*}{R}_m \|^2_2 = c_m(\overset{*}{R}_m)$ again by means of Theorem 3.13, where $\overset{*}{A}_m$ has now to be identified with $\overset{*}{R}_m$. So we get for $r \ge 3$

$$c_m(\overset{*}{R}_m) = \frac{2\mu + r - 2}{(r-2)\omega_{r-1}} \sum_{|n| \le \lfloor \frac{\mu}{2} \rfloor} a_{\mu-2|n|}^{\frac{r-2}{2}} \binom{\mu - 2|n|}{m - 2n} \binom{|n|}{n}^2 , \tag{4.56}$$

where we used the particular choice of Q_μ, (4.13), and (3.52). For $r = 2$ a similar equation holds.

Constructive Harmonic Extension of Spherical Polynomials

The $\overset{*}{R}_m$ are homogeneous and harmonic. Moreover, they agree on the unit sphere with R_m, i.e., they satisfy

$$\overset{*}{R}_m(x) = R_m(x) \text{ for all } x \in S^{r-1}.$$

Because of these properties, the following theorem holds.

Theorem 4.17 (Harmonic Extension of a Spherical Polynomial). *Let $r \in \mathbb{N} \setminus \{1\}$, $\mu \in \mathbb{N}_0$. Every $P \in \mathbb{P}_\mu^r$ has a uniquely determined representation*

$$P = \sum_{|m| \leq \mu} a_m R_m$$

with real coefficients a_m, and

$$\mathcal{H}(P) := \sum_{|m| \leq \mu} a_m \overset{*}{R}_m$$

is the harmonic extension of $P|_{S^{r-1}}$ to $\mathbb{R}^r$.

Proof. The R_m inherit the basis property from the A_m, see Theorem 3.12. Therefore, P can be represented uniquely as stated. $\mathcal{H}(P)$ is harmonic by construction, and for $x \in S^{r-1}$ we get

$$\mathcal{H}(P)(x) = \sum_{|m|=\mu} a_m \overset{*}{R}_m(x) = \sum_{|m|=\mu} a_m R_m(x) = P(x).$$

This finishes the proof. $\square$

Selection of a Basis

We mentioned already, for $\mu \geq 2$ the $\overset{*}{R}_m$ with $|m| = \mu$ are linearly dependent. So the question arises how a basis can be obtained by selection of a proper subsystem.

In view of (4.51) the right side of (4.50) is harmonic with respect to t. It follows that the $\overset{*}{R}_m(x)$ occur on the left side as the coefficients of a harmonic homogeneous polynomial of degree μ, which satisfy, as we know already, a difference equation. Actually, using Lemma 3.8 we get

$$\overset{*}{R}_m = (-1)^{\lfloor \frac{m_r}{2} \rfloor} \binom{\mu}{m} \sum_{|\bar{n}|=\lfloor \frac{m_r}{2} \rfloor} \binom{|\bar{n}|}{\bar{n}} \binom{\mu}{\bar{m} + 2\bar{n}, m_r - 2\lfloor \frac{m_r}{2} \rfloor}^{-1} \overset{*}{R}_{\bar{m}+2\bar{n}, m_r - 2\lfloor \frac{m_r}{2} \rfloor}.$$
$$(4.57)$$

Now let $H \in \overset{*}{\mathbb{H}}_\mu^r$ and $x \in S^{r-1}$ be arbitrary, again. Using the left side of (4.50) we obtain

$$H(x) = \langle G_\mu^r(x \cdot), H \rangle = \sum_{|m|=\mu} \langle M_m, H \rangle \overset{*}{R}_m(x),$$

and hence

$$H = \sum_{|m|=\mu} \langle M_m, H \rangle \overset{*}{R}_m. \qquad (4.58)$$

In view of (4.51) and of (4.57), this yields the right-side inclusion of

$$\text{span}\{\overset{*}{R}_m| m \in \mathcal{I}\} \subset \overset{*}{\mathbb{H}}{}^r_\mu \subset \text{span}\{\overset{*}{R}_m|m \in \mathcal{I}\},$$

while the left-side inclusion is obvious. For the definition of the index set $\mathcal{I} = \mathcal{I}(r, \mu)$ we recall (3.33). In other words,

$$\overset{*}{\mathbb{H}}{}^r_\mu = \text{span}\{\overset{*}{R}_m| m \in \mathcal{I}\} \tag{4.59}$$

is valid. This is, in essential, the statement of the following theorem.

Theorem 4.18 (Basis in $\overset{*}{\mathbb{H}}{}^r_\mu$). *Let* $r \in \mathbb{N}\backslash\{1\}$, $\mu \in \mathbb{N}_0$. *The system* $\{\overset{*}{R}_m\}_{m \in \mathcal{I}(r,\mu)}$ *is a basis of* $\overset{*}{\mathbb{H}}{}^r_\mu$.

Proof. In view of (3.33) and of Corollary 3.11 we get

$$|\mathcal{I}| = \dim \overset{*}{\mathbb{H}}{}^r_\mu.$$

Together with (4.59) this yields the statement.   □

Now recall once more (4.50), but we use only the equation

$$\sum_{|m|=\mu} \overset{*}{R}_m(x)t^m = \sum_{|m|=\mu} \overset{*}{R}_m(t)x^m. \tag{4.60}$$

If we represent the $\overset{*}{R}_m$, which occur on the right side, by the basis of Theorem 4.18, then this takes the form

$$\sum_{|m|=\mu} \overset{*}{R}_m(x)t^m = \sum_{m \in \mathcal{I}} \overset{*}{R}_m(t)\overset{*}{S}_m(x) \tag{4.61}$$

with some $\overset{*}{S}_m \in \overset{*}{\mathbb{P}}{}^r_\mu$. Now we apply the same procedure to the left side and obtain

$$\sum_{m \in \mathcal{I}} \overset{*}{R}_m(x)\overset{*}{S}_m(t) = \sum_{m \in \mathcal{I}} \overset{*}{R}_m(t)\overset{*}{S}_m(x) \tag{4.62}$$

for arbitrary $t, x \in \mathbb{R}^r$. By applying the Laplace operator with respect to the variable x we get

$$0 = \sum_{m \in \mathcal{I}} \left(\Delta\overset{*}{R}_m\right)(x)\overset{*}{S}_m(t) = \sum_{m \in \mathcal{I}} \overset{*}{R}_m(t)\left(\Delta\overset{*}{S}_m\right)(x),$$

and since the $\overset{*}{R}_m$, which occur on the right side, are linearly independent we conclude that $\Delta \overset{*}{S}_m = 0$ must hold. Together this yields

$$\overset{*}{S}_m \in \overset{*}{\mathbb{H}}^r_\mu$$

for $|m| = \mu$. So we derived another particular system of harmonic homogeneous polynomials, and actually it has the property wanted.

Theorem 4.19 (Biorthonormal Systems in $\overset{*}{\mathbb{H}}^r_\mu$). *Let* $r \in \mathbb{N} \setminus \{1\}$, $\mu \in \mathbb{N}_0$. *The systems* $\{\overset{*}{R}_m\}_{m \in \mathcal{I}(r,\mu)}$ *and* $\{\overset{*}{S}_m\}_{m \in \mathcal{I}(r,\mu)}$ *are biorthonormal bases in* $\overset{*}{\mathbb{H}}^r_\mu$, *i.e.,* $\overset{*}{R}_m, \overset{*}{S}_n \in \overset{*}{\mathbb{H}}^r_\mu$, *and*

$$\langle \overset{*}{R}_m, \overset{*}{S}_n \rangle = \delta_{m,n} \tag{4.63}$$

holds for $m, n \in \mathcal{I}(r,\mu)$.

Proof. By Theorem 4.18 the $\overset{*}{R}_m$, $m \in \mathcal{I} = \mathcal{I}(r,\mu)$, form a basis of $\overset{*}{\mathbb{H}}^r_\mu$. Moreover,

$$G^r_\mu(tx) = \sum_{n \in \mathcal{I}} \overset{*}{R}_n(t) \overset{*}{S}_n(x)$$

holds for $t, x \in S^{r-1}$. It follows that

$$\overset{*}{R}_m(t) = \langle G^r_\mu(t\,\cdot\,), \overset{*}{R}_m \rangle = \sum_{n \in \mathcal{I}} \langle \overset{*}{R}_m, \overset{*}{S}_n \rangle \overset{*}{R}_n(t),$$

and a comparison of the coefficients yields (4.63). From this it follows that the $\overset{*}{S}_n$, $n \in \mathcal{I}(r,\mu)$, are linearly independent. But their number is equal to the space dimension, so they form also a basis. □

Explicit Representation of the $\overset{*}{S}_{\bar{m},0}$.
The R_m share property (3.46) with the A_m. So R_m is even or odd with respect to the last argument, if m_r is even or odd, respectively. Moreover, from (4.62) and (4.61) we obtain

$$\sum_{m \in \mathcal{I}} \overset{*}{R}_m(x) \overset{*}{S}_m(t) = \sum_{|m|=\mu} \overset{*}{R}_m(x) t^m.$$

With respect to the variable x_r the even part is given by

$$\sum_{|\bar{m}|=\mu} \overset{*}{R}_{\bar{m},0}(x) \overset{*}{S}_{\bar{m},0}(t) = \sum_{\lambda=0}^{\lfloor \frac{\mu}{2} \rfloor} \sum_{|\bar{m}|=\mu-2\lambda} \overset{*}{R}_{\bar{m},2\lambda}(x) \bar{t}^{\bar{m}} t_r^{2\lambda},$$

where $\bar{t} = (t_1, \ldots, t_{r-1})'$, $\bar{m} = (m_1, \ldots, m_{r-1})$, and so on. We reduce the right side by means of (4.57), and using the substitution $\bar{l} = \bar{m} + 2\bar{n}$ we get

$$\sum_{|\bar{m}|=\mu} \overset{*}{R}_{\bar{m},0}(x)\overset{*}{S}_{\bar{m},0}(t)$$

$$= \sum_{\lambda=0}^{\lfloor\frac{\mu}{2}\rfloor} \sum_{|\bar{m}|=\mu-2\lambda} (-1)^{\lambda} \binom{\mu}{\bar{m},2\lambda} \sum_{|\bar{n}|=\lambda} \binom{|\bar{n}|}{\bar{n}}\binom{\mu}{\bar{m}+2\bar{n}}^{-1} \overset{*}{R}_{\bar{m}+2\bar{n},0}(x)\bar{t}^{\,\bar{m}}t_r^{2\lambda}$$

$$= \sum_{|\bar{l}|=\mu} \overset{*}{R}_{\bar{l},0}(x)\binom{\mu}{\bar{l}}^{-1} \sum_{|\bar{n}|\leq\lfloor\frac{\mu}{2}\rfloor} (-1)^{|\bar{n}|} \binom{\mu}{2|\bar{n}|}\binom{|\bar{n}|}{\bar{n}}\binom{\mu-2|\bar{n}|}{\bar{l}-2\bar{n}}\bar{t}^{\,\bar{l}-2\bar{n}}t_r^{2|\bar{n}|}.$$

Here we used the equation

$$\binom{\mu}{\bar{l}-2\bar{n},2|\bar{n}|} = \frac{\mu!}{(\bar{l}-2\bar{n})!(2|\bar{n}|)!} = \binom{\mu}{2|\bar{n}|}\binom{\mu-2|\bar{n}|}{\bar{l}-2\bar{n}}.$$

By Theorem 4.18 the $\overset{*}{R}_{\bar{m},0}$, $|\bar{m}| = \mu$, are linearly independent. So we get by a comparison of the coefficients, and replacing t by x,

$$\overset{*}{S}_{\bar{m},0}(x) = \binom{\mu}{\bar{m}}^{-1} \sum_{|\bar{n}|\leq\lfloor\frac{\mu}{2}\rfloor} (-1)^{|\bar{n}|} \binom{\mu}{2|\bar{n}|}\binom{|\bar{n}|}{\bar{n}}\binom{\mu-2|\bar{n}|}{\bar{m}-2\bar{n}}\bar{x}^{\,\bar{m}-2\bar{n}}x_r^{2|\bar{n}|}, \quad (4.64)$$

again for $|\bar{m}| = \mu$ and arbitrary $x \in \mathbb{R}^r$, where $\bar{x} = (x_1, \ldots, x_{r-1})'$.

To justify our denoting the coefficients on the right side of (4.61) by $\overset{*}{S}_m(x)$, let us consider the polynomial

$$S_{\bar{m},0}(x) := \binom{\mu}{\bar{m}}^{-1} \sum_{|\bar{n}|\leq\lfloor\frac{\mu}{2}\rfloor} (-1)^{|\bar{n}|} \binom{\mu}{2|\bar{n}|}\binom{|\bar{n}|}{\bar{n}}\binom{\mu-2|\bar{n}|}{\bar{m}-2\bar{n}}\bar{x}^{\,\bar{m}-2\bar{n}}(1-|\bar{x}|^2)^{|\bar{n}|}$$

$$(4.65)$$

for $\bar{m} \in \mathbb{N}_0^{r-1}$, $\mu = |\bar{m}|$, and $\bar{x}$ as above. Actually, due to this definition (4.64) can be obtained by a homogenisation of (4.65), i.e.,

$$\overset{*}{S}_{\bar{m},0}(x) = |x|^{\mu} S_{\bar{m},0}\left(\frac{x}{|x|}\right)$$

holds. Therefore we obtain from (4.63)

$$\langle R_{\bar{m},0}, S_{\bar{n},0} \rangle = \delta_{\bar{m},\bar{n}}, \quad (4.66)$$

first for $|\bar{m}| = |\bar{n}|$, only. But since $R_{\bar{m},0}$ and $S_{\bar{n},0}$ are in $\mathbb{H}_{|\bar{m}|}^r(S^{r-1})$ and in $\overset{*}{\mathbb{H}}_{|\bar{n}|}^r(S^{r-1})$, respectively, (4.66) is valid even for arbitrary $\bar{m}, \bar{n} \in \mathbb{N}_0^{r-1}$.

Moreover, from (4.65) we obtain the important equation

$$S_{\bar{m},0}(\bar{x},0) = \bar{x}^{\bar{m}} \tag{4.67}$$

for $\bar{x} \in S^{r-2}$ and $\bar{m} \in \mathbb{N}_0^{r-1}$.

Restrictions onto S^{k-1}

The $R_{\bar{m},0}$, $|\bar{m}| = \mu$, restricted to S^{r-1}, form a basis in the subspace of $\overset{*}{\mathbb{H}}{}_\mu^r(S^{r-1})$ where the elements are even with respect to x_r. The restrictions of the corresponding $S_{\bar{m},0}$ agree with the restrictions of the $\overset{*}{S}_{\bar{m},0}$, and so they are also contained in this subspace, and because of (4.66) they form also a basis. Moreover, using this biorthogonality relation again, we get the basis transforms

$$R_{\bar{m},0} = \sum_{|\bar{n}|=\mu} \langle R_{\bar{m},0}, R_{\bar{n},0} \rangle S_{\bar{n},0}, \tag{4.68}$$

$$S_{\bar{m},0} = \sum_{|\bar{n}|=\mu} \langle S_{\bar{m},0}, S_{\bar{n},0} \rangle R_{\bar{n},0}, \tag{4.69}$$

for $|\bar{m}| = \mu$. In particular, the transform matrices are inverse one to another, i.e., we obtain

$$\left(\langle R_{\bar{m},0}, R_{\bar{n},0} \rangle \right) = \left(\langle S_{\bar{m},0}, S_{\bar{n},0} \rangle \right)^{-1}. \tag{4.70}$$

Both matrices are Gram matrices and hence positive definite.

Next assume that $k \in \{1,2,\ldots,r-1\}$ is a fixed number. This includes the assumption $r \in \mathbb{N} \setminus \{1\}$, in what follows. The elements of $\mathbb{R}^k$ and the multiindices in $\mathbb{N}_0^k$ are denoted by $\bar{\bar{x}} = (x_1,\ldots,x_k)'$ and $\bar{\bar{m}} = (m_1,\ldots,m_k)$, respectively, and so on. We order the indices

$$m = (\bar{m},0) \in \mathbb{N}_0^{r-1} \times \mathbb{N}_0^1, \quad |\bar{m}| = \mu,$$

arbitrarily, but such that all of them which have the form

$$m = (\bar{\bar{m}},0) \in \mathbb{N}_0^k \times \mathbb{N}_0^{r-k}, \quad |\bar{\bar{m}}| = \mu,$$

and which are called to be of the *first kind*, precede all the other ones. Then (4.68) takes the form, with $\bar{\bar{x}} \in S^{k-1}$,

$$\begin{bmatrix} R_{\bar{m},0}(\bar{\bar{x}},0) \\ * \end{bmatrix} = \left(\langle R_{\bar{m},0}, R_{\bar{n},0} \rangle \right) \begin{bmatrix} \bar{\bar{x}}^{\bar{\bar{n}}} \\ 0 \end{bmatrix},$$

where the column on the right side is obtained by the help of (4.67).

Now we get because of the particular shape of the right side

$$R_{\bar{\bar{m}},0}(\bar{\bar{x}},*) = \sum_{|\bar{\bar{n}}|=\mu} \langle R_{\bar{\bar{m}},0}, R_{\bar{n},0} \rangle \bar{\bar{x}}^{\bar{\bar{n}}} \tag{4.71}$$

for $|\bar{\bar{m}}| = \mu$ and $\bar{\bar{x}} \in S^{k-1}$, and the following theorem is valid.

Theorem 4.20 (Basis in $\overset{*}{\mathbb{P}}{}^k_\mu(S^{r-1})$). *Let* $r \in \mathbb{N} \setminus \{1\}$, $k \in \{1, 2, \ldots, r-1\}$, *and* $\mu \in \mathbb{N}_0$. *The restrictions of the functions*

$$R_{\bar{\bar{m}},0}, \quad (\bar{\bar{m}}, 0) \in \mathbb{N}_0^k \times \mathbb{N}_0^{r-k}, \ |\bar{\bar{m}}| = \mu,$$

onto S^{k-1} *form a basis in* $\overset{*}{\mathbb{P}}{}^k_\mu(S^{r-1})$.

Proof. Because of (4.71) the restrictions are in $\overset{*}{\mathbb{P}}{}^k_\mu(S^{r-1})$. The matrix

$$R := \left(\langle R_{\bar{\bar{m}},0}, R_{\bar{\bar{n}},0} \rangle \right)$$

which occurs in this linear system is regular. So (4.71) is a basis transform. $\square$

The T-Kernels in $\overset{*}{\mathbb{H}}{}^r_\mu(S^{r-1})$

Let $1 \leq k \leq r - 2$, and assume $T = (t_1, \ldots, t_k)$ is an $r \times k$-matrix with orthogonal columns $t_1, \ldots, t_k \in S^{r-1}$. We want to describe the T-kernel $\mathbf{V}_T$ of the space

$$\mathbf{V} := \overset{*}{\mathbb{H}}{}^r_\mu(S^{r-1}).$$

First let us recall that T can be completed by columns $u_{k+1}, \ldots, u_r \in S^{r-1}$ such that

$$V := \left(T, U \right) = \left(t_1, \ldots, t_k, u_{k+1}, \ldots, u_r \right) \tag{4.72}$$

is in $\mathbf{A}^r$. With $E := (e_1, \ldots, e_k)$ we get $T = VE$, and it is easy to see that the equivalence

$$F \in \mathbf{V}_E \Longleftrightarrow F_{V^{-1}} \in \mathbf{V}_T \tag{4.73}$$

is true. So it suffices to study $\mathbf{V}_E$.

To this end let $F \in \overset{*}{\mathbb{H}}{}^r_\mu$ define an element of $\mathbf{V}_E$. We define $f \in \mathbb{P}^k_\mu$ by

$$f(x_1, \ldots, x_k) \ := \ \tfrac{1}{2}\Big(F(x_1, \ldots, x_k, 0, \ldots, 0, +\sqrt{1 - x_1^2 - \cdots - x_k^2}$$
$$+ F(x_1, \ldots, x_k, 0, \ldots, 0, -\sqrt{1 - x_1^2 - \cdots - x_k^2} \Big) \tag{4.74}$$

for $x_1^2 + \cdots + x_k^2 \leq 1$. Note that

$$f(-x_1, \ldots, -x_k) = (-1)^\mu f(x_1, \ldots, x_k) \tag{4.75}$$

holds. Now let $x \in S^{r-1}$ be arbitrary. Because $1 \leq k \leq r - 2$ there are matrices $A, \bar{A} \in \mathbf{A}^r_E$ such that

$$Ax = \left(x_1, \ldots, x_k, 0, \ldots, 0, +\sqrt{1 - x_1^2 - \cdots - x_k^2} \right)'$$

and

$$\bar{A}x = \left(x_1, \ldots, x_k, 0, \ldots, 0, -\sqrt{1 - x_1^2 - \cdots - x_k^2}\right)'$$

holds. And using $F \in V_E$ we get $F_A = F = F_{\bar{A}}$, and hence

$$F(x_1, \ldots, x_r) = \tfrac{1}{2}\left(F(Ax) + F(\bar{A}x)\right) = f(x_1, \ldots, x_k).$$

It follows that

$$F(x_1, \ldots, x_r) = |x|^\mu f\left(\frac{x_1}{|x|}, \ldots, \frac{x_k}{|x|}\right),$$

first for $x \in S^{r-1}$, only. But because of (4.75), homogeneous polynomials occur on both sides. So the equation is valid identically. The question arises now what does the condition $\Delta F = 0$ impose on the polynomial f.

Theorem 4.21. *Let* $r \in \mathbb{N} \setminus \{1, 2\}$, $k \in \{1, \ldots, r - 2\}$, $\mu \in \mathbb{N}_0$, $g \in \overset{*}{\mathbb{P}}{}^k_\mu$. *Then a uniquely determined* $h \in \mathbb{P}^k_{\mu-2}$ *with* $h(-x_1, \ldots, -x_k) = (-1)^\mu h(x_1, \ldots, x_k)$ *exists, such that* $f = g + h$ *satisfies*

$$\Delta |x|^\mu f\left(\tfrac{x_1}{|x|}, \ldots, \tfrac{x_k}{|x|}\right) = 0, \tag{4.76}$$

where $x = (x_1, \ldots, x_r)'$ *and where* Δ *is the Laplace operator in* $\mathbb{R}^r$.

Proof. Let f be a polynomial of the form $f = g + h$, this means of the form

$$f(\xi) = {\sum_m}' a_m \xi^m \quad , \quad \xi = (\xi_1, \ldots, \xi_k)' \in \mathbb{R}^k, \tag{4.77}$$

where the sum is extended over all $m \in \mathbb{N}_0^k$ with $|m| \leq \mu$ and $|m| \equiv \mu \bmod (2)$. By a simple, but lengthy calculation we find that (4.76) is equivalent to the equation

$$\sum_{\nu=1}^{k} \sum_{\kappa=1}^{k} f^{(\nu,\kappa)}\left[\delta_{\nu,\kappa} - \xi_\nu \xi_\kappa\right] - (r-1)\sum_{\nu=1}^{k} \xi_\nu f^{(\nu)} + \mu(\mu + r - 2)f = 0,$$

to hold for $\xi_1 = x_1/|x|, \ldots, \xi_k = x_k/|x|$ and $x \in S^{r-1}$, i.e., to hold for arbitrary $\xi \in B^k$, and hence for arbitrary $\xi \in \mathbb{R}^k$. Inserting (4.77) in this partial differential equation, we get the equivalent equation

$$\begin{aligned}
0 = {}& \sum_{\nu=1}^{k} {\sum_m}' a_m m_\nu (m_\nu - 1)\xi^{m-2e_\nu} - \sum_{\nu=1}^{k} \sum_{\kappa=1}^{k} {\sum_m}' a_m m_\nu m_\kappa \xi^m \\
& + \sum_{\nu=1}^{k} {\sum_m}' a_m m_\nu \xi^m - (r-1)\sum_{\nu=1}^{k} {\sum_m}' a_m m_\nu \xi^m + \mu(\mu + r - 2){\sum_m}' a_m \xi^m.
\end{aligned}$$

Comparing the coefficients we find that this equation is equivalent to

$$\left[|m|(|m| + r - 2) - \mu(\mu + r - 2)\right] a_m = \sum_{\nu=1}^{k} (m_\nu + 2)(m_\nu + 1)\, a_{m+2e_\nu}$$

holding for $|m| = \mu - 2, \mu - 4, \ldots$. For a given

$$g(\xi) = \sum_{|m|=\mu} a_m \xi^m,$$

this recurrence equation has a unique solution. So h is uniquely determined by g.
□

Corollary 4.22. *Let* $1 \leq k \leq r - 2$, $\mu \in \mathbb{N}_0$, T, E *and* $\mathbf{V} = \overset{*}{\mathbb{H}}{}^r_\mu(S^{r-1})$ *be as above. Then the following spaces are isomorphic:*

$$\mathbf{V}_T \cong \mathbf{V}_E \cong \overset{*}{\mathbb{P}}{}^k_\mu.$$

Proof. $\mathbf{V}_T \cong \mathbf{V}_E$ is valid because of (4.73). Next let $F \in \mathbf{V}_E$, and let $f \in \overset{*}{\mathbb{P}}{}^k_\mu$ be defined by (4.74). In view of (4.75) we can represent f in the form $f = g + h$ with $g \in \overset{*}{\mathbb{P}}{}^k_\mu$ and with $h \in \mathbb{P}^k_{\mu-2}$, which satisfies $h(-x_1, \ldots, -x_k) = (-1)^\mu h(x_1, \ldots, x_k)$, too, and where g is uniquely determined by f, and hence by F. This defines a map

$$\mathbf{V}_E \ni F \mapsto g \in \overset{*}{\mathbb{P}}{}^k_\mu.$$

Moreover, assume $g = 0$. By Theorem 4.21, h is uniquely determined by g, but in the present case, 0 is a solution, and therefore $h = 0$ must hold. So the kernel of our map $\mathbf{V}_E \to \overset{*}{\mathbb{P}}{}^k_\mu$ is the null-space, the map is bijective, and $\mathbf{V}_E \cong \overset{*}{\mathbb{P}}{}^k_\mu$ holds. □

Corollary 4.23. *Assumptions as in Corollary 4.22. Then the following holds:*

$$\dim \mathbf{V}_T = \binom{\mu + k - 1}{k - 1}.$$

Proof. The right side equals the dimension of $\overset{*}{\mathbb{P}}{}^k_\mu$, which is given by (3.9). So the statement follows from Corollary 4.22. □

We have prepared now the tools by which the following theorem can be proved.

Theorem 4.24 (T-Kernel Basis in $\overset{*}{\mathbb{H}}{}^r_\mu(S^{r-1})$). *Let* $1 \leq k \leq r - 2$, $\mu \in \mathbb{N}_0$, *and let* $T = (t_1, \ldots, t_k)$ *consist of orthogonal columns* $t_1, \ldots, t_k \in S^{r-1}$. *Then a basis for the T-kernel of* $\overset{*}{\mathbb{H}}{}^r_\mu(S^{r-1})$, $\mu \in \mathbb{N}_0$, *is given by the spherical harmonics with the representation*

$$R_{\bar{\bar{m}},0}(t_1 x, \ldots, t_k x, *, \cdots, *) \quad \text{for } x \in S^{r-1}, \tag{4.78}$$

where $(\bar{\bar{m}}, 0) \in \mathbb{N}_0^k \times \mathbb{N}_0^{r-k}$ *and* $|\bar{\bar{m}}| = \mu$.

Proof. Because of $R_{\bar{m},0}|_{S^{r-1}} = \overset{*}{R}_{\bar{m},0}|_{S^{r-1}}$ the functions in (4.78) are in $\mathbf{V} :=$
$\overset{*}{\mathbb{H}}{}^r_\mu(S^{r-1})$. So are the functions $R_{\bar{m},0}(V' \cdot)$ for arbitrary rotations $V \in \mathbf{A}^r$. Choosing V as in (4.72), we get

$$R_{\bar{m},0}(V' \cdot) = R_{\bar{m},0}\big((t_1 \cdot), \ldots, (t_k \cdot), *, \cdots, *\big) \in \mathbf{V}_T, \qquad (4.79)$$

since $R_{\bar{m},0}$ does not depend on the arguments notified by $*$, see (4.49). By Theorem 4.19 the $\overset{*}{R}_{\bar{m},0}$ are linearly independent. So are their restrictions onto S^{r-1} and hence the functions (4.79). By Corollary 4.23 their number equals the dimension of $\mathbf{V}_T$, which finishes the proof. $\qquad \square$

Remark In Section 1.4 we mentioned that $k = r - 1$ implies $\mathbf{V}_T = \mathbf{V}$. Besides, a comparison of the dimensions yields $\mathbf{V}_T \not\cong \overset{*}{\mathbb{P}}{}^{r-1}_\mu$ for $\mu \geq 1$. So the assumption $1 \leq k \leq r - 2$ is indispensable.

T-Kernel Projections in $\overset{*}{\mathbb{H}}{}^r_\mu(S^{r-1})$

Let $1 \leq k \leq r - 2$, $\mu \in \mathbb{N}_0$, $T = (t_1, \ldots, t_k)$ and

$$V = (T, U) = (t_1, \ldots, t_k, u_{k+1}, \ldots, u_r) \in \mathbf{A}^r,$$

as above. We want to determine

$$\Pi^r_T G^r_\mu(a' \cdot)$$

for a fixed $a \in S^{r-1}$, which is the image of the function $G^r_\mu(a' \cdot) \in \overset{*}{\mathbb{H}}{}^r_\mu(S^{r-1})$ under the projection Π^r_T onto the T-kernel of $\overset{*}{\mathbb{H}}{}^r_\mu(S^{r-1})$, see Theorem 1.9, (iv). From Theorem 4.24 we know that with some real coefficients $a_{\bar{m},0}$

$$[\Pi^r_T G^r_\mu(a' \cdot)](x) = \sum_{|\bar{m}|=\mu} a_{\bar{m},0} R_{\bar{m},0}(xt_1, \ldots, xt_k, *, \ldots, *) \qquad (4.80)$$

holds for $x \in S^{r-1}$, where $\bar{m} = (m_1, \ldots, m_k)$ varies in $\mathbb{N}_0^k$, again. The fixed points in S^{r-1} of the group $\mathbf{A}^r_T$ are exactly the points of the form

$$t = T\bar{\bar{s}}, \quad \bar{\bar{s}} \in S^{k-1}, \qquad (4.81)$$

and Theorem 1.9, (iii), yields

$$G^r_\mu(a'T\bar{\bar{s}}) = \sum_{|\bar{m}|=\mu} a_{\bar{m},0} R_{\bar{m},0}(s_1, \ldots, s_k, *, \ldots, *), \qquad (4.82)$$

which should enable us to determine the coefficients $a_{\bar{m},0}$. Note that

$$G^r_\mu(a'T\bar{\bar{s}}) = G^r_\mu\big((at_1)s_1 + \cdots + (at_k)s_k\big) = \sum_{|\bar{m}|=\mu} R_{\bar{m},0}(at_1, \ldots, at_k, *, \ldots, *) \cdot \bar{\bar{s}}^{\bar{m}}$$

holds, where we used that

$$s = (s_1, \ldots, s_k, 0, \ldots, 0)' \in S^{r-1}$$

and

$$(at_1, \ldots, at_k, au_{k+1}, \ldots, au_r)' \in S^{r-1}$$

is valid, while s^m vanishes if $m_\nu > 0$ holds for some $\nu \in \{k+1, \ldots, r\}$. Together with (4.82) this yields

$$\sum_{|\bar{m}|=\mu} a_{\bar{m},0} R_{\bar{m},0}(s_1, \ldots, s_k, *, \ldots, *) = \sum_{|\bar{m}|=\mu} R_{\bar{m},0}(at_1, \ldots, at_k, *, \ldots, *) \cdot \bar{s}^{\bar{m}}.$$

(4.83)

Actually, using this equation we get

Theorem 4.25 (T-Kernel Projection of $G_\mu^r(a' \cdot)$). *Let $r \in \mathbb{N} \setminus \{1, 2\}$, $\mu \in \mathbb{N}_0$, $k \in \{1, 2, \ldots, r-2\}$, $T = (t_1, \ldots, t_k)$ as above, and $a \in S^{r-1}$. Then*

$$\left[\Pi_T^r G_\mu^r(a' \cdot) \right](x) =$$

$$\sum_{|\bar{m}|=\mu} \sum_{|\bar{n}|=\mu} c_{\bar{m},\bar{n}}^{r,k} R_{\bar{m},0}(at_1, \ldots, at_k, *, \cdots, *) R_{\bar{n},0}(xt_1, \ldots, xt_k, *, \cdots, *) \qquad (4.84)$$

holds for $x \in S^{r-1}$, where the coefficients $c_{\bar{m},\bar{n}}^{r,k}$ are defined for $(\bar{m}, 0), (\bar{n}, 0) \in \mathbb{N}_0^k \times \mathbb{N}_0^{r-k}$ by

$$\left(c_{\bar{m},\bar{n}}^{r,k} \right) = \left(\langle R_{\bar{m},0}, R_{\bar{n},0} \rangle \right)^{-1}. \qquad (4.85)$$

Proof. Using (4.71) we get from (4.83) for $\bar{\bar{s}} \in S^{k-1}$,

$$\sum_{|\bar{m}|=\mu} a_{\bar{m},0} \sum_{|\bar{n}|=\mu} \langle R_{\bar{m},0}, R_{\bar{n},0} \rangle \bar{\bar{s}}^{\bar{n}} = \sum_{|\bar{m}|=\mu} R_{\bar{m},0}(at_1, \ldots, at_k, *, \ldots, *) \bar{\bar{s}}^{\bar{m}}.$$

A comparison of the coefficients which occur with $\bar{\bar{s}}^{\bar{m}}$ yields

$$\sum_{|\bar{n}|=\mu} \langle R_{\bar{m},0}, R_{\bar{n},0} \rangle a_{\bar{n},0} = R_{\bar{m},0}(at_1, \ldots, at_k, * \ldots, *)$$

for $|\bar{m}| = \mu$. This is a system of linear equations for the unknown $a_{\bar{n},0}$. The matrix on the left side is a Gram matrix, and hence regular. So we obtain, with the coefficients defined by (4.85),

$$a_{\bar{m},0} = \sum_{|\bar{n}|=\mu} c_{\bar{m},\bar{n}}^{r,k} \cdot R_{\bar{n},0}(at_1, \ldots, at_k, *, \ldots, *),$$

and inserting this in (4.80) we get the statement of the theorem. $\square$

Particular Assumptions on the Fixed Points

In the case $at_k = 0$ the representation (4.84) becomes redundant. To see this recall that every rotation in $\mathbf{A}_T^r$ has the form

$$B = V \begin{pmatrix} I & O \\ O & A \end{pmatrix} V', \quad A \in \mathbf{A}^{r-k},$$

where $V = \left(T, U\right) \in \mathbf{A}^r$, see (1.16). Now let us use the transform

$$\xi := V'x.$$

Then we obtain

$$
\begin{aligned}
a'Bx &= a'\left(T, U\right) \begin{pmatrix} I & O \\ O & A \end{pmatrix} \left(T, U\right)' x \\
&= \left(at_1, \ldots, at_{k-1}, 0, au_{k+1}, \ldots, au_r\right) \begin{pmatrix} I & O \\ O & A \end{pmatrix} \left(\xi_1, \ldots, \xi_r\right)',
\end{aligned}
$$

and it is obvious that $G_\mu^r(a'Bx)$ does not depend on ξ_k. Nor does the average over all $B \in \mathbf{A}_T^r$ depend on ξ_k. On the other hand, using (3.45) with $A_m = R_m$ we get from (4.80)

$$[\Pi_T^r G_\mu^r(a' \cdot)](x) = \alpha_\mu^r \sum_{|\bar{m}|=\mu} a_{\bar{m},0} \left(\frac{\mu}{\bar{m}}\right) \bar{\xi}^{\bar{m}} + TLD,$$

where $\alpha_\mu^r > 0$ is now the leading coefficient of $Q_\mu = G_\mu^r$, see (4.48). The values of $\bar{\xi} = (\xi_1, \ldots, \xi_k)'$ cover B^k while x varies in S^{r-1}. Together this implies

$$a_{\bar{m},0} = 0 \quad \text{for} \quad m_k = 0,$$

and (4.80), (4.82) change so far as $\bar{m}$ takes now the meaning of $(m_1, \ldots, m_{k-1}, 0)$. Theorem 4.25 including its proof remains valid literally, except that the inverse of the coefficient matrix is now the submatrix of (4.85) which belongs to the rows and columns with the indices $(m_1, \ldots, m_{k-1}, 0)$, $m_1 + \cdots + m_{k-1} = \mu$, only. Formally this means that k has to be replaced by $k - 1$.

In the case $at_k = 0$, $at_{k-1} = 0$ we repeat this procedure, and so on. This yields finally

Corollary 4.26 (Particular Assumptions). *Assumptions as in Theorem 4.25. Moreover, for some $\kappa \in \{0, 1, \ldots, k - 1\}$ let $E_{r,\kappa} := span\{e_{r-\nu} | \nu = 0, 1, \ldots, \kappa - 1\}$, where $E_{r,\kappa}$ has to be put to $[0]$ for $\kappa = 0$. And assume that the last κ fixed points are in $E_{r,\kappa}$,*

$$t_{k-\nu} \in E_{r,\kappa} \quad for \quad \nu = 0, 1, \ldots, \kappa - 1,$$

while the remaining $k - \kappa$ fixed points satisfy

$$t_{k-\nu} \in E_{r,\kappa}^{\perp} \quad for \quad \nu = \kappa, \ldots, k - 1.$$

Finally let $a \in E_{r,\kappa}^{\perp}$. Then

$$\left[\Pi_T^r G_\mu^r (a' \cdot) \right](x) =$$
$$\sum_{|\bar{m}|=\mu} \sum_{|\bar{n}|=\mu} c_{\bar{m},\bar{n}}^{r,k-\kappa} R_{\bar{m},0}(at_1, \ldots, at_{k-\kappa}, *, \ldots, *) R_{\bar{n},0}(xt_1, \ldots, xt_{k-\kappa}, *, \ldots, *)$$

holds for $x \in S^{r-1}$, where $(\bar{m}, 0)$, $(\bar{n}, 0) \in \mathbb{N}_0^{k-\kappa} \times \mathbb{N}_0^{r-k+\kappa}$.

Remark. For $\kappa = 0$ no further assumption is put upon the fixed points, and the statements of Corollary 4.26 and of Theorem 4.25 agree.

4.3 Biorthonormal Systems on the Ball

In the beginning of this section we assume $r \in \mathbb{N} \setminus \{1, 2\}$. The inner product on the left side of (4.66) is defined by an integral over S^{r-1}. In view of (4.49) and of (4.65), the integrand does not depend on x_r, so formula (1.24) can be used to reduce the integration domain to B^{r-1}. The result is a biorthogonal relation on the ball B^{r-1}. But a further reduction, say by formula (1.25), to a ball B^{r-s-1}, where s is positive, is impossible since $S_{\bar{n},0}$ depends on $|x|^2 = x_1^2 + \cdots + x_{r-1}^2$, i.e., on all of the remaining variables.

For this reason we change our strategy. We assume $s \in \{1, \ldots, r - 2\}$ and

$$(\bar{m}, 0), (\bar{n}, 0), \ldots \in \mathbb{N}_0^{r-s-1} \times \mathbb{N}_0^{s+1}.$$

For these indices (4.66) takes more than ever the form

$$\delta_{\bar{m},\bar{n}} = \int_{S^{r-1}} R_{\bar{m},0}(\bar{x}, *) \overset{*}{S}_{\bar{n},0}(\bar{x}, x_r) d\omega(x).$$

Now the integrand does not depend on the variables $x_{r-s}, \ldots, x_{r-1}$, see (4.49) and (4.64), and we get by an application of (1.25) with respect to these s nonoccuring arguments

$$\delta_{\bar{m},\bar{n}} = \int_{S^{r-1}} R_{\bar{m},0}(\bar{\bar{x}}, *, *) \overset{*}{S}_{\bar{n},0}(\bar{\bar{x}}, *, x_r) \, d\omega(x)$$

$$= \omega_{s-1} \int_{B^{r-s}} R_{\bar{m},0}(\bar{\bar{x}}, *, *) \overset{*}{S}_{\bar{n},0}(\bar{\bar{x}}, *, x_r) \left(1 - \bar{\bar{x}}^2 - x_r^2\right)^{\frac{s-2}{2}} d(\bar{\bar{x}}, x_r)$$

$$= \omega_{s-1} \int_{B^{r-s-1}} R_{\bar{m},0}(\bar{\bar{x}}, *, *) \left[\int_{-\sqrt{1-\bar{\bar{x}}^2}}^{\sqrt{1-\bar{\bar{x}}^2}} \overset{*}{S}_{\bar{n},0}(\bar{\bar{x}}, *, x_r) \left(1 - \bar{\bar{x}}^2 - x_r^2\right)^{\frac{s-2}{2}} dx_r \right] d\bar{\bar{x}},$$

where we used the notation $\bar{\bar{x}} = (x_1, \ldots, x_{r-s-1})'$. Substituting $x_r = \sqrt{1 - \bar{\bar{x}}^2} \cdot \xi$ we bring this to the form

$$\delta_{\bar{\bar{m}},\bar{\bar{n}}} = \omega_{s-1} \int\limits_{B^{r-s-1}} R_{\bar{m},0}(\bar{\bar{x}}, *, *) \left[\int\limits_{-1}^{1} \overset{*}{S}_{\bar{n},0}(\bar{\bar{x}}, *, \sqrt{1 - \bar{\bar{x}}^2} \cdot \xi) (1 - \xi^2)^{\frac{s-2}{2}} d\xi \right] (1 - \bar{\bar{x}}^2)^{\frac{s-1}{2}} d\bar{\bar{x}}.$$

Here we arrive at a point where we can derive the biorthogonality of the Appell systems from the biorthogonality of the systems $\{\overset{*}{R}_{\bar{m},0}\}$ and $\{\overset{*}{S}_{\bar{n},0}\}$.

First we get by a comparison of (4.48) and (3.51), together with (4.13),

$$R_m = \frac{2\mu + r - 2}{(r-2)\omega_{r-1}} \cdot V_m^{r,-1} \quad \text{for } m \in \mathbb{N}_0^r, \ |m| = \mu.$$

In the case of our particular indices this yields, in view of (3.58),

$$R_{\bar{m},0} = \frac{2\mu + r - 2}{(r-2)\omega_{r-1}} \cdot V_{\bar{m}}^{r-s-1,s}, \tag{4.86}$$

which is still valid for $s = 0$.

Next we introduce the polynomials $\hat{U}_{\bar{n}}^{r-s-1,s}$ by

$$\hat{U}_{\bar{n}}^{r-s-1,s}(\bar{\bar{x}}) := \frac{2\mu + r - 2}{r - 2} \cdot \frac{\omega_{s-1}}{\omega_{r-1}} \cdot \int\limits_{-1}^{1} \overset{*}{S}_{\bar{n},0}(\bar{\bar{x}}, *, \sqrt{1 - \bar{\bar{x}}^2} \cdot \xi)(1 - \xi^2)^{\frac{s-2}{2}} d\xi, \tag{4.87}$$

but in the beginning for $s \in \{1, \ldots, r-2\}$, only. Inserting this above we get the result wanted,

$$\int\limits_{B^{r-s-1}} V_{\bar{m}}^{r-s-1,s}(\bar{\bar{x}}) \hat{U}_{\bar{n}}^{r-s-1,s}(\bar{\bar{x}})(1 - \bar{\bar{x}}^2)^{\frac{s-1}{2}} d\bar{\bar{x}} = \delta_{\bar{\bar{m}},\bar{\bar{n}}}.$$

For $s = 0$ we obtain the same result from (4.66) by an application of (1.24), provided we define $\hat{U}_{\bar{n}}^{r-1,0}$ by

$$\hat{U}_{\bar{n}}^{r-1,0}(\bar{x}) := \frac{2\mu + r - 2}{(r-2)\omega_{r-1}} \cdot \frac{1}{2} \cdot S_{\bar{n},0}(\bar{x}, *)$$

for $(\bar{n}, 0) = (\bar{n}, 0) \in \mathbb{N}_0^{r-1} \times \mathbb{N}_0$ and with $\bar{\bar{x}} = \bar{x} = (x_1, \ldots, x_{r-1})'$, in this case.

Our result takes a more handsome form if we replace $r - s - 1$ by r. So we get finally the following theorem.

Theorem 4.27 (Biorthonormal Systems on B^r, Appell and Kampé de Feriét). *For $r \in \mathbb{N}$ and $s \in \mathbb{N}_0$, the polynomials $V_m^{r,s}$ and $\hat{U}_n^{r,s}$, $m, n \in \mathbb{N}_0^r$, have the following properties:*

$$(i) \qquad V_m^{r,s} \in \mathbb{P}_m^r, \quad \hat{U}_n^{r,s} \in \mathbb{P}_\mu^r,$$

$$(ii) \qquad [V_m^{r,s}, \hat{U}_n^{r,s}]_{r,s} = \delta_{m,n}, \tag{4.88}$$

where the inner product is defined by

$$[E, F]_{r,s} := \int_{B^r} E(x)F(x)(1 - x^2)^{\frac{s-1}{2}} dx \qquad (4.89)$$

for $E, F \in C(B^r)$. Each of the systems $V_m^{r,s}$, $|m| \leq \mu$, and $\hat{U}_m^{r,s}$, $|m| \leq \mu$, is a basis in $\mathbb{P}_\mu^r$. For $|m| = \mu$, the $V_m^{r,s}$ and $\hat{U}_m^{r,s}$ are orthogonal to $\mathbb{P}_{|m|-1}^r$.

Proof. With $\bar{r} := r - s - 1$ the assumptions $r \in \mathbb{N} \setminus \{1, 2\}$ and $s \in \{0, 1, \ldots, r - 2\}$ are together equivalent to $1 \leq \bar{r} \leq \bar{r} + s$, and hence with $\bar{r} \in \mathbb{N}$, only. For (4.88) we refer to (3.53) and our result from above. Moreover, because of Theorem 3.12, the $V_m^{r-s-1,s}$ with $|m| \leq \mu$ form a basis in $\mathbb{P}_\mu^r$. Together with (4.88) this implies that the corresponding $\hat{U}_n^{r-s-1,s}$ do the same. $\qquad \square$

The $V_m^{r,s}$ are the polynomials of Appell and Kampé de Feriét, as introduced in Section 3.1. Theorem 4.27 says that the $\hat{U}_n^{r,s}$ are biorthogonal to the $V_m^{r,s}$, as the polynomials $U_n^{r,s}$ of Appell and Kampé de Feriét are, see [2], p.260. Therefore

$$U_m^{r,s} = const \cdot \hat{U}_m^{r,s}$$

must hold, where the constant is obtained from

$$[V_m^{r,s}, U_m^{r,s}]_{r,s} = const \cdot [V_m^{r,s}, \hat{U}_m^{r,s}]_{r,s} = const.$$

This yields

$$U_m^{r,s} = [V_m^{r,s}, U_m^{r,s}]_{r,s} \cdot \hat{U}_m^{r,s} \qquad (4.90)$$

for $m \in \mathbb{N}_0^r$.

We should remark that for $r = 1$ the definition of the inner product $[\,\cdot\,, \,\cdot\,]_{r,s}$, see (4.89), is consistent with Definition 2.1, where $B^r = [-1, 1]$.

In order to get an explicit expression for $\hat{U}_m^{r,s}$ we insert (4.64) in (4.87), replacing $r - s - 1$ by r, $\bar{m}$ by m and $\bar{x}$ by x. So we get, for $r \in \mathbb{N} \setminus \{1, 2\}$, $s \in \mathbb{N}_0$, $\mu = |m|$,

$$\hat{U}_m^{r,s}(x) = \frac{2\mu + r + s - 1}{r + s - 1} \cdot \frac{\omega_{s-1}}{\omega_{r+s}} \cdot \binom{\mu}{m}^{-1} \times$$
$$\sum_{|n| \leq \lfloor \frac{\mu}{2} \rfloor} (-1)^{|n|} \binom{\mu}{2|n|} \binom{|n|}{n} \binom{\mu - 2|n|}{m - 2n} B\left(|n| + \tfrac{1}{2}, \tfrac{s}{2}\right) \cdot x^{m-2n}(1 - |x|^2)^{|n|},$$

where we used the equation

$$\int_{-1}^{1} \xi^{2|n|}(1 - \xi^2)^{\frac{s-2}{2}} d\xi = B\left(|n| + \tfrac{1}{2}, \tfrac{s}{2}\right),$$

which follows from (1.4) by the substitution of $\xi^2 = \tau$. Using (1.8), (1.4) and *Legendre's formula*

$$\Gamma(2|n| + 1) = \frac{2^{2|n|}}{\sqrt{\pi}} \Gamma\left(|n| + \frac{1}{2}\right) \Gamma(|n| + 1),$$

we get by some further calculation the representation

$$\hat{U}_m^{r,s}(x) = \left(\mu + \tfrac{r+s-1}{2}\right)\Gamma\left(\tfrac{r+s-1}{2}\right)\tfrac{m!}{\pi^{\frac{r}{2}}} \sum_{|n|\leq\lfloor\frac{\mu}{2}\rfloor} (-1)^{|n|} \frac{x^{m-2n}(1-|x|^2)^{|n|}}{2^{2|n|}n!(m-2n)!\Gamma(|n|+\tfrac{s+1}{2})}.$$

(4.91)

For $x \in S^{r-1}$ this formula takes the simple form

$$\hat{U}_m^{r,s}(x) = \left(\mu + \tfrac{r+s-1}{2}\right)\Gamma\left(\tfrac{r+s-1}{2}\right)\frac{1}{\pi^{\frac{r}{2}}\Gamma(\tfrac{s+1}{2})} \cdot x^m.$$

Inserting this in (4.90) and comparing the result with (3.62), we obtain together with (2.13) the important formula

$$[V_m^{r,s}, U_m^{r,s}]_{r,s} = \frac{\Gamma(\tfrac{s+1}{2})}{\Gamma(\tfrac{r+s-1}{2})} \cdot \frac{\pi^{\frac{r}{2}}}{\mu + \tfrac{r+s-1}{2}} \cdot \frac{(s)_\mu}{m!},$$

(4.92)

but for $r \in \mathbb{N} \setminus \{1\}$, only. For $r = 1$ the equation (2.13) has to be replaced by $C_\mu^0(1) = 1$, and a similar formula holds.

Decomposition of the Space $\mathbb{P}_\mu^r(\mathbf{B}^r)$

In what follows let $r \in \mathbb{N}$, $s \in \mathbb{N}_0$ and $\mu \in \mathbb{N}_0$, again. We identify $\mathbb{P}_\mu^r = \mathbb{P}_\mu^r(B^r)$ by isomorphy, but the elements of these spaces are always treated as functions $B^r \to \mathbb{R}$.

Theorem 4.27 suggests the following definition.

Definition 4.2 (Appell Spaces $\mathbf{V}_\mu^{r,s}$). *The Appell space $\mathbf{V}_\mu^{r,s}$ with respect to the inner product $[\cdot, \cdot]_{r,s}$ is the orthogonal complement of $\mathbb{P}_{\mu-1}^r$ in $\mathbb{P}_\mu^r$. The reproducing kernel of $\mathbf{V}_\mu^{r,s}$ is denoted by $G_\mu^{r,s}(\cdot, \cdot)$.*

Recall that $\mathbb{P}_{-1}^r = [0]$ holds by definition.

It follows from Theorem 4.27 that $\mathbf{V}_\mu^{r,s}$ has the following basis representations,

$$span\{\hat{U}_m^{r,s} : |m| = \mu\} = \mathbf{V}_\mu^{r,s} = span\{V_m^{r,s} : |m| = \mu\}.$$

(4.93)

Moreover, every $F \in \mathbf{V}_\mu^{r,s}$ can be represented in the form

$$F = \sum_{|m|=\mu} a_m \hat{U}_m^{r,s}$$

with real coefficients a_m, and using (4.88) we get for $|n| = \mu$

$$[F, V_n^{r,s}]_{r,s} = \sum_{|m|=\mu} a_m [\hat{U}_m^{r,s}, V_n]_{r,s} = a_n.$$

Inserting this above we obtain

$$F = \sum_{|n|=\mu} [F, V_n^{r,s}]_{r,s} \hat{U}_n^{r,s}.$$

(4.94)

Likewise we get

$$F = \sum_{|n|=\mu} [F, \hat{U}_n^{r,s}]_{r,s} V_n^{r,s}. \tag{4.95}$$

In (4.94) and (4.95) we identify $F = V_m^{r,s}$ and $F = \hat{U}_m^{r,s}$, respectively, and obtain the following result.

Basis Transform in $\mathbf{V}_\mu^{r,s}$

$$\hat{U}_m^{r,s} = \sum_{|n|=\mu} [\hat{U}_m^{r,s}, \hat{U}_n^{r,s}]_{r,s} V_n^{r,s}, \tag{4.96}$$

$$V_m^{r,s} = \sum_{|n|=\mu} [V_m^{r,s}, V_n^{r,s}]_{r,s} \hat{U}_n^{r,s} \tag{4.97}$$

holds for $|m| = \mu$. It follows that the Gram matrices which occur in these systems are inverse to another,

$$\left([V_m^{r,s}, V_n^{r,s}]_{r,s} \right)^{-1}_{m,n\in\mathcal{M}} = \left([\hat{U}_m^{r,s}, \hat{U}_n^{r,s}]_{r,s} \right)_{m,n\in\mathcal{M}}. \tag{4.98}$$

These results enable us to prove the following theorem.

Theorem 4.28 (Orthogonal Decomposition of $\mathbb{P}_\mu^r$, Reproducing Kernel of $\mathbf{V}_\kappa^{r,s}$).
Let $r \in \mathbb{N} \setminus \{1, 2\}$, $s \in \mathbb{N}_0$. Then

$$\mathbb{P}_\mu^r = \bigoplus_{\kappa=0}^{\mu} \mathbf{V}_\kappa^{r,s} \tag{4.99}$$

holds, and this is an orthogonal decomposition with respect to the inner product $[\cdot, \cdot]_{r,s}$. The spaces $\mathbf{V}_\kappa^{r,s}$, $\kappa \in \mathbb{N}_0$, are rotations-invariant. For $x, y \in B^r$ the reproducing kernel of $\mathbf{V}_\mu^{r,s}$ has the representations

$$\sum_{|m|=\mu} \hat{U}_m^{r,s}(x) V_m^{r,s}(y) = G_\mu^{r,s}(x,y) = \sum_{|m|=\mu} \hat{U}_m^{r,s}(y) V_m^{r,s}(x). \tag{4.100}$$

Proof. The $V_m^{r,s}$ have the basis property, so every $F \in \mathbb{P}_\mu^r$ has a unique representation

$$F = \sum_{\nu=0}^{\mu} \sum_{|n|=\nu} a_n V_n^{r,s}$$

with real coefficients a_n. This corresponds to the decomposition (4.99). Next let $\nu, \kappa \in \mathbb{N}_0$, where $\nu \neq \kappa$, and use (4.93) to write

$$\mathbf{V}_\nu^{r,s} = span\left\{ V_n^{r,s} : |n| = \nu \right\},$$

$$\mathbf{V}_\kappa^{r,s} = span\left\{ U_k^{r,s} : |k| = \kappa \right\}.$$

Then (4.88) implies immediately

$$\mathbf{V}_\nu^{r,s} \perp \mathbf{V}_\kappa^{r,s},$$

so the decomposition is orthogonal.

Moreover, the inner product satisfies $[E_A, F_A]_{r,s} = [E, F]_{r,s}$ for $E, F \in C(B^r)$ and $A \in \mathbf{A}^r$. So it follows from Definition 4.2 that the $\mathbf{V}_\kappa^{r,s}$ are rotation-invariant.

Next we define the polynomial function $H_\mu^{r,s}$ by

$$H_\mu^{r,s}(x,y) := \sum_{|m|=\mu} \hat{U}_m^{r,s}(x) V_m^{r,s}(y)$$

for $x, y \in B^r$. Because of (4.93) we get

(i) $H_\mu^{r,s}(x, \cdot) \in \mathbf{V}_\mu^{r,s}$.

Moreover, every $F \in \mathbf{V}_\mu^{r,s}$ can be represented in the form (4.94), and this implies

(iii) $[H_\mu^{r,s}(x, \cdot), F]_{r,s} = F(x)$.

Finally we want to prove symmetry, i.e.,

(ii) $H_\mu^{r,s}(x,y) = H_\mu^{r,s}(y,x)$.

To this end we introduce the vectors

$$\hat{U}(x) := \left[\hat{U}_m^{r,s}(x)\right], \quad V(y) := \left[V_m^{r,s}(y)\right],$$

again for $x, y \in B^r$, where the row index m runs in $\mathcal{M}(r, \mu)$. Likewise we define the matrix

$$A := \left([\hat{U}_m^{r,s}, \hat{U}_n^{r,s}] \right)$$

with row and column indices $m, n \in \mathcal{M}(r, \mu)$. Note that $A = A'$ holds.

With this notation we get from (4.96) and (4.97), respectively, together with (4.98),

$$\hat{U}(x) = AV(x), \quad V(y) = A^{-1}\hat{U}(y),$$

and this yields

$$\begin{aligned}
H_\mu^{r,s}(x,y) &= \hat{U}'(x)V(y) \\
&= V'(x)A'A^{-1}\hat{U}(y) \\
&= V'(x)\hat{U}(y) &= H_\mu^{r,s}(y,x),
\end{aligned}$$

for arbitrary $x, y \in B^r$. So (ii) is also valid. By Definition 1.1 and Theorem 1.1 $H_\mu^{r,s}(x, y)$ is the uniquely determined reproducing kernel, and we get

$$H_\mu^{r,s}(x,y) = G_\mu^{r,s}(x,y) = H_\mu^{r,s}(y,x),$$

which is equation (4.100) in short notation. This finishes the proof.   □

In the following section we use our results to give Theorem 4.25 and Corollary 4.26 a definite form.

4.4   The Image of $G_\mu^{r+\kappa}(a' \cdot)$ under T-Kernel Projections

In Section 7.3 we prove that the spaces $\mathbf{V}_\mu^{r,s}$ can be represented, additionally to (4.93), in the form

$$\mathbf{V}_\mu^{r,s} = span\{G_\mu^{r+s+1}(a' \cdot)|\, a \in S^{r-1}\},$$

see Theorem 7.6. So it is important to know the action of the T-kernel projections onto the functions $G_\mu^r(a' \cdot)$, and more generally, onto $G_\mu^{r+\kappa}(a' \cdot)$, where $a \in S^{r-1}$ and $\kappa \in \mathbb{N}$. We begin with the first case.

Let $r \in \mathbb{N} \setminus \{1,2\}$, $k \in \{1,\dots,r-1\}$ and put $s := r-1-k$. Then (4.86) takes the form

$$R_{\bar{\bar{m}},0} = \tfrac{2\mu+r-2}{(r-2)\omega_{r-1}} \cdot V_{\bar{\bar{m}}}^{k,s} \tag{4.101}$$

where $\mu = |\bar{\bar{m}}|$, and for $(\bar{\bar{m}},0), (\bar{\bar{n}},0) \in \mathbb{N}_0^k \times \mathbb{N}_0^{r-k}$, $|\bar{\bar{m}}| = \mu = |\bar{\bar{n}}|$, we get

$$\begin{aligned}
\langle R_{\bar{\bar{m}},0}, R_{\bar{\bar{n}},0}\rangle_r &= \left(\tfrac{2\mu+r-2}{(r-2)\omega_{r-1}}\right)^2 \int_{S^{r-1}} V_{\bar{\bar{m}}}^{k,s}(x)V_{\bar{\bar{n}}}^{k,s}(x)\, d\omega(x) \\
&= \left(\tfrac{2\mu+r-2}{(r-2)\omega_{r-1}}\right)^2 \omega_s \cdot \left[V_{\bar{\bar{m}}}^{k,s}, V_{\bar{\bar{n}}}^{k,s}\right]_{k,s},
\end{aligned}$$

where we used formula (1.25) again. Because of (4.98) this yields

$$\left(\langle R_{\bar{\bar{m}},0}, R_{\bar{\bar{n}},0}\rangle_r\right)^{-1}_{\bar{\bar{m}},\bar{\bar{n}}\in\mathcal{M}} = \left(\tfrac{(r-2)\omega_{r-1}}{2\mu+r-2}\right)^2 \cdot \tfrac{1}{\omega_s} \cdot \left(\left[\hat{U}_{\bar{\bar{m}}}^{k,s}, \hat{U}_{\bar{\bar{n}}}^{k,s}\right]_{k,s}\right)_{\bar{\bar{m}},\bar{\bar{n}}\in\mathcal{M}}, \tag{4.102}$$

where $\mathcal{M} = \mathcal{M}(k,\mu)$. In view of (4.85) this means that we obtained the $c_{\bar{\bar{m}},\bar{\bar{n}}}^{r,k}$, $\bar{\bar{m}}, \bar{\bar{n}} \in \mathcal{M}(k,\mu)$, in an explicit form. Inserting this in (4.84) we get, together with (4.101),

$$\left(\Pi_T^r G_\mu^r(a' \cdot)\right)(x) =$$

$$\tfrac{1}{\omega_s} \sum_{|m|=\mu} \sum_{|n|=\mu} \left[\hat{U}_m^{k,s}, \hat{U}_n^{k,s}\right]_{k,s} \cdot V_m^{k,s}(at_1,\dots,at_k)\, V_n^{k,s}(xt_1,\dots,xt_k). \tag{4.103}$$

A more compact formulation is contained in the following theorem.

Theorem 4.29 (T-Kernel Projection of $G_\mu^r(a' \cdot)$, Final Form.). *Let* $r \in \mathbb{N} \setminus \{1,2\}$, $k \in \{1,\dots,r-1\}$, $\mu \in \mathbb{N}_0$, *and* $a \in S^{r-1}$. *Let* $T = (t_1,\dots,t_k)$ *consist of the orthogonal columns* $t_1,\dots,t_k \in S^{r-1}$, *and put* $s := r-1-k$. *Then*

$$\left(\Pi_T^r G_\mu^r(a' \cdot)\right)(x) = \tfrac{1}{\omega_s} G_\mu^{k,s}(a'T, x'T) \tag{4.104}$$

holds for arbitrary $x \in S^{r-1}$.

Proof. Using (4.96) and (4.100) we get from (4.103)

$$\left(\Pi_T^r G_\mu^r(a' \cdot)\right)(x) = \tfrac{1}{\omega_s} \sum_{|m|=\mu} V_m^{k,s}(at_1, \dots, at_k)\, \hat{U}_m^{k,s}(xt_1, \dots, xt_k)$$
$$= \tfrac{1}{\omega_s} G_\mu^{k,s}(a'T, x'T),$$

as claimed. □

Particular Assumptions

In the particular situation of Corollary 4.26 we have to determine the coefficients $c_{\bar{m},\bar{n}}^{r,k-\kappa}$ for $(\bar{m},0),(\bar{n},0) \in \mathbb{N}_0^{k-\kappa} \times \mathbb{N}_0^{r-k+\kappa}$, $|\bar{m}| = \mu = |\bar{n}|$, from

$$\left(c_{\bar{m},\bar{n}}^{r,k-\kappa}\right)^{-1} = \left(\langle R_{\bar{m},0}, R_{\bar{n},0}\rangle\right),$$

where (4.101) has to be replaced by

$$R_{\bar{m},0} = \tfrac{2\mu+r-2}{(r-2)\omega_{r-1}} \cdot V_{\bar{m}}^{k-\kappa,s},$$

with $s = r - k + \kappa - 1$. It follows, similar to above, that

$$\left(\Pi_T^r G_\mu^r(a \cdot)\right)(x) =$$

$$\tfrac{1}{\omega_s} \sum_{|m|=\mu} \sum_{|n|=\mu} \left[\hat{U}_m^{k-\kappa,s}, \hat{U}_n^{k-\kappa,s}\right]_{k-\kappa,s} \cdot V_m^{k-\kappa,s}(at_1, \dots, at_{k-\kappa})\, V_n^{k-\kappa,s}(xt_1, \dots, xt_{k-\kappa})$$

$$(4.105)$$

holds for $x \in S^{r-1}$, now, instead of (4.103). By the assumptions of Corollary 4.26, a, x and $t_1, \dots, t_{k-\kappa}$ have the particular form

$$a = \begin{pmatrix} \bar{a} \\ 0 \end{pmatrix}, \quad x = \begin{pmatrix} \bar{x} \\ * \end{pmatrix}, \quad t_\nu = \begin{pmatrix} \bar{t}_\nu \\ 0 \end{pmatrix}$$

for $\nu = 1, \dots, k - \kappa$, where $\bar{a}, \bar{t} \in S^{r-\kappa-1}$, while $\bar{x}$ varies in $B^{r-\kappa}$ for $x \in S^{r-1}$. So we are invited to view the problem from beneath, i.e., from the $\mathbb{R}^{r-\kappa}$–level. For this reason we transform it by the help of the following definitions. First let

$$\bar{r} := r - \kappa,$$
$$\bar{k} := k - \kappa,$$
$$\bar{s} := \bar{r} + \kappa - 1 - \bar{k},$$
$$\bar{T} := (\bar{t}_1, \dots, \bar{t}_{k-\kappa}),$$

which implies $\bar{s} = s$. Hence T can be written in the form

$$T = \begin{pmatrix} \bar{T} & O \\ O & I \end{pmatrix}$$

with the $\kappa \times \kappa$ unity matrix I. Next we want to define the action of Π_T^r onto functions $\bar{F} \in C(B^{r-\kappa})$. Obviously, we may identify $\bar{F}$ with $F \in C(S^{r-1})$ defined by

$$F(x_1, \ldots, x_r) := \bar{F}(x_1, \ldots, x_{r-\kappa}) \text{ for } x \in S^{r-1}.$$

Because of the particular form of T the function F_B depends for $B \in \mathbf{A}_T^r$ on the arguments $x_1, \ldots, x_{r-\kappa}$, only. So does the average over all B. It follows that $(\Pi_T^r F)(x)$ depends on these arguments, only. This allows us the additional definition.

Definition 4.3. *Let T be as above. The* average operator

$$\Pi_{\bar{T}}^{r-\kappa,\kappa} : C(B^{r-\kappa}) \to C(B^{r-\kappa})$$

is defined by $\left(\Pi_{\bar{T}}^{r-\kappa,\kappa} \bar{F}\right)(\bar{x}) := (\Pi_T^r F)(x)$ *for* $x = \binom{\bar{x}}{*} \in S^{r-1}$, *i.e.,* $\bar{x} \in B^{r-\kappa}$.

Using the transform from above and Definition 4.3 we formulate Corollary 4.26 now in the following form.

Corollary 4.30 (Particular Assumptions). *Let $r \in \mathbb{N} \setminus \{1\}$, $k \in \{1, \ldots, r-1\}$, $\kappa \in \mathbb{N}$, $\mu \in \mathbb{N}_0$, and define $s := r + \kappa - 1 - k$. Assume that $T = (t_1, \ldots, t_k)$ consists of orthogonal columns $t_1, \ldots, t_k \in S^{r-1}$, and let $a \in S^{r-1}$. Then*

$$\left(\Pi_T^{r,\kappa} G_\mu^{r+\kappa}(a' \cdot)\right)(x) = \tfrac{1}{\omega_s} G_\mu^{k,s}(a'T, x'T) \tag{4.106}$$

holds for $x \in B^r$.

Proof. First we replace in formula (4.105) all terms by the corresponding barred terms. After that we omit the bars. So we get the explicit representation

$$\left(\Pi_T^{r,\kappa} G_\mu^{r+\kappa}(a' \cdot)\right)(x) = \tfrac{1}{\omega_s} \sum_{|m|=\mu} \sum_{|n|=\mu} \left[\hat{U}_m^{k,s}, \hat{U}_n^{k,s}\right]_{k,s} V_m^{k,s}(a'T) V_n^{k,s}(x'T).$$

Finally we use (4.96) and (4.100), again, to obtain (4.106). □

For an application in *tomography* we add the following corollary, which is just Corollary 4.30 under the assumption $k = 1$.

Corollary 4.31 (Particular assumptions, and $k = 1$). *Assumptions as in Corollary 4.30, but $k := 1$, $t := t_1 \in S^{r-1}$. Then*

$$\left(\Pi_T^{r,\kappa} G_\mu^{r+\kappa}(a' \cdot)\right)(x) = \tilde{G}_\mu^{r+\kappa}(at) G_\mu^{r+\kappa}(tx)$$

holds for $x \in B^r$, where

$$\tilde{G}_\mu^{r+\kappa} := G_\mu^{r+\kappa} / G_\mu^{r+\kappa}(1) = \tilde{C}_\mu^{\frac{r+\kappa-2}{2}}.$$

Proof. For $k = 1$ we get $s = r + \kappa - 2$, and in view of (4.100), formula (4.106) takes the simple form

$$\left(\Pi_T^{r,s} G_\mu^{r+\kappa}(a' \cdot)\right)(x) = const \cdot U_\mu^{1,s}(at) V_\mu^{1,s}(tx).$$

From (3.51) and (3.59), respectively, we get with $t := e_1$,

$$V_\mu^{1,s} = C_\mu^{\frac{s}{2}} = U_\mu^{1,s}. \tag{4.107}$$

In view of (4.12) this implies

$$\left(\Pi_T^{r,s} G_\mu^{r+\kappa}(a' \cdot)\right)(x) = const \cdot G_\mu^{r+\kappa}(at) G_\mu^{r+\kappa}(tx),$$

again with some constant. We determine the constant by inserting the fixed point $x = t$, which yields

$$G_\mu^{r+\kappa}(at) = \left(\Pi_T^{r,s} G_\mu^{r+\kappa}(a' \cdot)\right)(t) = const \cdot G_\mu^{r+\kappa}(at) G_\mu^{r+\kappa}(1)$$

and hence $const = 1/G_\mu^{r+\kappa}(1)$, which finishes the proof. □

Remark 1. It follows from Theorem 1.9, (i), that

$$\Pi_T^{r,s} 1 = 1 \tag{4.108}$$

is valid again.

Remark 2. In the case $k = 1$ the right side of (4.104) takes the form

$$\frac{1}{\omega_{r-2}} G_\mu^{1,r-2}(at, xt) = \frac{1}{\omega_{r-2}} \hat{U}_\mu^{1,r-2}(at) V_\mu^{1,r-2}(xt)$$

where we put $t := t_1$ and used (4.100). It follows as above that

$$\left(\Pi_T^r G_\mu^r(a' \cdot)\right)(x) = \tilde{G}_\mu^r(at) G_\mu^r(xt).$$

This result agrees formally with the result of Corollary 4.31 if κ is put to zero.

Remark 3. For $k \geq 2$ the spaces $\mathbf{V}_\mu^{k,s}$ are rotation-invariant, see Theorem 4.28, and by Theorem 1.2 the reproducing kernel can be represented in the form

$$G_\mu^{k,s}(x, y) = h_\mu^{k,s}(|x|, |y|, xy)$$

for $x \neq 0$, $y \neq 0$. It follows from (4.104) that $\left(\Pi_T^r G_\mu^r(a' \cdot)\right)(x)$ is a function of the expressions

$$|a'T| = \sqrt{a'TT'a}, \quad |x'T| = \sqrt{x'TT'x}, \quad a'TT'x,$$

only. Now let T$:= span\{t_1, \ldots, t_k\}$. Then

$$\bar{a} := TT'a = \sum_{\nu=0}^{k} (at_\nu) t_\nu, \quad \bar{x} := TT'x = \sum_{\nu=0}^{k} (xt_\nu) t_\nu,$$

are the orthogonal projections onto T of a and x, respectively. It follows that

$$|a'T| = |\bar{a}|, \quad |x'T| = |\bar{x}|, \quad a'TT'x = \bar{a}\bar{x}.$$

This implies that

$$\left(\Pi_T^r G_\mu^r(a' \cdot)\right)(x) = \frac{1}{\omega_s} \cdot h_\mu^{k,s}(|\bar{a}|, |\bar{x}|, \bar{a}\bar{x})$$

depends on the space T only, not on the choice of the orthonormal basis $\{t_1, \ldots, t_k\}$ within this space.

4.5 Problems

Problem 4.1. Prove that (4.10) and (4.11) are equivalent.

Problem 4.2. Let $r \geq 2$, $\mu \in \mathbb{N}$. Prove $|G_\mu^r(\xi)| \leq G_\mu^r(1)$ for $-1 \leq \xi \leq +1$.

Problem 4.3. Let $\mathbf{V}$ be a rotation-invariant subspace of $C(S^{r-1})$, $r \in \mathbb{N} \setminus \{1\}$, $N := \dim \mathbf{V} \in \mathbb{N}$, and let $\| \cdot \|_2$ be the norm induced by $\langle \cdot, \cdot \rangle$. Then the estimate

$$\|F\|_\infty \leq \sqrt{N} \|F\|_2$$

holds for $F \in \mathbf{V}$.

Problem 4.4. Prove

$$\frac{1}{(1 - 2xt + x^2t^2)^{\frac{r+s-1}{2}}} = \sum_{\mu=0}^{\infty} \sum_{|m|=\mu} \overset{*}{V}_m^{r,s}(x) \, t^m$$

for $x, t \in B^r$, $|t| < 1$, $r + s > 1$. For which values of $\lambda > 0$ is the generating function on the left side harmonic with respect to x? Interpret the result for the polynomials $\overset{*}{V}_m^{r,s}(x)$.

Problem 4.5. For $(x_1, \ldots, x_{r+s+1})' \in \mathbb{R}^{r+s+1}$, $r, s \in \mathbb{N}$, define $x \in \mathbb{R}^r$ by $x := (x_1, \ldots, x_r)'$. And put

$$H_m^{r,s}(x_1, \ldots, x_{r+s+1}) := \left[x_1^2 + \cdots + x_{r+s+1}^2\right]^{\frac{\mu}{2}} U_m^{r,s}\left(\frac{x}{[x_1^2 + \cdots + x_{r+s+1}^2]^{\frac{1}{2}}}\right)$$

for $m \in \mathbb{N}_0^r$, $|m| = \mu$. Prove

$$\frac{1}{\left[(1 - xt)^2 + (x_{r+1} + \cdots + x_{r+s+1}^2)t^2\right]^{\frac{s}{2}}} = \sum_{\mu=0}^{\infty} \sum_{|m|=\mu} H_m^{r,s}(x_1, \ldots, x_{r+s+1}) \, t^m,$$

and $H_m^{r,s} \in \overset{*}{\mathbb{H}}_\mu^{r+s+1}$.

Part III

Multivariate Approximation

Chapter 5

Approximation Methods

5.1 Bounded Linear Operators

We investigate polynomial approximations to multivariate functions which are defined by linear operators. The corresponding theory is ruled by some important principles and theorems, which we present in the beginning.

Let X and Y denote real linear normed spaces with norm $\| \cdot \| = \| \cdot \|_X$ and $\| \cdot \| = \| \cdot \|_Y$, respectively. A linear map, or *operator*,

$$F : X \to Y$$

is called *bounded* if

$$\sup\{\|Fx\|_Y : x \in X, \|x\|_X \leq 1\} < \infty$$

holds. It is easy to show that the set $\mathcal{L}(X, Y)$ of all bounded linear operators $F : X \to Y$ is a linear space which can be provided with the norm $\| \cdot \| = \| \cdot \|_{X,Y}$ defined by

$$\|F\|_{X,Y} := \sup\{\|Fx\|_Y : x \in X, \|x\|_X \leq 1\}. \tag{5.1}$$

Note that $F \in \mathcal{L}(X, Y)$, $G \in \mathcal{L}(Y, Z)$, and $x \in X$ implies

$$\|Fx\| \leq \|F\| \cdot \|x\| \quad and \quad \|G \circ F\| \leq \|G\| \cdot \|F\|. \tag{5.2}$$

A linear operator $F : X \to Y$ is *continuous* if and only if it is bounded, which means, if it belongs to $\mathcal{L}(X, Y)$. It may happen that a sequence of elements $F_n \in \mathcal{L}(X, Y)$, $n \in \mathbb{N}$, converges in the norm, i.e., that

$$\lim_{n \to \infty} \|F_n - F\| = 0$$

holds for some $F \in \mathcal{L}(X, Y)$. Using (5.2) we get in this case

$$\|F_n x - Fx\| \leq \|F_n - F\| \cdot \|x\|$$

for $x \in X$ and $n \in \mathbb{N}$, and hence

$$\lim_{n \to \infty} \|F_n x - F x\| = 0 \qquad (5.3)$$

for arbitrary $x \in X$. However, operator sequences which converge in the norm are rare. For this reason it is usual to use the following weaker definition of convergence.

Definition 5.1 (Pointwise Convergent). *A sequence of operators* $F_n \in \mathcal{L}(X, Y)$, $n \in \mathbb{N}$, *is called* pointwise convergent *to* $F \in \mathcal{L}(X, Y)$ *if*

$$\lim_{n \to \infty} F_n x = F x$$

holds for all $x \in X$.

It is important to know an exact criterion about pointwise convergence. This is given by the *Theorem of Banach–Steinhaus*, which we prove in three steps.

Theorem 5.1 (Baire). *Let* (X, d) *denote a complete metric space and assume that the union*

$$A := \bigcup_{n \in \mathbb{N}} A_n$$

of countably many subsets $A_n \subset X$, $n \in \mathbb{N}$, *contains a nonempty open subset. Then* $k \in \mathbb{N}$ *exists such that the closure* $\bar{A}_k$ *of* A_k *contains a nonempty open set.*

Proof. We define the open balls

$$B(x, \rho) := \left\{ y \mid y \in X, \, d(y, x) < \rho \right\}$$

for $x \in X$ and $\rho \in \mathbb{R}$, $\rho > 0$, and we assume that the statement of Theorem 5.1 is false, which means that no $\bar{A}_k$ contains a nonempty open set.

Since A contains a nonempty open set there exists a ball $B_0 = B(x_0, \rho_0)$, $\rho_0 > 0$, such that $B_0 \subset A$ is valid. B_0 is not contained in $\bar{A}_1$, which we write in the form

$$B_0 \cap \mathcal{C}\bar{A}_1 \neq \emptyset.$$

This set is open, hence we can find a ball $B_1 = B(x_1, \rho_1)$, $0 < \rho_1 < \frac{1}{2}\rho_0$, such that

$$\bar{B}_1 \subset B_0 \cap \mathcal{C}\bar{A}_1.$$

$B_1 \cap \mathcal{C}\bar{A}_2$ is again nonempty and open, hence we can find a ball $B_2 = B(x_2, \rho_2)$, $0 < \rho_2 < \frac{1}{2}\rho_1$, such that

$$\bar{B}_2 \subset B_1 \cap \mathcal{C}\bar{A}_2.$$

Continuing this procedure we find a whole sequence of balls $B_j = B(x_j, \rho_j)$, $j \in \mathbb{N}$, satisfying

$$\bar{B}_j \subset B_{j-1} \cap \mathcal{C}\bar{A}_j, \quad \rho_j < 2^{-j}\rho_0,$$

for $j = 1, 2, \ldots$. By construction, $x_j \in B_k$ holds for $k \in \mathbb{N}_0$ and $k < j$. This implies

$$d(x_j, x_k) < \rho_k < 2^{-k}\rho_0.$$

So the x_j form a Cauchy sequence which is convergent with the limit

$$x := \lim_{j \to \infty} x_j,$$

where $x \in \bar{B}_k \subset B_{k-1} \subset \cdots \subset B_0 \subset A$ is valid for arbitrary $k \in \mathbb{N}$. From $x \in A$ it follows that $n \in \mathbb{N}$ exists such that $x \in A_n$ holds, while, simultaneously,

$$x \in \bar{B}_{n+1} \subset B_n \cap C\bar{A}_n \subset CA_n$$

must also be valid, which is a contradiction. So the statement of Theorem 5.1 is not false.   □

Theorem 5.2 (Uniform Boundedness). *Let X be a Banach-space, Y a normed linear space, and assume that the sequence of operators $F_n \in \mathcal{L}(X, Y)$, $n \in \mathbb{N}$, satisfies the condition*

$$\sup\{\|F_n x\| : n \in \mathbb{N}\} < \infty$$

pointwise for all $x \in X$. Then the operator norms are uniformly bounded, i.e.,

$$\sup\{\|F_n\| : n \in \mathbb{N}\} < \infty$$

holds.

Proof. Let

$$\Phi(x) := \sup\{\|F_n x\| : n \in \mathbb{N}\} \text{ for } x \in X,$$
$$A_k := \{x \in X | \ \Phi(x) \le k\} \text{ for } k \in \mathbb{N}.$$

First we claim that every set A_k is closed. Actually, in the case $A_k = X$ nothing has to be proved. Next assume $A_k \ne X$ and choose $y \in CA_k$. Then

$$\Phi(y) > k$$

is valid, and there exists some $n \in \mathbb{N}$ such that even $\|F_n y\| > k$ holds. F_n is bounded and hence continuous, so a neighbourhood $\mathcal{U}$ of y exists such that

$$\|F_n z\| > k$$

remains valid for all $z \in \mathcal{U}$. Hence CA_k is open and A_k is closed, as claimed. It follows that

$$A_k = \bar{A}_k \text{ for } k \in \mathbb{N}.$$

Next we use that

$$X = \bigcup_{k \in \mathbb{N}} A_k$$

holds, since for every $x \in X$ there exists a $k \in \mathbb{N}$ with $k \geq \Phi(x)$. X is nonempty and open itself. So Theorem 5.1 says that $k \in \mathbb{N}$ exists such that A_k contains a nonempty open subset, say an open ball K with center x_0 and radius $\rho > 0$. Since A_k is closed, even the closed ball

$$\bar{K} = \{z \in X : \|z - x_0\| \leq \rho\}$$

is contained in A_k, i.e., we get $\bar{K} \subset A_k$. By the definitions of A_k and of $\Phi(x)$ it follows that $z \in \bar{K}$ implies $z \in A_k$ and hence

$$\|F_n z\| \leq \Phi(z) \leq k \text{ for all } n \in \mathbb{N}.$$

This allows us to estimate $\|F_n\|$ as follows.

Let $x \in X$, $\|x\| \leq 1$, $n \in \mathbb{N}$. The points x_0 and $z := x_0 + \rho x$ are contained in $\bar{K}$, and from

$$F_n x = F_n\left(\frac{z - x_0}{\rho}\right)$$

we get

$$\|F_n x\| \leq \frac{1}{\rho}\left(\|F_n z\| + \|F_n x_0\|\right) \leq \frac{2k}{\rho}.$$

This holds for arbitrary $x \in X$ with $\|x\| \leq 1$, so we get

$$\|F_n\| \leq \frac{2k}{\rho}$$

for arbitrary $n \in \mathbb{N}$, which finishes the proof. $\qquad\square$

Theorem 5.3 (Banach–Steinhaus). *Let X be a Banach-space, Y a normed linear space, and E a subset of X such that $\operatorname{span}(E)$ is dense in X. Then a sequence of operators $F_n \in \mathcal{L}(X,Y)$, $n \in \mathbb{N}$, converges pointwise to $F \in \mathcal{L}(X,Y)$ if and only if the following conditions are satisfied.*

$$(i) \quad \sup\{\|F_n\| : n \in \mathbb{N}\} < \infty,$$

$$(ii) \quad \lim_{n \to \infty} F_n x = F x \text{ for all } x \in E.$$

Proof. If the sequence of operators F_n converges pointwise to F, then (ii) is obvious, while (i) is valid by Theorem 5.2.

Next we assume that (i) and (ii) are valid. We choose $A > 0$ such that

$$\|F\| \leq A \quad and \quad \|F_n\| \leq A \text{ for all } n \in \mathbb{N} \tag{5.4}$$

holds. Next let $x \in X$ be fixed, and choose an $\epsilon > 0$. Then there is an element $y \in \operatorname{span}(E)$, say

$$y = \sum_{j=1}^{k} c_j x_j, \quad x_j \in E, \quad j = 1, \dots, k,$$

which satisfies

$$\|x - y\| < \frac{\epsilon}{3A}. \tag{5.5}$$

By the triangular inequality we obtain

$$\|F_n x - Fx\| \leq \|F_n x - F_n y\| + \|F_n y - Fy\| + \|Fy - Fx\|.$$

Moreover,

$$F_n y - Fy = \sum_{j=1}^{k} c_j \left(F_n x_j - Fx_j \right)$$

tends to zero for $n \to \infty$. So there is an $N \in \mathbb{N}$ such that

$$\|F_n y - Fy\| < \frac{\epsilon}{3}$$

holds for $n \geq N$, and together with (5.4) and (5.5), we get

$$\|F_n x - Fx\| < \|F_n\| \cdot \|x - y\| + \frac{\epsilon}{3} + \|F\| \cdot \|x - y\| \leq \epsilon,$$

again for $n \geq N$. Since $\epsilon > 0$ was arbitrary, this yields

$$\lim_{n \to \infty} F_n x = Fx,$$

and since x was arbitrary, the theorem is proved. $\qquad\square$

Remark. In the context of Theorem 5.3 the operator norms $\|F_n\|$ are called *Lebesgue constants*.

The application of Theorem 5.3 requires the knowledge of a suitable subset E of X. In the important case $X = C(D)$, D a nonempty compact subset of $\mathbb{R}^r$, the polynomial space $\mathbb{P}^r(D)$ is dense by the *Theorem of Weierstrass*, which we prove later. The most elegant proofs use the setting of *positive linear operators*.

Definition 5.2 (Positive Linear Operator). *In $C(D)$ the semi-order $\geq$ is defined as follows. $F \geq 0$ holds for a function $F \in C(D)$ if and only if $F(x) \geq 0$ is valid for arbitrary $x \in D$. A linear operator $L : C(D) \to C(D)$ is called* positive *if $F \in C(D)$, $F \geq 0$, implies $LF \geq 0$.*

We remark that $F \geq G$ is the same as $F - G \geq 0$, which is often also written as $G \leq F$.

As usual $C(D)$ is provided with the maximum norm $\| \cdot \| = \| \cdot \|_\infty$. If $L : C(D) \to C(D)$ is a positive linear operator and if $F \in C(D)$ satisfies $\|F\| \leq 1$, then we get $1 \pm F \geq 0$ and hence $L1 \pm LF \geq 0$, which is the same as $|(LF)(x)| \leq (L1)(x)$ for arbitrary $x \in D$. It follows that

$$\|LF\| \leq \|L1\|,$$

which means that a positive linear operator is always *bounded*. So we can determine the pointwise convergence of a sequence of positive linear operators $L_n : C(D) \to C(D)$ with the help of Theorem 5.3. However, often it is easier to use the *Theorem of Bohman and Korovkin*.

Definition 5.3 (Korovkin Set). *A finite set of nonzero functions $F_0, \ldots, F_k \in C(D)$ is called a* Korovkin set *in $C(D)$ if there are functions $G_0, \ldots, G_k \in C(D)$ such that*

$$\sum_{j=0}^{k} F_j(x) G_j(y) \geq 0$$

holds for all $x, y \in D$, where equality holds if and only if $x = y$.

The following theorem describes some Korovkin sets for the sphere S^{r-1} and for the simplex Σ^{r-1} (see Section 1.1).

Theorem 5.4 (Korovkin Sets in $C(S^{r-1})$ and in $C(\Sigma^{r-1})$). *Let $r \in N \setminus \{1\}$, $D \in \{S^{r-1}, \Sigma^{r-1}\}$, and define the functions $F_j \in C(D)$, $j = 0, \ldots, r$, by*

$$F_j(x) := \begin{cases} 1 & , \quad \text{for } j = 0, \\ x_j & , \quad \text{for } j = 1, \ldots, r, \text{ if } D = S^{r-1}, \\ \sqrt{x_j} & , \quad \text{for } j = 1, \ldots, r, \text{ if } D = \Sigma^{r-1}. \end{cases}$$

Then $F_0, \ldots, F_r$ form a Korovkin set in $C(D)$. In both cases the functions $1, x_1, \ldots, x_r, x_1^2, \ldots, x_r^2$ form another Korovkin set in $C(D)$.

Proof. For the first statement let us define $G_0 := 1$, $G_j := -F_j$ for $j = 1, \ldots, r$. Then we obtain

$$\sum_{j=0}^{r} F_j(x) G_j(y) = \begin{cases} 1 - xy \geq 0 & , \text{ if } D = S^{r-1}, \\ 1 - \sum_{j=1}^{r} \sqrt{x_j y_j} \geq 0 & , \text{ if } D = \Sigma^{r-1}, \end{cases}$$

for $x, y \in D$, in both cases, where equality holds exactly for $x = y$.

The second statement can be proved as follows. For arbitrary $x, y \in D$ we get

$$\sum_{j=1}^{r} \left(x_j^2 - 2x_j y_j + y_j^2 \right) = |x - y|^2 \geq 0,$$

where equality holds exactly for $x = y$. Writing the left side in the form

$$1 \cdot \left(\sum_{k=1}^{r} y_k^2 \right) + \sum_{j=1}^{r} x_j \cdot (-2y_j) + \sum_{j=1}^{r} \left(x_j^2 \cdot 1 \right)$$

we see immediately that the functions considered form also a Korovkin set. $\square$

Theorem 5.5 (Bohman and Korovkin). *Let $D \subset \mathbb{R}^r$ be nonempty and compact and let $F_0, \ldots, F_k$ form a Korovkin set in $C(D)$, which is provided with the uniform norm. And assume that L_n, $n \in \mathbb{N}$, is a sequence of positive linear operators $C(D) \to C(D)$. Then*

$$\lim_{n \to \infty} \|L_n F - F\| = 0$$

is valid for all functions $F \in C(D)$ if and only if

$$\lim_{n \to \infty} \|L_n F_j - F_j\| = 0$$

holds for $j = 0, 1, \ldots, k$.

Remark. In view of Theorem 5.5 the finitely many functions $F_0, \ldots, F_k$ are also called a *Korovkin test family*.

Proof. The necessary condition is obvious. So it suffices to consider the case where

$$\lim_{n \to \infty} \|L_n F_j - F_j\| = 0$$

holds for $j = 0, 1, \ldots, k$.

We begin with the — most noninteresting — case where $D = \{x\}$ consists of a single point. Then the functions $F \in C(D)$ are all constants. In particular we get

$$\lim_{n \to \infty} L_n F_0 = F_0 \neq 0,$$

and for arbitrary $F \in C(D)$ we obtain

$$\lim_{n \to \infty} L_n F = \lim_{n \to \infty} L_n \left(\frac{F}{F_0} \cdot F_0 \right) = \frac{F}{F_0} \cdot L_n F_0 = F,$$

as claimed.

Next we assume that D contains more than one point, say y_1 and $y_2 \neq y_1$. By Definition 5.3 there are $G_0, \ldots, G_k \in C(D)$ such that

$$\Phi(x, y) := \sum_{j=0}^{k} F_j(x) G_j(y) \geq 0 \qquad (5.6)$$

is valid for $x, y \in D$, where $\Phi(x, y) = 0$ holds exactly for $x = y$. Note that $\Phi \in C(D \times D)$ holds, and that $D \times D$ is compact, again. Our first aim is to show that the functions $L_n 1$ are uniformly bounded to above, and this even if 1 does not belong to the Korovkin set. To this end we use that the function

$$\Phi(\cdot, y_1) + \Phi(\cdot, y_2) \in C(D)$$

is strictly positive, such that a constant a exists such that

$$0 < a \leq \Phi(x, y_1) + \Phi(x, y_2) \qquad (5.7)$$

holds for arbitrary $x \in D$. As L_n is positive it follows that

$$a \ L_n 1 \leq L_n \Big(\Phi(\cdot, y_1) + \Phi(\cdot, y_2) \Big) = \sum_{j=0}^{k} \Big(G_j(y_1) + G_j(y_2) \Big) L_n F_j.$$

The right side converges uniformly to a function of $C(D)$. This is bounded above. So a constant A exists, such that

$$L_n 1 \leq A \tag{5.8}$$

holds. Moreover, for fixed $y \in D$ we get

$$\|L_n \Phi(\,\cdot\,, y) - \Phi(\,\cdot\,, y)\| = \|\sum_{j=0}^{k} G_j(y)\left(L_n F_j - F_j\right)\| \leq \sum_{j=0}^{k} \|G_j\| \cdot \|L_n F_j - F_j\|.$$

This implies

$$\lim_{n \to \infty} \|L_n \Phi(\,\cdot\,, y) - \Phi(\,\cdot\,, y)\| = 0, \tag{5.9}$$

uniformly for all $y \in D$.

Our next aim is to estimate arbitrary functions $\Psi \in C(D \times D)$, which satisfy $\Psi(x, x) = 0$ for $x \in D$, with the help of Φ in order to understand their behaviour near the diagonal $x = y$. So assume Ψ is such a function. Then a constant M exists such that

$$|\Psi(x, y)| \leq M$$

holds for all $(x, y) \in D \times D$. And Ψ is uniformly continuous. So let $\epsilon > 0$ be an arbitrary number. Then $\delta > 0$ exists such that, in view of $\Psi(x, x) = 0$,

$$|\Psi(x, y)| < \epsilon$$

holds for all $(x, y) \in D \times D$, which satisfy $|x - y| < \delta$. Moreover, since $\Phi(x, y)$ does not vanish for $|x - y| \geq \delta$, a constant m exists such that

$$0 < m \leq \Phi(x, y)$$

holds on the compact set defined by $(x, y) \in D \times D$, $|x - y| \geq \delta$. Together this implies, for both signs, that

$$\pm \Psi(x, y) < \epsilon + \tfrac{M}{m}\,\Phi(x, y)$$

is valid for arbitrary $(x, y) \in D \times D$. Especially we get for fixed $y \in D$ by the positivity of the operator

$$\pm L_n \Psi(\,\cdot\,, y) \leq \epsilon\, L_n 1 + \tfrac{M}{m}\, L_n \Phi(\,\cdot\,, y).$$

But in view of (5.9) a number $n(\epsilon)$ exists such that

$$L_n \Phi(\,\cdot\,, y) - \Phi(\,\cdot\,, y) \leq \epsilon$$

is valid for $n \geq n(\epsilon)$ and arbitrary $y \in D$. Together with (5.8) this yields

$$\pm L_n \Psi(\,\cdot\,, y) \leq \epsilon\, A + \tfrac{M}{m}\left(\epsilon + \Phi(\,\cdot\,, y)\right)$$

for $n \geq n(\epsilon)$ and arbitrary $y \in D$. Here we may insert y as the argument, and we obtain, in view of $\Phi(y, y) = 0$,

$$\left| \left(L_n \Psi(\cdot, y) \right)(y) \right| \leq \epsilon (A + \tfrac{M}{m})$$

and hence

$$\lim_{n \to \infty} \left(L_n \Psi(\cdot, y) \right)(y) = 0, \tag{5.10}$$

uniformly for $y \in D$.

Now it is easy to finish the proof. For, let $F \in C(D)$ be an arbitrary function. Define Ψ by

$$\Psi(x, y) := F(x) - F(y) \cdot \frac{\Phi(x, y_1) + \Phi(x, y_2)}{\Phi(y, y_1) + \Phi(y, y_2)}$$

for $(x, y) \in D \times D$, which is admissable because of $\Psi(x, x) = 0$. Then it follows from (5.10) and (5.9) that

$$\lim_{n \to \infty} \left(L_n F \right)(y) - F(y) = 0$$

holds uniformly for all $y \in D$, and the theorem is proved. $\qquad\square$

5.2 Bernstein Polynomials and the Theorem of Weierstrass

Throughout this section, $C(D)$ is provided with the uniform norm $\| \cdot \| = \| \cdot \|_D$.

The Homogeneous Bernstein Polynomials on Σ^{r-1}

Let $r \in \mathbb{N} \setminus \{1\}$, $\mu \in \mathbb{N}$, $F \in C(\Sigma^{r-1})$. The corresponding *homogeneous Bernstein polynomial* of degree μ is defined by

$$\left(\overset{*}{B}_\mu F \right)(x) := \sum_{|m| = \mu} F(\tfrac{m}{\mu}) \binom{\mu}{m} x^m. \tag{5.11}$$

Actually, this is a homogeneous polynomial of degree μ. By restriction of the polynomial onto Σ^{r-1} a map

$$\overset{*}{B}_\mu : C(\Sigma^{r-1}) \to \overset{*}{\mathbb{P}}{}^r_\mu(\Sigma^{r-1}) \tag{5.12}$$

is defined which is the corresponding *homogeneous Bernstein operator*. This operator is linear and positive such that Theorem 5.5 is applicable.

Lemma 5.6 (Convergence of the Homogeneous Bernstein Operators). *Let $r \in \mathbb{N} \setminus \{1\}$. Then*

$$\lim_{\mu \to \infty} \| \overset{*}{B}_\mu F - F \| = 0$$

holds for arbitrary $F \in C(\Sigma^{r-1})$.

Proof. By Theorem 5.4 the set $K := \{1, x_1, \ldots, x_r, x_1^2, \ldots, x_r^2\}$ is a Korovkin set in $C(\Sigma^{r-1})$. In view of (3.10) we get, for $x \in \Sigma^{r-1}$,

$$\left(\overset{*}{B}_\mu 1\right)(x) = \sum_{|m|=\mu} \binom{\mu}{m} x^m = (x_1 + \cdots + x_r)^\mu = 1. \tag{5.13}$$

Likewise we obtain for $j \in \{1, \ldots, r\}$, substituting $n = m - e_j$,

$$\left(\overset{*}{B}_\mu x_j\right)(x) = \sum_{|m|=\mu} \frac{m_j}{\mu}\binom{\mu}{m} x^m = x_j \sum_{|n|=\mu-1} \binom{\mu-1}{n} x^n = x_j, \tag{5.14}$$

and for $\mu \geq 2$ we get

$$\left(\overset{*}{B}_\mu x_j^2\right)(x) = \sum_{|m|=\mu} (\tfrac{m_j}{\mu})^2 \binom{\mu}{m} x^m = x_j^2 + \tfrac{1}{\mu} x_j (1 - x_j). \tag{5.15}$$

Hence $\overset{*}{B}_\mu F$ converges to F for all $F \in K$. The rest of the argument follows from Theorem 5.5. $\qquad\square$

Lemma 5.6 says that $\mathbb{P}^r(D)$ is dense in $C(D)$ in the particular case where $D = \Sigma^{r-1}$. This is a subset of measure zero in $\mathbb{R}^r$. But in spite of this the Theorem of Weierstrass can be derived from Lemma 5.6 in its full extent. This takes the following two steps.

The Bernstein Polynomials on E^r

Let $r \in \mathbb{N}$, and note that Σ^r is now a subset of $\mathbb{R}^{r+1}$. The map $E^r \to \Sigma^r$ defined by

$$E^r \ni (x_1, \ldots, x_r)' = x \mapsto \bar{x} = (x_1, \ldots, x_r, 1 - x_1 - \cdots - x_r)' \in \Sigma^r$$

is bijective and defines an *isometry*

$$C(E^r) \ni F \mapsto \bar{F} \in C(\Sigma^r) \tag{5.16}$$

by $\bar{F}(\bar{x}) := F(x)$ for $x \in E^r$. This map is affine-linear and maps polynomials onto polynomials of the same degree.

Now let $F \in C(E^r)$ be given. We write the homogeneous Bernstein polynomial of $\bar{F}$ for $\mu \in \mathbb{N}$ in the form

$$\left(\overset{*}{B}_\mu \bar{F}\right)(\bar{x}) = \left(B_\mu F\right)(x) \tag{5.17}$$

where

$$\left(B_\mu F\right)(x) = \sum_{\nu=0}^{\mu} \sum_{|m|=\mu-\nu} F(\tfrac{m}{\mu})\binom{\mu}{m,\nu} x^m (1 - x_1 - \cdots - x_r)^\nu. \tag{5.18}$$

Note that $B_\mu F$ is now an element of $\mathbb{P}^r_\mu$, and in general no longer homogeneous. It is called the *Bernstein polynomial* to F of degree μ. The map

$$B_\mu : C(E^r) \to \mathbb{P}^r_\mu(E^r) \tag{5.19}$$

defined by $F \mapsto B_\mu F$ is the corresponding *Bernstein operator*. By construction the following lemma is valid.

Lemma 5.7 (Convergence of the Bernstein Operators). *Let $r \in \mathbb{N}$. Then*

$$\lim_{\mu \to \infty} \|B_\mu F - F\| = 0$$

holds for arbitrary $F \in C(E^r)$.

Proof. Using (5.17) and the isomorphy (5.16) we get for $\mu \in \mathbb{N}$

$$\|B_\mu F - F\|_{\Sigma^r} = \|\overset{*}{B}_\mu \bar{F} - \bar{F}\|_{E^r},$$

where the right side converges to zero by Lemma 5.6. $\qquad\square$

The Theorem of Weierstrass

Now we are ready to perform the final step, which means to prove the Theorem of Weierstrass in its full extent.

Theorem 5.8 (Weierstrass). *Let $r \in \mathbb{N}$ and let D be a nonempty compact set in $\mathbb{R}^r$. Then $\mathbb{P}^r(D)$ is dense in $C(D)$.*

Proof. From Lemma 5.7 it follows that $\mathbb{P}^r(E^r)$ is dense in $C(E^r)$. Next assume D is arbitrary, but satisfying the assumptions. Since the affine-linear transforms map polynomials onto polynomials of the same degree, we may assume $D \subset E^r$ without restriction of generality. Now let $F \in C(D)$. By a theorem of *Tietze and Urysohn*, see [62], e.g., F is the restriction onto D of a function $G \in C(E^r)$, and we can approximate F by the restrictions F_μ of the Bernstein polynomials $B_\mu G$ onto D. From

$$\|F - F_\mu\|_D \le \|G - B_\mu G\|_{E^r}$$

it follows now in view of Lemma 5.7 that

$$\lim_{\mu \to \infty} \|F - F_\mu\|_D = 0$$

holds. This finishes the proof. $\qquad\square$

Remark. Lemma 5.6 and hence the Theorem of Weierstrass can be proved without the help of the Theorem of Bohman and Korovkin. In the univariate case this is well known. For the multivariate case see [48]. It is essential that the points $\frac{m}{\mu}$, which occur in the definition of the homogeneous Bernstein polynomials (5.11), are the peaks, i.e., the maximum points of the monomials x^m on Σ^{r-1}. On the sphere another direct proof exists, which uses the spherical modulus of continuity. See Section 6.9, Corollary 6.37.

5.3   Best Approximation and Projections

A fundamental question is to ask how an arbitrary continuous function, as complicated it may be, can be approximated by a simpler one, say by a polynomial of a
certain degree. Bernstein polynomials are an answer to this question, though not
always the best possible one. We investigate this question in a more systematic
way.

Best Approximation

Let X be a normed linear space, V a finite-dimensional subspace of X, and $x \in X$.
By a compactness argument it can be proved that

$$\min\{\|x - v\| : v \in V\}$$

exists for arbitrary $x \in X$, see, e.g., [13], Theorem 7.4.1. Therefore the following
definition is possible.

Definition 5.4 (Minimal Deviation, Best Approximation). *Let V be a finite-dimensional subspace of the normed linear space X. For $x \in X$*

$$E(x, V) := \min\{\|x - v\| : v \in V\}$$

is called the minimal deviation *of x in V. Every $\overset{*}{v} \in V$ for which the minimum is
attained, i.e., with*

$$\|x - \overset{*}{v}\| = E(x, V),$$

is called a best approximation *to x in V.*

Note that by our remark from above a best approximation always exists.

Projections

Definition 5.5 (Projection). *An operator $L \in \mathcal{L}(X, V)$, $[0] \neq V \subset X$, is called a
projection if the following holds.*

$$(i) \quad L \text{ is surjective,}$$
$$(ii) \quad L \circ L = L.$$

Note that the assumption $V \neq [0]$ implies $L \neq 0$.

Theorem 5.9. *Let $L : X \to V$ be a projection. Then $L|_V = id_V$ holds.*

Proof. Let $v \in V$. As L is surjective, there is an $x \in X$ such that $v = Lx$. It follows
that $Lv = L(Lx) = Lx = v$.   □

Theorem 5.10. *Let X be a normed linear space and $L : X \to V$ a projection.
Then*

$$\|L\| \geq 1. \tag{5.20}$$

Proof. In view of (5.2) we obtain $\|L\| = \|L \circ L\| \leq \|L\|^2$, where $\|L\| \neq 0$ holds by the remark from above. This implies (5.20). □

Projections define reasonable approximations in the following sense.

Theorem 5.11. *Let $L \in \mathcal{L}(X, V)$ be a projection onto the finite-dimensional subspace V. Then*

$$\|x - Lx\| \leq (1 + \|L\|) \cdot E(x, V)$$

holds for arbitrary $x \in X$.

Proof. Let $\overset{*}{v}$ be a best approximation to $x \in X$ in V. Using $L\overset{*}{v} = \overset{*}{v}$ we obtain

$$
\begin{aligned}
\|x - Lx\| &= \|x - \overset{*}{v} + L(\overset{*}{v} - x)\| \\
&\leq \|x - \overset{*}{v}\| + \|L\| \cdot \|\overset{*}{v} - x\| = (1 + \|L\|) \cdot E(x, V),
\end{aligned}
$$

as claimed. □

The result can be interpreted as follows. If x is well approximable in V, then $E(x, V)$ is small, and $\|x - Lx\|$ is also small. The advantage is that it is often easier to calculate Lx than to calculate a best approximation $\overset{*}{v}$ to x.

Orthogonal Projections

The theory of best approximation attains a very handsome form if the norm of X is induced by an *inner product* $(\cdot, \cdot)$ and V is a finite-dimensional subspace of X.

Definition 5.6 (Orthogonal Projection). *Let X be provided with the inner product $(\cdot, \cdot)$ and let V be a subspace of X. A linear map $\Pi : X \to V$ is called* orthogonal projection *onto V if for all $x \in X$ the equation*

$$(x - \Pi x, v) = 0 \tag{5.21}$$

holds for all $v \in V$.

The definition is justified in view of the following theorem.

Theorem 5.12 (Orthogonal Projection). *Let X be provided with the inner product $(\cdot, \cdot)$ and let V be a subspace of finite dimension. Then a uniquely determined orthogonal projection $\Pi : X \to V$ exists. Π is a projection with $\|\Pi\| = 1$. For all $x \in X$ the element $\overset{*}{v} := \Pi x$ is the uniquely determined best approximation to x in V.*

Proof. *Uniqueness.* Assume Π_1 and Π_2 are orthogonal projections. Taking the difference of the corresponding equations (5.21) we get

$$(\Pi_1 x - \Pi_2 x, v) = 0$$

for arbitrary $v \in V$. Putting $v := \Pi_1 x - \Pi_2 x$ we get

$$\|\Pi_1 x - \Pi_2 x\| = 0$$

and hence $\Pi_1 x = \Pi_2 x$ for arbitrary $x \in X$. It follows that $\Pi_1 = \Pi_2$.

Existence. Let $v_1, \ldots, v_N$ form an orthonormal basis in V and define $\Pi : X \to V$ by

$$\Pi x := \sum_{\nu=1}^{N} (x, v_\nu) v_\nu. \tag{5.22}$$

Then we get

$$(x - \Pi x, v_\kappa) = (x, v_\kappa) - (x, v_\kappa) = 0$$

for $\kappa = 1, \ldots, N$, and by linear combination of the equalities we find that (5.21) holds for all $v \in V$.

Projection. The uniquely determined orthogonal projection Π can be represented in the form (5.22). It follows that

$$\Pi v_\kappa = v_\kappa$$

holds for $\kappa = 1, \ldots, N$, and by linear combination of these equalities we find that Π is surjective. Moreover, for arbitrary $x \in X$ we obtain

$$\Pi(\Pi x) = \sum_{\nu=1}^{N} (x, v_\nu) \Pi v_\nu = \sum_{\nu=1}^{N} (x, v_\nu) v_\nu = \Pi x.$$

So Π is a projection.

Norm. Inserting $v := \Pi x$ in (5.21) we obtain

$$\|\Pi x\|^2 = (x, \Pi x),$$

and by the help of the inequality of Schwarz we get for arbitrary $x \in X$ with $\|x\| \le 1$,

$$\|\Pi x\|^2 \le \|x\| \cdot \|\Pi x\| \le \|\Pi x\|,$$

and hence $\|\Pi x\| \le 1$. This implies $\|\Pi\| \le 1$. But Π is a projection, so $\|\Pi\| \ge 1$ holds because of Theorem 5.10. Together this implies $\|\Pi\| = 1$.

Best Approximation. Let $x \in X$ be fixed. For $x \in V$ the minimal deviation is zero, i.e., $\overset{*}{v} = x = \Pi x$ is the unique best appoximation. Next assume $x \notin V$, and let

$$v = \sum_{\nu=1}^{N} a_\nu v_\nu,$$

$a_1, \ldots, a_N \in \mathbb{R}$, be an arbitrary element in V. Then we get

$$0 < \|x - v\|^2 = \|x\|^2 - 2\sum_{\nu=1}^{N} a_\nu (x, v_\nu) + \sum_{\nu=1}^{N} a_\nu^2.$$

As a function of $a_1, \ldots, a_N$ the right side is a positive quadratic form. So there is a minimum point. In all minimum points the gradient vanishes, which is equivalent to

$$a_\kappa = (x, v_\kappa)$$

for $\kappa = 1, \ldots, N$. Hence the minimum is attained at a unique point, which corresponds to $v = \Pi x$, and the theorem is proved. $\square$

Remark. In view of (5.17), but not quite correctly, we also say that Πx is the orthogonal projection of x in V.

Representation of Orthogonal Projections

If the inner product space is a function space of the form $X = C(D)$ and if the finite-dimensional subspace V has the *reproducing kernel* $G(x, y)$, then

$$G(x, y) = \sum_{\nu=0}^{N} v_\nu(x) v_\nu(y)$$

holds for $x, y \in D$, see (1.5), and in view of (5.22) the orthogonal projection can be written in the form

$$(\Pi F)(x) = \Big(F, G(\,\cdot\,, x) \Big) \tag{5.23}$$

for $F \in C(D)$ and $x \in D$.

Minimal Projections

Assume that X is again an arbitrary normed linear space and that V is a subspace. In view of Theorem 5.11 it is most desirable to know projections $L \in \mathcal{L}(X, V)$ with $\|L\|$ as small as possible.

Definition 5.7 (Minimal Projection). *A projection* $L^* \in \mathcal{L}(X, V)$ *is called a* minimal projection *if*

$$\|L^*\| \leq \|L\|$$

holds for all projections $L \in \mathcal{L}(X, V)$.

Remark. The operator norm depends on the choice of the norm in X. Hence a minimal projection which belongs to a certain norm need not be a minimal projection with respect to another norm. It is known that if V is finite-dimensional, then a minimal projection must exist. See [26], or, for an outline of the proof, [10]. We shall not need this theoretical knowlege, since in the cases which we consider, existence will always be evident. A first example is the following theorem.

Theorem 5.13 (Minimal Projection in Inner Product Spaces). *Let X be provided with the inner product $(\,\cdot\,, \cdot\,)$ and let V be a finite-dimensional subspace of X. Then the orthogonal projection $\Pi : X \to V$ is a minimal projection onto V. The norm of every minimal projection equals 1.*

Proof. From Theorem 5.12 and Theorem 5.10 we get for arbitrary projections L

$$\|\Pi\| = 1 \le \|L\|,$$

with equality if L is a minimal projection.   □

Interpolatory Projections

Except for inner product spaces, it is often difficult to calculate a best approximation or a minimal projection. For this reason, interpolation, which is a very constructive principle, plays an important role in numerical analysis. The defining linear operators are projections, so they belong to this section. Though interpolation, which tries to understand a function from a finite number of its function values, is a quite inadequate tool if the function is discontinuous, we consider it in the beginning in a very general setting. After that, the necessary restriction to continuous functions will be the topic of the following section.

Assume that V is a linear space of functions $D \to \mathbb{R}$ where D is a nonempty set. Let the dimension of V be finite.

Fundamental Systems

Definition 5.8 (Fundamental System). *Assume* $N := \dim V \in \mathbb{N}$. *A family* $T = \{t_1, \ldots, t_N\}$ *of nodes* $t_k \in D$ *is a* fundamental system *for* V, *if the evaluation functionals*

$$V \ni F \mapsto F(t_k) \in \mathbb{R},$$

$k = 1, \ldots, N$, *are linearly independent.*

By this definition T is a fundamental system if

$$\sum_{k=1}^{N} a_k F(t_k) = 0 \text{ for all } F \in V$$

implies $a_1 = \cdots = a_N = 0$, or if for any basis $\{F_1, \ldots, F_N\}$ in V the linear system of equations

$$\sum_{k=1}^{N} F_j(t_k) a_k = 0, \quad j = 1, \ldots, N,$$

has the trivial solution, only. So we are led to

Theorem 5.14 (Existence and Characterisation of Fundamental Systems). *Assume* $N := \dim V \in \mathbb{N}$. *Then the following holds.*
(*i*) *There exists a fundamental system for* V.
(*ii*) *A node system* $T = \{t_1, \ldots, t_N\}$ *is a fundamental system for* V *if and only if*

$$\det\left(F_j(t_k)\right)_{j,k=1,\ldots,N} \ne 0$$

is valid for any basis $\{F_1, \ldots, F_N\}$ *of* V.

Proof. (ii) follows immediately from our consideration from above. For (i) assume that $F_1 \ldots, F_N$ is a basis in V. Since

$$\det\left(F_1(x)\right) = F_1(x)$$

is not the null function, there is an $x_1 \in D$ such that

$$\det\left(F_1(x_1)\right) \neq 0.$$

Next let us introduce the determinant

$$\det\begin{pmatrix} F_1(x_1) & F_1(x) \\ F_2(x_1) & F_2(x) \end{pmatrix},$$

which is, since the first main minor determinant does not vanish, a nontrivial linear combination of $F_1(x)$ and $F_2(x)$. It is not the null function, so there is an $x_2 \in D$ such that

$$\det\begin{pmatrix} F_1(x_1) & F_1(x_2) \\ F_2(x_1) & F_2(x_2) \end{pmatrix} \neq 0$$

is valid. So we proceed. Finally the determinant

$$\det\begin{pmatrix} F_1(x_1) & F_1(x_2) & \cdots & F_1(x) \\ F_2(x_1) & F_2(x_2) & \cdots & F_2(x) \\ \hdotsfor{4} \\ F_N(x_1) & F_N(x_2) & \cdots & F_N(x) \end{pmatrix}$$

is a nontrivial linear combination of $F_1(x), \ldots, F_N(x)$, such that an $x_N \in D$ exists with

$$\det\begin{pmatrix} F_1(x_1) & F_1(x_2) & \cdots & F_1(x_N) \\ F_2(x_1) & F_2(x_2) & \cdots & F_2(x_N) \\ \hdotsfor{4} \\ F_N(x_1) & F_N(x_2) & \cdots & F_N(x) \end{pmatrix} \neq 0,$$

which finishes the proof. □

We point out that Theorem 5.14 does not contain an assumption on the quality of the space V or on the domain D.

Example 1. To give an example, let $V := \mathbb{P}^1_\mu([a,b])$, where $a < b$ and $N = \mu + 1$. It is well known that in this case every family of N different points in $[a,b]$ is a fundamental system.

Example 2. Example 1 is not typical for what happens in the multivariate case. We explain this in what follows.

Let $r \geq 2$ and assume that V is a subspace of $C(D)$ of dimension $N \geq 3$, where $D \subset \mathbb{R}^r$ contains an interior point a_0. Hence there is an open ball $B_0 := B(a_0, \rho)$ with center a_0 and radius $\rho > 0$ contained in D, and we can choose N additional points $a_1, \ldots, a_N \in D$ such that $a_0, a_1, \ldots, a_N$ are, altogether, different in pairs while no three of them are located on a line. And we define the line segments $s_\nu := conv\{a_0, a_\nu\}$ for $\nu = 1, 2, 3$.

In the beginning we choose the nodes $x_1, \ldots, x_N \in B_0$ in the position

$$x_1 = a_1, \ x_2 = a_2, \ x_3 = a_3, \ldots, x_N = a_N.$$

Next we move x_1 on $s_1 \cup s_3$ continuously to the midpoint of s_3. After that we move x_2 on $s_2 \cup s_1$ to the position of a_1, and finally x_1 on $s_3 \cup s_2$ to the position of a_2. Note that all the time the nodes were different in pairs, and that they reached finally the position

$$x_1 = a_2, \ x_2 = a_1, \ x_3 = a_3, \ldots, x_N = a_N$$

by a continuous transform. Since x_1 and x_2 have changed now their position, the determinant

$$\det \begin{pmatrix} F_1(x_1) & F_1(x_2) & \cdots & F_1(x_N) \\ F_2(x_1) & F_2(x_2) & \cdots & F_2(x_N) \\ \hdotsfor{4} \\ F_N(x_1) & F_N(x_2) & \cdots & F_N(x) \end{pmatrix}$$

must have changed its sign. Therefore, a node configuration appeared between, where the determinant vanishes. In other words, a configuration of N pairwise different nodes exists which do not form a fundamental system.

Lagrange Elements

Assume that $T = \{t_1, \ldots, t_N\}$ is a fundamental system of V, and assume that $F_1, \ldots, F_N$ form a basis in V. Then the following holds.

Let $F = \sum_{j=1}^N a_j F_j$ be an arbitrary element in V, and choose $y_1, \ldots, y_n \in \mathbb{R}$ also arbitrarily. Then the system

$$F(t_k) = y_k, \quad k = 1, \ldots, N, \tag{5.24}$$

is equivalent to the system of linear equations

$$\sum_{j=1}^N F_j(t_k) a_j = y_k, \quad k = 1, \ldots, N,$$

and because of Theorem 5.14 it has a uniquely determined solution. In other words, F is uniquely determined by its function values $F(t_1), \ldots, F(t_N)$. Hence we can define the functions $L_j \in V$, $j = 1 \ldots, N$, by the equations

$$L_j(t_k) = \delta_{j,k}, \quad k = 1, \ldots, N. \tag{5.25}$$

Next let $F \in V$ be arbitrary again and define $G \in V$ by

$$G := \sum_{j=1}^N F(t_j) L_j.$$

Obviously $G(t_k) = F(t_k)$ holds for $k = 1, \ldots, N$, and this implies $G = F$. This means that every element F in V has the representation

$$F = \sum_{j=1}^{N} F(t_j) L_j. \tag{5.26}$$

In particular, $L_1, \ldots, L_N$ are a basis in V.

Definition 5.9 (Lagrange Elements). *The elements L_j in V, $j = 1, \ldots, N$, defined by (5.25) are the* Lagrange elements *belonging to the fundamental system T. The system $\{L_1, \ldots, L_N\}$ is called a* Lagrange basis.

Now we define, with a basis $F_1, \ldots, F_N \in V$, the determinant

$$\Delta(x_1, \ldots, x_N) := \det \begin{pmatrix} F_1(x_1) & F_1(x_2) & \cdots & F_1(x_N) \\ F_2(x_1) & F_2(x_2) & \cdots & F_2(x_N) \\ \cdots\cdots\cdots\cdots\cdots\cdots\cdots\cdots\cdots\cdots\cdots \\ F_N(x_1) & F_N(x_2) & \cdots & F_N(x_N) \end{pmatrix}$$

for arbitrary $x_1, \ldots, x_N \in D$. By Theorem 5.14 we get $\Delta(t_1, \ldots, t_N) \neq 0$. We claim that

$$L_j(x) = \frac{\Delta(t_1, \ldots, x, \ldots, t_N)}{\Delta(t_1, \ldots, t_j, \ldots, t_N)} \tag{5.27}$$

holds for $x \in D$. Actually, expanding $\Delta(t_1, \ldots, x, \ldots, t_N)$ with respect to its j-th column we see that, as a function in x, this determinant and hence the right side of (5.27) is an element of V. Both sides attain the same function values at the points $t_1, \ldots, t_N$. Together this proves the identity.

The Interpolation Operator

In what follows we assume that V is a subspace of a normed linear space X of functions $F : D \to \mathbb{R}$, and that $T = \{t_1, \ldots, t_N\}$ is a fundamental system of V. Then a linear map

$$\Lambda : X \to V$$

is defined by

$$\Lambda F := \sum_{j=1}^{N} F(t_j) L_j. \tag{5.28}$$

It follows from (5.26) that Λ is *surjective*. Besides we get $(\Lambda F)(t_j) = F(t_j)$ for arbitrary $F \in X$, which implies

$$\Lambda \circ \Lambda F = \sum_{j=1}^{N} \left(\Lambda F\right)(t_j) L_j = \sum_{j=1}^{N} F(t_j) L_j = \Lambda F.$$

So Λ is a projection onto V.

Definition 5.10 (Interpolatory Projection and Norm). *The linear map* $\Lambda : X \to V$ *defined by (5.28) is called the* interpolatory projection *which belongs to the fundamental system* T. $\|\Lambda\|$ *is the corresponding* interpolation norm *or* interpolatory Lebesgue constant.

5.4   Interpolatory Projections in C(D)

In what follows, $\mathbf{V}$ is a finite-dimensional subspace of $C(D)$, where D is a nonempty and compact subset of $\mathbb{R}^r$, $r \in \mathbb{R}$. As in Chapter 1, $C(D)$ is provided with the *uniform norm* $\| \cdot \|_D = \| \cdot \|_\infty$. Moreover, we assume that it is furnished also with the *inner product* $\langle \cdot, \cdot \rangle$, which induces the norm $\| \cdot \|$. The corresponding reproducing kernel of $\mathbf{V}$ is again denoted by $G(x, y)$.

Kernel Functions

Let $N := \dim \mathbf{V}$ and assume $T = \{t_1, \ldots, t_N\}$ is an arbitrary system of nodes $t_j \in D$. We define the *kernel functions* $K_j \in \mathbf{V}$ for $j = 1, \ldots, N$ by

$$K_j(x) := G(t_j, x) \text{ for } x \in D. \tag{5.29}$$

Their linear independence is the subject of the following theorem.

Theorem 5.15 (Kernel-Basis). *Let* $\mathbf{V}$ *be a subspace of* $C(D)$ *of dimension* N *with the reproducing kernel* $G(x, y)$. *And assume* $T = \{t_1, \ldots, t_N\}$, $N = \dim \mathbf{V}$, *is a system of nodes* $t_j \in D$. *Then the kernel functions* $K_1, \ldots, K_N$ *form a basis in* $\mathbf{V}$ *if and only if* T *is a fundamental system.*

Proof. Let $\{S_1, \ldots, S_N\}$ be an orthonormal system in $\mathbf{V}$, such that

$$G(x, y) = \sum_{k=1}^{N} S_k(x) S_k(y)$$

holds for $x, y \in D$, see (1.5). Then we obtain

$$K_j = \sum_{k=1}^{N} S_k(t_j) S_k$$

for $j = 1, \ldots, N$. Because of Theorem 5.14 this is a basis transform if and only if T is a fundamental system. $\square$

Theorem 5.15 leads us to use in Theorem 5.14 the functions $F_1 = K_1, \ldots, F_N = K_N$ in the characterisation of fundamental systems. However we do not know in advance whether these functions form a basis. Nevertheless, the following version of Theorem 5.14 is actually true.

Theorem 5.16 (Kernel-Bases and Fundamental Systems). *Assumptions as in Theorem 5.15. The following statements are equivalent.*

(i) $t_1, \ldots, t_N$ *form a fundamental system for* $\mathbf{V}$,

(ii) $K_1, \ldots, K_N$ *form a basis in* $\mathbf{V}$,

(iii) $\det \left(G(t_j, t_k) \right)_{j,k=1,\ldots,N} \neq 0.$

Proof. (i) implies (ii), see Theorem 5.15. Identifying $F_j = K_j$ in Theorem 5.14, we see that (ii) implies (iii). Finally, it follows from (iii) that the evaluation functionals are linearly independent, such that (i) holds. This finishes the proof. □

Condition (iii) has the advantage that it can be evaluated without the knowledge of a concrete basis, provided the reproducing kernel is known, as is the case if $\mathbf{V}$ is a rotation-invariant subspace of $\mathbb{P}^r(S^{r-1})$ of finite dimension, see Theorem 4.11.

Moreover, in this case the evaluation of the kernel functions is very easy, since it requires the calculation of a euclidean inner product and the following evaluation of a univariate orthogonal polynomial, only. This evaluation itself can be arranged at low cost and high stability by the use of the corresponding three-term recurrence relation

In view of (5.29) we obtain, in the general case, $\langle K_j, K_k \rangle = G(t_j, t_k)$, where we used the reproducing property of the kernel. This yields

$$\left(G(t_j, t_k) \right)_{j,k=1,\ldots,N} = \left(\langle K_j, K_k \rangle \right)_{j,k=1,\ldots,N}. \tag{5.30}$$

So the matrix on the left side proves to be a *Gram matrix* and hence to be positive definite, with all the useful and well-known numerical advantages which are included by this fact. No doubt, reasonable constructive multivariate polynomial approximation is only possible thanks to the two basic facts just mentioned.

Kernel Functions and Lagrangians are Biorthogonal

We assume again that $T = \{t_1, \ldots, t_N\}$ is a fundamental system for $\mathbf{V}$. Then the corresponding Lagrange elements L_j and kernel functions K_k are tightly related.

As a basic fact, (5.25) can be written in the form $\langle L_j, G(\cdot, t_k) \rangle = \delta_{j,k}$, where we use the reproducing kernel property of G, again. This is equivalent to

$$\langle L_j, K_k \rangle = \delta_{j,k} \tag{5.31}$$

for $j, k = 1, \ldots, N$. In other words, the Lagrange basis $\{L_1, \ldots, L_N\}$ and the kernel basis $\{K_1, \ldots, K_N\}$ are *biorthonormal*. It follows that the corresponding basis tranforms take the form

$$L_j = \sum_{k=1}^{N} \langle L_j, L_k \rangle K_k, \tag{5.32}$$

$$K_k = \sum_{j=1}^{N} \langle K_k, K_j \rangle L_j, \tag{5.33}$$

for $j = 1, \ldots, N$ and $k = 1, \ldots, N$, respectively. Moreover,

$$\Big(\langle L_j, L_k \rangle \Big) = \Big(\langle K_k, K_l \rangle \Big)^{-1} \tag{5.34}$$

holds, where we do not notify that the row and the column indices are varying in $\{1, \ldots, N\}$.

The matrix $\Big(G(t_j, t_k) \Big) = \Big(\langle K_j, K_k \rangle \Big)$, which in view of (5.34) and of (5.30) rules the basis transforms, is called the *fundamental matrix*.

Remark. The fundamental matrix can be calculated straightly forward, provided the kernel is known, where even parallel procession is possible. Formula (5.34) enables us to calculate the matrix $\Big(\langle L_j, L_k \rangle \Big)$, which rules the basic transform (5.32), after that by a numerical inversion of a positive definite matrix, for instance by *Cholesky's method*.

The Square Sum of the Lagrangians

Again we assume that T is a fundamental system. If we consider $G(x, y)$ for fixed $x \in D$ as a function in y, then we get in view of (5.26)

$$G(x, y) = \sum_{k=1}^{N} G(x, t_k) L_k(y)$$

for $y \in D$. The right side can be treated likewise as a function of x, and we obtain the remarkable formula

$$G(x, y) = \sum_{j=1}^{N} \sum_{k=1}^{N} L_j(x) G(t_j, t_k) L_k(y) \tag{5.35}$$

for arbitrary $x, y \in D$. Now assume that the matrix $\big(G(t_j, t_k) \big)$, which is positive definite, has the eigenvalues

$$0 < \lambda_{\min} := \lambda_1 \le \lambda_2 \le \ldots \le \lambda_N =: \lambda_{\max}.$$

Identifying $x = y$ in (5.35) we get the 'quadratic form'

$$G(x, x) = \sum_{j=1}^{N} \sum_{k=1}^{N} L_j(x) G(t_j, t_k) L_k(x)$$

and hence

$$\lambda_{\min} \cdot \sum_{j=1}^{N} L_j^2(x) \leq G(x,x) \leq \sum_{j=1}^{N} L_j^2(x) \cdot \lambda_{\max}.$$

This is equivalent to the statement of the following theorem.

Theorem 5.17 (Lagrange Square Sum). *Assumptions as above. For arbitrary $x \in D$ the following inequalities hold,*

$$\lambda_{\max}^{-1} G(x,x) \leq \sum_{j} L_j^2(x) \leq G(x,x) \lambda_{\min}^{-1}. \tag{5.36}$$

There is a particularly interesting application of this theorem.

Corollary 5.18 (Rotation-Invariant Subspaces of $C(S^{r-1})$). *Assumptions as above, but assume that V is a rotation-invariant subspace of $C(S^{r-1})$, $r \geq 2$,, where $G(x,y) = K(xy)$ is its reproducing kernel. Then the following is valid for arbitrary $x \in S^{r-1}$,*

$$\lambda_{\max}^{-1} K(1) \leq \sum_{j} L_j^2(x) \leq K(1) \lambda_{\min}^{-1}, \tag{5.37}$$

$$\lambda_{\min} = \lambda_{\max} \ implies \ \sum_{j=1}^{N} L_j^2(x) = 1. \tag{5.38}$$

Proof. Obviously, the reproducing kernel takes the particular form by Theorem 1.11. (5.37) follows from (5.36), immediately. Moreover, the fundamental matrix has the form

$$\left(G(t_j, t_k)\right) = \begin{pmatrix} K(1) & & & \\ & K(1) & & * \\ & & \ddots & \\ & * & & K(1) \end{pmatrix}.$$

Therefore $\lambda_{\min} = \lambda_{\max}$ implies $\lambda_{\min} = \lambda_{\max} = K(1)$, and inserting this in (5.37) we get $\sum_{j=1}^{N} L_j^2(x) = 1$, as claimed. $\qquad\square$

Lebesgue Function and Interpolation Norm

We do not change our assumptions. In view of the Theorem of Banach and Steinhaus it is essential to know the uniform norm of the interpolatory projection

$$\Lambda F = \sum_{j=1}^{N} F(t_j) L_j,$$

which in view of (5.1) has now to be defined by

$$\|\Lambda\|_\infty = \max\{\|\Lambda F\|_\infty : F \in C(D), \|F\|_\infty \le 1\}.$$

So let $F \in C(D)$ and assume $\|F\|_\infty \le 1$. For arbitrary $x \in D$ we obtain

$$|(\Lambda F)(x)| \le \lambda(x)$$

if $\lambda \in C(D)$ is defined by

$$\lambda(x) := \sum_{j=1}^{N} |L_j(x)|$$

for $x \in D$. This function is called the *Lebesgue function* belonging to the fundamental system T. It is obvious that the following upper bound holds,

$$\|\Lambda\|_\infty \le \|\lambda\|_\infty, \tag{5.39}$$

and there is a point $x_0 \in D$ such that

$$\|\lambda\|_\infty = \lambda(x_0)$$

is valid. This will enable us to determine a lower bound. To this end let

$$\epsilon_j := sgn\, L_j(x_0)$$

for $j = 1, \ldots, N$, where the usual definition of the sign-function is completed by $sgn(0) := 0$. Together this yields

$$\lambda(x_0) = \sum_{j=1}^{N} \epsilon_j L_j(x_0).$$

Now we choose a such that

$$0 < a < \tfrac{1}{2} \min\left\{|t_j - t_k| : j, k \in \{1, \ldots, N\}, j \ne k\right\}$$

is valid. After that we define the function $E \in C(D)$ by putting

$$E(x) := \begin{cases} \epsilon_j(1 - \tfrac{1}{a}|x - t_j|), & for\ |x - t_j| < a,\ j \in \{1, \ldots, N\}, \\ 0 & else, \end{cases} \tag{5.40}$$

for $x \in D$. The definition is admissable since there is always at most one t_j satisfying $|x - t_j| < a$.

Obviously, E interpolates the values

$$E(t_j) = \epsilon_j \quad for\ j = 1, \ldots, N,$$

and satisfies $\|E\|_\infty = 1$, such that we obtain

$$\|\Lambda\|_\infty \geq \|\Lambda E\|_\infty \geq \left|(\Lambda E)(x_0)\right| = \sum_{j=1}^N \epsilon_j L_j(x_0) = \lambda(x_0),$$

and hence

$$\|\Lambda\|_\infty \geq \|\lambda\|_\infty.$$

Together with (5.39) this yields $\|\Lambda\|_\infty = \|\lambda\|_\infty$, which is the same as

$$\|\Lambda\|_\infty = \max\left\{ \sum_{j=1}^N |L_j(x)| : x \in D \right\}. \tag{5.41}$$

The evaluation of the interpolation norm from this formula is difficult. It is easier to invert the fundamental matrix and to estimate it by the help of the following theorem.

Theorem 5.19 (Estimate of the Interpolation Norm). *Let $D \subset \mathbb{R}^r$, $r \in \mathbb{R}$, be nonempty and compact, and assume that the subspace $\mathbf{V}$ of $C(D)$ is provided with the reproducing kernel $G(x,y)$ and the fundamental system $T = \{t_1, \ldots, t_N\}$. Then the uniform norm of the corresponding interpolatory projection satisfies*

$$\|\Lambda\|_\infty^2 \leq N \cdot \lambda_{\min}^{-1} \cdot \max\{G(x,x) : x \in D\}.$$

Proof. Using the inequality of Cauchy–Schwarz and Theorem 5.17 we get for $x \in D$

$$\left(\sum_{j=1}^N 1 \cdot |L_j(x)| \right)^2 \leq N \cdot \sum_{j=1}^N L_j^2(x) \leq N \cdot \lambda_{\min}^{-1} \cdot G(x,x),$$

and the statement follows in view of (5.41). $\qquad\square$

Corollary 5.20 (Estimate in Rotation-Invariant Subspaces of $C(S^{r-1})$). *Let $\mathbf{V}$ be a rotation-invariant subspace of $C(S^{r-1})$, $r \geq 2$, with the fundamental system $T = \{t_1, \ldots, t_N\}$. Assume that the inner product is induced by the surface integral, such that the reproducing kernel takes the form $G(x,y) = K(xy)$. Then the interpolatory projection Λ belonging to T satisfies*

$$\|\Lambda\|_\infty^2 \leq N \cdot K(1) \cdot \lambda_{\min}^{-1}.$$

If the eigenvalues of the fundamental matrix are all equal, i.e., if $\lambda_1 = \lambda_2 = \cdots = \lambda_N$ holds, then

$$\|\Lambda\|_\infty \leq \sqrt{N}$$

is valid.

Proof. For the form of the reproducing kernel we refer to Theorem 1.11. The first statement follows immediately from Theorem 5.19. For the second statement we refer to the proof of Corollary 5.18, by which $\lambda_{\min} = K(1)$ holds.   □

Remark. For every $\mu \in \mathbb{N}$ let us choose a fundamental system $T = T_\mu$ for the space $\mathbb{P}_\mu^r(S^{r-1})$. Then the convergence problem for the corresponding interpolatory projections $\Lambda = \Lambda_\mu$ is governed by the Theorem of Banach–Steinhaus. So boundedness of the interpolation norms is required, while Corollary 5.20 ensures us in view of (4.4) even in a rather promising situation of a rate of growth

$$\|\Lambda_\mu\|_\infty = \mathcal{O}(\mu^{\frac{r-1}{2}}) \quad as \ \mu \to \infty, \tag{5.42}$$

only. This is poor, however boundedness is unattainable, as will be proved in Section 6.2. Therefore convergence $\Lambda_\mu F \to F$ cannot hold for arbitrary $F \in C(S^{r-1})$, whatever the choice of the fundamental systems may be. This does not make interpolation useless. For instance, from Theorem 5.11 we obtain

$$\|F - \Lambda_\mu F\|_\infty \leq (1 + \|\Lambda_\mu\|_\infty)E_\mu(F),$$

where $E_\mu(F)$ denotes the *algebraic minimal deviation*, which is defined by

$$E_\mu(F) := \min\{\|F - P\|_\infty : P \in \mathbb{P}_\mu^r(S^{r-1})\}. \tag{5.43}$$

This inequality says that in spite of the divergence of the interpolation norms, convergence takes place if the interpolated function F is sufficiently well approximable by spherical polynomials.

We recall that the growth of the interpolation norms depends on the choice of the fundamental systems. The question for a best possible choice is far from being solved. A good choice would be *extremal fundamental systems*, which we present in the following section.

5.5   Extremal Bases and Extremal Fundamental Systems

In this section we discuss the question of what a 'good' basis might be in a real normed linear space X of finite dimension $N \in \mathbb{N}$. What is wanted is that there is a tight relation between the space elements and their coefficient vectors with respect to the basis which is chosen. A reasonable and always possible choice is *extremal bases*.

Definition 5.11 (Extremal Basis). *A basis $\{x_1, \ldots, x_N\}$ in X is called* extremal, *if for every $j \in \{1, \ldots, N\}$ the basis element x_j is best approximated by zero in the complementary space*

$$X_j := span\{x_k \mid k \in \{1, \ldots, N\}, \ k \neq j\}.$$

Note that the definition requires that for all $j \in \{1, \ldots, N\}$,

$$\left\| x_j - \sum_{k \neq j} a_k x_k \right\| \geq \|x_j\|$$

holds for all choices of the coefficients $a_k \in \mathbb{R}$, so making sure, in some sense, that the basis elements are 'as linearly independent, as possible'.

To give an example, we assume that the norm $\| \cdot \|$ is induced by the inner product $\langle \, , \cdot \, \rangle$. It is obvious that in this case every orthogonal basis is extremal. This example is far from being 'trivial', since it makes apparent a very essential property of orthogonal bases.

If the norm is uniform or even arbitrary, an extremal basis can be obtained by a rather general principle, where an *extremal Lagrange basis* is constructed. In the beginning this is possible only in function spaces. But since every finite-dimensional normed space can be identified with a normed function space, the results are transferrable to arbitrary normed spaces.

Theorem 5.21 (Existence of an Extremal Basis). *In every finite-dimensional real normed linear space an extremal basis exists.*

Proof. First let X be a finite-dimensional real linear space of continuous functions on D where D is a nonempty compact metric space, and let X be provided as usual with the uniform norm

$$\|F\|_\infty := \max\{|F(x)| : x \in D\}$$

for $F \in X$. Then an extremal basis can be constructed as follows.

Let $F_1, \ldots, F_N$ form a basis in X and define

$$\Delta(t_1, \ldots, t_N) := \det \big(F_j(t_k) \big)$$

for $t_1, \ldots t_N \in D$. This is a continuous function on the compact set D^N and attains a maximum in absolute value, say for $\bar{t}_1, \ldots, \bar{t}_N \in D$. The maximum is positive, since a fundamental system exists, see Theorem 5.14. So we obtain

$$\Delta\big(\bar{t}_1, \ldots, \bar{t}_N\big) \neq 0,$$

and we may define the functions $L_j \in X$ for $j = 1, \ldots, N$ by

$$L_j(x) := \frac{\Delta(\bar{t}_1, \ldots, x, \ldots, \bar{t}_N)}{\Delta(\bar{t}_1, \ldots, \bar{t}_j, \ldots, \bar{t}_N)} \quad \text{for } x \in D. \tag{5.44}$$

Expanding the nominator determinant with respect to its *j-th* column we see that the definition is correct, i.e., that the L_j are linear combinations of the F_k, and hence in X. And as the denominator is maximum in absolute value, we obtain $|L_j(x)| \leq 1$ for arbitrary $x \in D$. Together with

$$L_j(\bar{t}_k) = \delta_{j,k}, \quad \text{for } j, k = 1, \ldots, N, \tag{5.45}$$

this yields
$$\|L_j\|_\infty = 1 \quad \text{for} \quad j = 1, \ldots, N. \tag{5.46}$$

Finally we get for arbitrary $a_k \in \mathbb{R}$, using (5.44)–(5.46),

$$\left\|L_j - \sum_{k \neq j} a_k L_k\right\|_\infty \geq \left(L_j - \sum_{k \neq j} a_k L_k\right)(\bar{t}_j) = L_j(\bar{t}_j) = 1 = \|L_j\|_\infty,$$

again for $j = 1, \ldots, N$. This means that $L_1, \ldots, L_N$ form an extremal Lagrange basis.

We transfer this result to the arbitrary case. So, let X be an arbitrary finite-dimensional real normed linear space, provided with the norm $\| \cdot \|$. The dual space X^* (algebraic or continuous, which is in the present case the same) consists of the linear functionals $\phi \in X^*$,

$$X \ni x \mapsto \phi(x) \in \mathbb{R}.$$

For arbitrary $x \in X$ a map $\Phi_x \in X^{**}$ is defined by

$$X^* \ni \phi \mapsto \Phi_x(\phi) := \phi(x) \in \mathbb{R}.$$

Together this defines a map $\Lambda : X \to X^{**}$ which maps $x \in X$ onto $\Phi_x \in X^{**}$, where $\Phi_x = 0$ implies $\phi(x) = 0$ for all $\phi \in X^*$, and hence $x = 0$. So we get $\ker \Lambda = 0$, and hence

$$X \cong X^{**}.$$

By this isomorphy, X can be understood to be a function space. We provide X^* and X^{**} with the following norms:

$$\|\phi\| := \max\{|\phi(x)| : x \in X, \|x\| \leq 1\}, \quad \text{for } \phi \in X^*,$$
$$\|\Phi\| := \max\{|\Phi(\phi)| : \phi \in X^*, \|\phi\| \leq 1\}, \quad \text{for } \Phi \in X^{**}.$$

Then we get, in particular,

$$\|\Phi_x\| = \max\left\{|\phi(x)| : \phi \in X^*, \|\phi\| \leq 1\right\} \leq \|x\|. \tag{5.47}$$

We claim that even $\|\Phi_x\| = \|x\|$ holds for arbitrary $x \in X$.

For $x = 0$ this is obvious. In what follows let $x \neq 0$. We consider the linear functional $\psi : span\{x\} \to \mathbb{R}$ defined by

$$\psi(\alpha x) := \alpha\|x\| \quad \text{for } \alpha \in \mathbb{R}.$$

Its norm is given by

$$\|\psi\| = \max\left\{|\alpha\|x\|| : \|\alpha x\| \leq 1\right\} = 1.$$

By the *Theorem of Hahn–Banach* ψ can be continued by a functional $\psi_x \in X^*$ such that

$$\|\psi_x\| = \|\psi\| = 1$$

holds, where

$$\psi_x(x) = \psi(x) = \|x\|$$

is valid. This implies

$$\|\Phi_x\| \geq \|\psi_x\| \geq |\psi_x(x)| = \|x\|.$$

Together with (5.47) this yields

$$\|\Phi_x\| = \|x\|,$$

which holds also for $x = 0$. It follows that X and X^{**} are isometrically isomorphic under Λ.

We finish our proof by considering the unit ball in X^*,

$$D := \{\phi \in X^* : \|\phi\| \leq 1\}.$$

Note that $\dim X^* = \dim X = N$ is finite. Therefore, D is compact, and of course it is not empty. Moreover, it follows directly from the definition of the X^{**}-norm that the map

$$X^{**} \ni \Phi \mapsto \Phi\big|_D \in X^{**}(D)$$

defines an isometric isomorphism between X^{**} and $X^{**}(D)$. So we get finally an isometric isomorphism

$$X \cong X^{**}(D).$$

But the elements of X^{**} are bounded linear functionals and hence continuous. So are their restrictions onto D. In other words, X is isometrically isomorphic with the space $X^{**}(D)$, which is now a space of continuous real functions on D which is provided with the uniform norm. This means that $X^{**}(D)$ is a space just as we considered in the first part of the proof. It follows that $X^{**}(D)$ is furnished with an extremal (Lagrange) basis, whose elements are, by isometry, the images of the elements of an extremal basis in X itself. This finishes the proof. □

Remarks. The earliest version of Theorem 5.21 seems to be due to A. E. Taylor [74], but is also known under the name of Auerbach. It holds as well in complex spaces, which are, however, without interest in our setting. Note also that the determinant $\det\big(F_j(t_k)\big)$, which occurred above, is maximum in absolute value independently of the choice of the basis $F_1, \ldots, F_N$. Unfortunately we cannot conclude that, vice versa, the fundamental determinant is maximum in absolute value if the Lagrange basis is extremal, though we shall be able to prove, later, that it defines a local extremum, at least. Finally, it is convenient to introduce the following definition.

Definition 5.12 (Extremal Fundamental System). *A fundamental system is called extremal if* $\|L_1\|_\infty = \ldots = \|L_N\|_\infty = 1$ *holds.*

Let us remark that an extremal Lagrange basis is always obtained from a maximum fundamental determinant, but the converse need not be true. We do not even know whether every extremal Lagrange basis is supported by an extremal fundamental system. In spite of this we are able to present an algorithm which calculates extremal fundamental systems. But first of all we discuss the quality of extremal bases.

Condition of Extremal Bases

Let $\{x_1, \ldots, x_N\}$ be a basis of the real linear space X, which is provided with the norm $\| \cdot \|_X$. Every $x \in X$ has a uniquely determined representation

$$x = \sum_{j=1}^{N} a_j x_j,$$

with a coefficient vector $a(x) := (a_1, \ldots, a_N)' \in \mathbb{R}^N$. Now let $\mathbb{R}^N$ be furnished with the norm $\| \cdot \|_{R^N}$, and define, as usual, the operator norms

$$\|a\| := \max\{\|a(x)\|_{R^N} : \|x\|_X \leq 1\},$$

$$\|a^{-1}\| := \max\{\|x\|_X : \|a(x)\|_{R^N} \leq 1\}.$$

A measure for the stability of each of these maps against small changes in the argument, for instance by numerical round-off, is their norm. But as we want to recover the elements $x \in X$ from their images $a(x) \in \mathbb{R}^N$, this means from the identity $x = a^{-1}(a(x))$, it is desirable to have, simultaneously, $\|a\|$ and $\|a^{-1}\|$ as small as possible. This is a question of the choice of the basis. As a measure for its quality we introduce the condition of the map a.

Definition 5.13 (Condition of a Basis). *Let X and $\mathbb{R}^N$ be normed as above. The condition of the basis $x_1, \ldots, x_N$ of X is defined by*

$$cond(a) := \|a\| \cdot \|a^{-1}\|,$$

where $a : X \to \mathbb{R}^N$ is the map onto the coefficient vectors.

Example. Assume that X is a subspace of $C(D)$, D a nonempty and compact metric space, and that $\{L_1, \ldots, L_N\}$ is an extremal Lagrange basis, which belongs to the (extremal) fundamental system $\{t_1, \ldots, t_N\}$. And assume that $\mathbb{R}^N$ is furnished with $\|x\| := \max\{|x_j| : j = 1, \ldots, N\}$ for $x \in \mathbb{R}^N$. Then every $F \in X$ has the representation

$$F = \sum_{j=1}^{N} a_j L_j,$$

where $a_j = F(t_j)$. This yields

$$\|a\| \leq 1,$$

where equality holds if the constant function 1 is contained in X. Besides we get

$$\|F\| \le \sum_{j=1}^{N} \|a\| \cdot \|L_j\| \le \|a\| \cdot N \le N. \tag{5.48}$$

Together this yields

$$cond(a) \le N.$$

In other words, our Lagrange basis has a 'reasonable' condition in comparison with the space dimension.

Construction of Extremal Fundamental Systems

In what follows, $\mathbf{V}$ is a subspace of $C(D)$, where D is a nonempty compact subset of $\mathbb{R}^r$, $r \in \mathbb{N}$, and where $\dim \mathbf{V} = N$, $N \in \mathbb{N}$. Moreover, we assume that $\mathbf{V}$ is provided with the inner product $\langle \cdot, \cdot \rangle$ and with the reproducing kernel $G(x, y)$. Every system of nodes $\{x_1, \ldots, x_N\}$ is identified with the corresponding matrix $X = (x_1, \ldots, x_N) \in D^N$, whose columns are the nodes.

For arbitrary node systems $X, Y \in D^N$, we define the matrix

$$G(X; Y) := \begin{pmatrix} G(x_1, y_1) & G(x_1, y_2) & \ldots & G(x_1, y_N) \\ G(x_2, y_1) & G(x_2, y_2) & \ldots & G(x_2, y_N) \\ \cdots\cdots\cdots\cdots\cdots\cdots\cdots\cdots\cdots\cdots \\ G(x_N, y_1) & G(x_N, y_2) & \ldots & G(x_N, y_N) \end{pmatrix}. \tag{5.49}$$

For $X = Y = T$ this is the matrix which occurs in (5.30) and in Theorem 5.16. Moreover, let us assume that the functions $F_1 = S_1, \ldots, F_N = S_N$ form an orthonormal basis in $\mathbf{V}$, and let us define the matrix

$$S(X) := \begin{pmatrix} S_1(x_1) & S_1(x_2) & \ldots & S_1(x_N) \\ S_2(x_1) & S_2(x_2) & \ldots & S_2(x_N) \\ \cdots\cdots\cdots\cdots\cdots\cdots\cdots\cdots\cdots \\ S_N(x_1) & S_N(x_2) & \ldots & S_N(x_N) \end{pmatrix},$$

again for $X \in D^N$. Recalling (1.5), we obtain

$$G(x_j, y_k) = \sum_{\nu=1}^{N} S_\nu(x_j) S_\nu(y_k)$$

for $j, k = 1, \ldots, N$. Inserting this in (5.49) we get

$$G(X; Y) = S(X)' S(Y). \tag{5.50}$$

Besides, $|\det S(\cdot)|$ is a continuous function on the compact set D^N. So it attains a maximum value, say at $T_{\max} \in D^N$,

$$|\det S(T_{\max})| = \max \{|\det S(X)| : X \in D^N\} > 0, \tag{5.51}$$

where the maximum is positive as a fundamental system exists, see Theorem 5.14. It follows from (5.50) that

$$0 \leq \det G(X; X) \leq |\det G(X; T_{\max})| \leq \det G(T_{\max}; T_{\max}) \qquad (5.52)$$

holds for arbitrary node systems $X \in D^N$, where $\det S(X) \neq 0$ and hence

$$\det G(X; X) > 0 \qquad (5.53)$$

is valid if and only if X is a fundamental system, again because of Theorem 5.14.

Next let

$$\Delta(X) := \det G(X; X) \qquad (5.54)$$

for $X \in D^N$. For fundamental systems X this definition is consistent with the definition used in (5.27) and in the proof of Theorem 5.21, except that the basis functions are now the kernel functions $F_j := K_j = G(x_j, \cdot)$, $j = 1, \ldots, N$. It follows from (5.52) that

$$\Delta(T_{\max}) = \max\{\Delta(X) : X \in D^N\} > 0 \qquad (5.55)$$

holds, and from (5.27) that the Lagrange elements belonging to $T_{\max}$ satisfy

$$\|L_1\|_\infty = \cdots = \|L_N\|_\infty = 1.$$

This says that $T_{\max}$ is an extremal fundamental system.

The essence is now that no particular basis functions occur in $\Delta(X)$. Seemingly, for the truth is that the kernel functions belonging to X take their part, with the advantage that if the maximum (5.55) is evaluated by some optimisation procedure, the basis functions change, step by step, thus improving the condition of the basis, in some sense.

Unfortunately, for complexity reasons the construction of an extremal fundamental system by an evaluation of the maximum (5.55) is, in general, an unsolvable task, since it requires solving a complicated nonlinear optimisation problem in the $r \cdot N^2$-dimensional euclidean space, where N is large in the interesting cases. So it is important to know that extremal fundamental systems can be generated also by an exchange algorithm which acts in the $r \cdot N$-dimensional space, only. We present it below.

The exchange algorithm can be understood from a particularly dense representation formula for the squares of the Lagrangians, which is of interest of itself. To derive it, let $T \in D^N$ be a fundamental system, again. If an orthonormal system is used as the basis, (5.27) takes the form

$$L_j(x) = \frac{\det S(\ldots, x, \ldots)}{\det S(\ldots, t_j, \ldots)}$$

for $x, y \in D$ and $j = 1, \ldots, N$. Together with (5.50) this yields

$$L_j(x)L_j(y) = \frac{1}{\Delta(T)} \det \mathbf{G}(\ldots, x, \ldots; \ldots, y, \ldots)$$

for $x, y \in D$ and $j = 1, \ldots, N$, again. We write this formula in the explicit form

$$L_j(x)L_j(y) = \frac{1}{\Delta(T)} \begin{vmatrix} G(t_1, t_1) & \ldots & G(t_1, x) & \ldots & G(t_1, t_N) \\ \vdots & & \vdots & & \vdots \\ G(y, t_1) & \ldots & G(y, x) & \ldots & G(y, t_N) \\ \vdots & & \vdots & & \vdots \\ G(t_N, t_1) & \ldots & G(t_N, x) & \ldots & G(t_N, t_N) \end{vmatrix}, \quad (5.56)$$

where

$$\Delta(T) = \begin{vmatrix} G(t_1, t_1) & \ldots & G(t_1, t_j) & \ldots & G(t_1, t_N) \\ \vdots & & \vdots & & \vdots \\ G(t_j, t_1) & \ldots & G(t_j, t_j) & \ldots & G(t_j, t_N) \\ \vdots & & \vdots & & \vdots \\ G(t_N, t_1) & \ldots & G(t_N, t_j) & \ldots & G(t_N, t_N) \end{vmatrix}. \quad (5.57)$$

In particular, identifying $X = Y$ we obtain the formula wanted,

$$L_j^2(x) = \frac{1}{\Delta(T)} \cdot \Delta(t_1, \ldots, t_{j-1}, x, t_{j+1}, \ldots, t_N) \quad (5.58)$$

for $x \in D$ and $j = 1, \ldots, N$. It is the base of the following algorithm.

Exchange Algorithm (Reimer–Sündermann)

Step 1: Choose an arbitrary fundamental system $T^{(0)} = \{t_1^{(0)}, \ldots, t_N^{(0)}\} \in D^N$, and choose $\epsilon > 0$. Put n:=0.

Step 2: Let $L_1^{(n)}, \ldots, L_N^{(n)}$ be the Lagrange elements belonging to $T^{(n)}$. Calculate
$$M_n := \max\{\|L_1^{(n)}\|_\infty, \ldots, \|L_N^{(n)}\|_\infty\}$$
with the help of (5.32), and after the inversion of (5.34).

Step 3: If $M_n \leq 1 + \epsilon$ **then** STOP **else** some $j \in \{1, \ldots, N\}$ and some $x \in D$ are known such that $|L_j^{(n)}(x)| > 1 + \epsilon$ holds. Put
$$T^{(n+1)} := T^{(n)} \setminus \{t_j^{(n)}\} \cup \{x\}, \quad n := n+1,$$
go to Step 2.

Theorem 5.22 (Exchange Algorithm). *Let $\epsilon > 0$. The Exchange Algorithm is finite and ends with STOP after m steps with the result*

$$\max\{\|L_1^{(m)}\|_\infty, \ldots, \|L_N^{(m)}\|_\infty\} \leq 1 + \epsilon,$$

where

$$m \leq \frac{1}{2} \log \frac{\Delta(T_{\max})}{\Delta(T^{(0)})} \cdot \frac{1}{\log(1+\epsilon)}.$$

Proof. First we remark that, if the algorithm did not stop, a new fundamental system is produced in Step 3, which satisfies, in view of (5.58),

$$\Delta(T^{(n+1)}) > (1+\epsilon)^2 \Delta(T^{(n)}),$$

and we obtain, in the beginning of Step 3,

$$\Delta(T_{\max}) \geq \Delta(T^{(n)}) \geq (1+\epsilon)^{2n} \cdot \Delta(T^{(0)}) > 0.$$

Therefore the algorithm must be finite, this means it ends with STOP after m steps, where

$$\Delta(T_{\max}) \geq (1+\epsilon)^{2m} \cdot \Delta(T^{(0)}) > 0$$

is valid. This furnishes the bound on m. The remaining follows from $M_m \leq 1+\epsilon$. $\square$

Undisturbed Performance of the Exchange Algorithm

The Exchange Algorithm takes into account numerical disturbances. But it is also worthwhile to investigate how it acts in case of an exact performance. So let us put $\epsilon := 0$ and assume, in Step 3, that j and $x = t_j^{(n+1)}$ are determined such that

$$|L_j^{(n)}(x)| = \|L_j^{(n)}\|_\infty = M_n$$

holds, exactly. Then the algorithm either ends with $M_m = 1$ for some $m \in \mathbb{N}_0$, or it is infinite and produces a sequence of fundamental systems $T^{(n)}$, where

$$\Delta(T^{(0)}) < \cdots < \Delta(T^{(n)}) < \Delta(T^{(n+1)}) < \cdots < \Delta(T_{\max})$$

holds for $n = 1, 2, \ldots.$ Because of (5.58) we obtain

$$M_n^2 = |L_j^{(n)}(t_j^{(n+1)})|^2 = \frac{\Delta(T^{(n+1)})}{\Delta(T^{(n)})},$$

where the products

$$\prod_{\nu=0}^{n-1} \frac{\Delta(T^{(\nu+1)})}{\Delta(T^{(\nu)})} = \frac{\Delta(T^{(n)})}{\Delta(T^{(0)})}$$

are monotonically increasing and bounded above by $\Delta(T_{\max})/\Delta(T^{(0)})$. So the factors converge to unity, and we obtain

$$\lim_{n\to\infty} M_n^2 = \lim_{n\to\infty} \frac{\Delta(T^{(n+1)})}{\Delta(T^{(n)})} = 1,$$

and hence

$$\lim_{n\to\infty} \max\{\|L_1^{(n)}\|_\infty, \ldots, \|L_N^{(n)}\|_\infty\} = 1.$$

In this sense, the Exchange Algorithm is convergent, though the fundamental systems $T^{(n)}$ themselves need not converge. But since D^N is compact, a convergent subsequence exists. Moreover, if $T \in D^N$ is any limit point of the $T^{(n)}$, then the corresponding Lagrange elements L_j satisfy

$$\|L_1\|_\infty = \cdots = \|L_N\|_\infty = 1.$$

In other words, every limit point is an extremal fundamental system. Moreover, in view of (5.58) we get

$$0 \leq \Delta(t_1, \ldots, t_{j-1}, x, t_{j+1}, \ldots, t_N) \leq \Delta(t_1, \ldots, t_{j-1}, t_j, t_{j+1}, \ldots, t_N)$$

for arbitrary $x \in D$ and $j \in \{1, \ldots, N\}$, saying that $\Delta(\cdot)$ attains at T a *relative maximum* with respect to each of the arguments. However, the *absolute maximum* can fail to hold, as recent numerical investigations of Sloan and Womersley [65] with respect to some of the spaces $\mathbb{P}_\mu^3(S^2)$ prove.

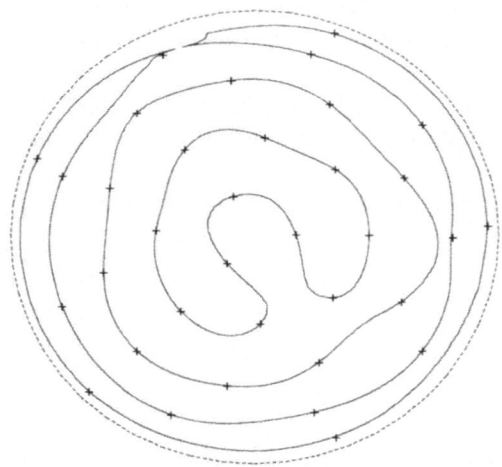

Figure 5.1. Extremal Fundamental System for $\mathbb{P}_5^3(S^2)$ and the Zero Set of L_{36} under an Area Preserving Map $S^2 \to 2B^2$. The Pivot $t_{36} = -e_3$ is Mapped onto the Circle $2S^1$.

In the general case the Exchange Algorithm is very complex and rather slow, where m is large of the order ϵ^{-1}. The performance is much easier if the space and hence the reproducing kernel have a particular structure, as is the case if **V** is a rotation-invariant subspace of $C(S^{r-1})$, where the reproducing kernel takes the form $\mathcal{G}(xy) = K(xy)$, $K \in C[-1, 1]$, see Theorem 1.11. In particular, if K is a differentiable function, Newton's method is helpful in the evaluation of the norms

$\|L_j\|_\infty$, which is the most expensive task within the algorithm. Moreover, in the limit we get $\|L_j\|_\infty = 1$. So the L_j must satisfy Lagrange's maximality conditions

$$grad\, L_j(t_j) \,=\, \lambda_j t_j \tag{5.59}$$

for $j = 1, \ldots, N$. They can be evaluated easily, since, in the present case, the kernel functions have the form $K_k(\,\cdot\,) = K(t_k\,\cdot\,)$, such that (5.32) implies

$$grad\, L_j(x) \,=\, \sum_{k=1}^{N} \langle L_j, L_k \rangle K'(t_k x) t_k \tag{5.60}$$

for $x \in S^{r-1}$, where the matrix $\left(\langle L_j, L_k \rangle \right)$ is known from (5.34) and (5.30).

Starting the Exchange Algorithm

The efficiency of the Exchange Algorithm depends crucially on the choice of the initial fundamental system $T^{(0)}$, where $\Delta(T^{(0)})$ should exhaust the value of $\Delta(T_{\max})$ considerably. If there is no better guess, $T^{(0)}$ has to be chosen at random. This is a promising method only if the expected value of the determinants $\Delta(X)$, $X \in D^N$, has a reasonable size. In what follows we discuss this question in a more general setting.

Let $A(D)$ be a *commutative algebra* of real functions on D with *unity* $1 \in A(D)$. $C(D)$ can serve us as an example, but we drop the continuity assumption, and assume $D \neq \emptyset$, only.

Next we assume that I is a positive linear functional on $A(D)$. Such a functional is called an *integral*. Without restriction of generality we assume that the integral I satisfies

$$I1 \,=\, 1. \tag{5.61}$$

It induces an inner product $[\,\cdot\,,\,\cdot\,]$ by the definition

$$[F_1, F_2] \,:=\, I(F_1 F_2) \quad \text{for} \quad F_1, F_2 \in A(D). \tag{5.62}$$

In what follows, V is a linear subspace of $A(D)$ with finite dimension $N \in \mathbb{N}$. The reproducing kernel of V is denoted again by $G(x,y)$. Recalling (1.5) we see that $G(\,\cdot\,,\,\cdot\,) \in A(D)$ holds, and get

$$IG(\,\cdot\,,\,\cdot\,) \,=\, N. \tag{5.63}$$

We are interested in the average size of the determinants $\Delta(X) = \det \mathrm{G}(X; X)$ for $X \in D^N$. In view of (5.61) this average is defined by

$$average\,(\Delta) \,:=\, I_N \cdots I_2 I_1 \Delta(x_1, \ldots, x_N),$$

where the index of I_i indicates, for $i \in \{1, \ldots, N\}$, that the integral I is to be applied to $\Delta(\ldots, x_i, \ldots)$ as a function of x_i. It is easy to see that this function belongs to $A(D)$, as the functions $G(\,\cdot\,,\,\cdot\,)$ and $G(\,\cdot\,,y)$, $y \in D$, are in $A(D)$.

Theorem 5.23 (Average of Δ). *For $X \in D^N$ let $\Delta(X) = \det G(X; X)$. Then*

$$average(\Delta) = N! \qquad (5.64)$$

holds.

Proof. For $i_1, k_1, i_2, k_2, \ldots \in \{1, \ldots, N\}$ let $\Delta_{k_1, k_2, \ldots}^{i_1, i_2, \ldots}$ denote the adjoint of Δ which belongs to the row indices $i_1, i_2, \ldots$ and to the column indices $k_1, k_2, \ldots$. Expanding $\Delta = \Delta(X)$ with respect to its first column, and expanding the adjoints Δ_1^i, $i = 2, \ldots, N$, after that with respect to their first row, we get

$$\Delta(x_1, \ldots, x_N) = G(x_1, x_1) \cdot \Delta_1^1 - \sum_{i=2}^{N} \sum_{k=2}^{N} G(x_i, x_1) \cdot \Delta_{1,k}^{1,i} \cdot G(x_1, x_k).$$

We recall that, with respect to the variable x_1, this is a function of $A(D)$, but we note also that none of the adjoints which occur on the right side depends on x_1. Therefore we are able to evaluate the integral $I_1 \Delta(x_1, \ldots, x_N)$ with the help of (5.63) and of the reproducing property of the kernel, and we obtain, for fixed $x_2, \ldots, x_N \in D$,

$$I_1 \Delta(x_1, \ldots, x_N) = N \cdot \Delta_1^1 - \sum_{i=2}^{N} \sum_{k=2}^{N} \Delta_{1k}^{1i} \cdot G(x_i, x_k).$$

For arbitrary $i \in \{2, \ldots, N\}$, the inner sum is the expansion of Δ_1^1 with respect to its row with the index i. So the values of the N–1 inner sums are all equal to Δ_1^1, which implies

$$I_1 \Delta(x_1, \ldots, x_N) = 1 \cdot \Delta_1^1.$$

$\Delta_1^1 = \Delta_1^1(x_2, \ldots, x_N)$ depends on $x_2, \ldots, x_N$ and can be expanded likewise. We obtain

$$\Delta_1^1(x_2, \ldots, x_N) = G(x_2, x_2) \cdot \Delta_{12}^{12} - \sum_{i=3}^{N} \sum_{k=3}^{N} G(x_i, x_2) \cdot \Delta_{12k}^{12i} \cdot G(x_2, x_k),$$

where the adjoints which now occur do not depend on x_2. This yields

$$I_2 \Delta_1^1(x_2, \ldots, x_N) = N \cdot \Delta_{12}^{12} - \sum_{i=3}^{N} \sum_{k=3}^{N} \Delta_{12k}^{12i} \cdot G(x_i, x_k),$$

where the N-2 inner sums have the value of Δ_{12}^{12}, which implies

$$I_2 \Delta_1^1(x_2, \ldots, x_N) = 2 \cdot \Delta_{12}^{12}.$$

Repeating this procedure we get finally

$$\Delta_{1,2,\ldots,N-1}^{1,2,\ldots,N-1} = \Delta_{1,2,\ldots,N-1}^{1,2,\ldots,N-1}(x_N) = G(x_N, x_N),$$

and hence

$$I_N \Delta^{1,2,\ldots,N-1}_{1,2,\ldots,N-1}(x_N) = N,$$

where we used (5.63), again. Inserting these results one into another we get

$$average\,(\Delta) = 1 \cdot 2 \cdots \cdot N = N!,$$

as claimed.   □

Theorem 5.23 says that if we choose X at random, the expected value of $\Delta(X)$ is $N!$. This value has to be compared with $\Delta_{\max} = \Delta(T_{\max})$. Such a comparison is possible if $\mathbf{V}$ is a rotation-invariant subspace of $C(S^{r-1})$, $r \in \mathbb{N} \setminus \{1\}$, and if the integral is defined by

$$IF := \frac{1}{\omega_{r-1}} \int\limits_{S^{r-1}} F(x)d\omega(x) \tag{5.65}$$

for $F \in C(S^{r-1})$. For, in this case the kernel has the form

$$G(x,y) = K(xy)$$

for $x, y \in S^{r-1}$, where $K \in C[-1, +1]$, see Theorem 1.11. In view of (5.61) and of (5.63) this implies $K(1) = N$, such that the determinant $\Delta = \Delta(X)$ takes the form

$$\Delta = \begin{vmatrix} N & & & \\ & N & & * \\ & & \ddots & \\ & * & & N \end{vmatrix},$$

independently of the choice of the node system X. Obviously, if $\lambda_1, \cdots, \lambda_N$ are the eigenvalues of $G(X; X)$, then we have

$$\Delta = \lambda_1 \cdots \lambda_N,$$

where $\lambda_1, \ldots, \lambda_N$ satisfy the side conditions

$$\lambda_1 \geq 0, \ldots, \lambda_N \geq 0,$$

$$\lambda_1 + \cdots + \lambda_N = N^2.$$

It is easy to prove with the help of Lagrange's maximality condition that the maximum value of the product $\lambda_1 \cdots \lambda_N$ under these side conditions is attained for $\lambda_1 = \cdots = \lambda_N = N$. The maximum has the value N^N, but this value need not be attained by eigenvalues originating from a matrix $G(X; X)$. However,

$$\Delta_{\max} \leq N^N \tag{5.66}$$

is always valid.

Theorem 5.24. *Let* $r \in \mathbb{N} \setminus \{1\}$. *There is an absolute constant* γ_r *with the following property. If* **V** *is a rotation-invariant subspace of* $C(S^{r-1})$ *with dimension* $N \in \mathbb{N}$, *and if the integral is defined by (5.65), then*

$$\sqrt[N]{\frac{average(\Delta)}{\Delta_{\max}}} \geq \gamma_r > 0$$

holds, independently of the value of N.

Proof. From (5.64) and (5.66) we obtain

$$\sqrt[N]{\frac{average(\Delta)}{\Delta_{\max}}} \geq \sqrt[N]{\frac{N!}{N^N}} \sim \frac{1}{e},$$

where we use Stirling's formula to get the right side. From this it follows that a proper constant γ_r exists. $\square$

Actually, by Theorem 5.23, and even more by Theorem 5.24 we are encouraged to start the Exchange Algorithm by a random choice of $T^{(0)}$. But in particular spaces, random selection can be replaced by a specific construction method, see Section 6.3 and Section 7.3.

5.6 Quadrature

In this section we are concerned with the space $C(D)$ of continuous real functions on D, where D is a nonempty compact set in $\mathbb{R}^r$, $r \in \mathbb{N}$. As usual, $C(D)$ is provided with the uniform norm

$$\| \cdot \| = \| \cdot \|_\infty,$$

while the dual space of bounded linear functionals λ is furnished with the norm

$$\|\lambda\| := \sup \left\{ |\lambda F| : F \in C(D), \|F\| \leq 1 \right\}.$$

Note that $F \in C(D)$, $\|F\| \leq 1$ implies $1 \pm F \geq 0$ such that we obtain for arbitrary positive linear functionals I,

$$I1 \pm IF = I(1 \pm F) \geq 0,$$

and hence $|IF| \leq I1$. The bound is attained for $F = 1$. This means that every positive linear functional is bounded and satisfies

$$\|I\| = I1. \tag{5.67}$$

Definition 5.14 (Integral). *A positive linear functional* I *is called* integral *if* $F \in C(D)$, $F \geq 0$ *and* $IF = 0$ *implies* $F = 0$.

In what follows let $\hat{I}$ denote a linear functional of the form

$$\hat{I}F = \sum_{j=1}^{M} A_j F(t_j)$$

for $F \in C(D)$, with $M \in \mathbb{N}$, *weights* $A_1, \ldots, A_M \in \mathbb{R}$, and with pairwise different *nodes* $t_1, \ldots, t_M \in D$, which are said to *support* the functional. Introducing the *evaluation functionals* $\hat{E}_j$ by the definition

$$\hat{E}_j F := F(t_j)$$

for $F \in C(D)$ and $j = 1, \ldots, M$, we can write $\hat{I}$ in the form

$$\hat{I} = \sum_{j=1}^{M} A_j \hat{E}_j. \tag{5.68}$$

Definition 5.15 (Quadrature). *Let I be an integral and let $\mathbf{V}$ be a subspace of $C(D)$ with $\dim \mathbf{V} \in \mathbb{N}$. $\hat{I}$ is a* quadrature *on $\mathbf{V}$ with respect to I, if $\hat{I}F = IF$ holds for all $F \in \mathbf{V}$. In this case we say also that $\hat{I}$ is* exact *on $\mathbf{V}$. A quadrature which is exact on $\mathbb{P}_\mu^r(D)$ is said to be* exact of degree μ.

Note that the definition of a quadrature depends on the choice of the integral and of the subspace.

The Norm of an Integral and of a Quadrature

An integral I is a particular positive linear functional. So it is bounded, and its norm is given by (5.67).

Next let $\hat{I}$ be a linear functional of the form (5.68). For arbitrary $F \in C(D)$ with $\|F\| \leq 1$ we get

$$|\hat{I}F| \leq \sum_{j=1}^{M} |A_j|.$$

Therefore $\hat{I}$ is also bounded, and we want to calculate its norm. It is obvious that

$$\|\hat{I}\| \leq \sum_{j=1}^{M} |A_j|$$

holds. To obtain a lower bound, we define the numbers

$$\epsilon_j := sgn(A_j)$$

for $j = 1, \ldots, M$, and after that the function $E \in C(D)$ with $\|E\| = 1$ by (5.40), again. It follows that

$$\|\hat{I}\| \geq \hat{I}E = \sum_{j=1}^{M} A_j \cdot sgn(A_j) = \sum_{j=1}^{M} |A_j|$$

is also valid. Together this yields

$$\|\hat{I}\| = \sum_{j=1}^{M} |A_j|. \tag{5.69}$$

In theory and applications, positive quadratures play an important role. They are characterized by the following theorem.

Theorem 5.25 (Weights of a Positive Quadrature). *A quadrature $\hat{I}$ is positive if and only if $A_j \geq 0$ holds for $j = 1, \ldots, M$.*

Proof. Obviously, if the weights are nonnegative, then the quadrature is positive. Vice versa, assume $\hat{I}$ is positive. Then (5.67) is valid with $\hat{I}$ instead of I. Comparing this with (5.69) we obtain

$$\sum_{j=1}^{M} |A_j| = \sum_{j=1}^{M} A_j,$$

and this implies $A_j \geq 0$ for $j = 1, \ldots, M$, as claimed. □

The Quadrature Error

In general it is difficult to evaluate an integral. But it is quite easy to evaluate a quadrature. Therefore the question arises, how integrals can be approximated by quadratures. The following theorem gives a first and rather general, though by no means exhaustive answer.

Theorem 5.26 (Quadrature Error). *Let I be an integral on $C(D)$ and let $\hat{I}$ be a quadrature which is exact on the subspace $\mathbf{V}$ of $C(D)$. Then*

$$\|IF - \hat{I}F\| \leq \|I - \hat{I}\| \cdot E(F, \mathbf{V})$$

holds for arbitrary $F \in C(D)$.

Proof. Assume that F is a function of $C(D)$ with minimal deviation $E(F, \mathbf{V})$ in $\mathbf{V}$. Let V^* be a best approximation to F in $\mathbf{V}$, such that $\|F - V^*\| = E(F, \mathbf{V})$ holds. $\hat{I}$ is exact on $\mathbf{V}$, so we get $(I - \hat{I})V^* = 0$ and hence

$$|(I - \hat{I})F| = |(I - \hat{I})(F - V^*)| \leq \|I - \hat{I}\| \cdot \|F - V^*\| = \|I - \hat{I}\| \cdot E(F, \mathbf{V}),$$

as claimed. □

A given quadrature approximates a fixed integral with some, but in general not arbitrary, precision. So we are forced to use a whole sequence of quadratures $\hat{I}_k$, $k \in \mathbb{N}_0$, each of which is exact on some subspace $\mathbf{V}_k$ of $C(D)$. The convergence of such a sequence is ruled by the Theorem of Banach–Steinhaus (Theorem 5.3, necessary and sufficient conditions). However, in the particular case where all quadratures are positive, the following version of this theorem is more satisfactory.

Theorem 5.27 (Convergence of Positive Quadratures). *Let I be a given integral on $C(D)$. Assume that $\{\hat{I}_k\}_{k\in\mathbb{N}_0}$ is a sequence of positive quadratures, where, for all $k \in \mathbb{N}_0$, $\hat{I}_k$ is exact with respect to I on the finite-dimensional subspace $\mathbf{V}_k$ of $C(D)$. Moreover assume that*

$$1 \in \mathbf{V}_0 \subset \mathbf{V}_1 \subset \cdots$$

is valid, where $E := \bigcup_{k=0}^{\infty} \mathbf{V}_k$ is dense in $C(D)$. Then

$$\lim_{k\to\infty} \hat{I}_k F = IF$$

holds for all $F \in C(D)$.

Proof. We want to apply Theorem 5.3. Because of $span(E) = E$ the assumption on E is satisfied. Next let $k \in \mathbb{N}_0$. Because of $1 \in \mathbf{V}_k$ we obtain $\hat{I}_k 1 = I1$. And since I and $\hat{I}_k$ are positive, we can apply (5.67) to both functionals to get

$$\|\hat{I}_k\| = \hat{I}1 = I1 = \|I\|.$$

So the $\hat{I}_k$ satisfy the assumption (i) of Theorem 5.3 on the F_k. Next let $F \in E$, say $F \in \mathbf{V}_j$ where $j \in \mathbb{N}_0$. Then, by the inclusion property of the subspaces, $F \in \mathbf{V}_k$ is also valid for all $k \in \mathbb{N}_0$, $k \geq j$. Therefore we get $\hat{I}_k F = IF$ for $k \geq j$, and hence

$$\lim_{k\to\infty} \hat{I}_k F = IF \text{ for all } F \in E.$$

This means that assumption (ii) is also satisfied, and the statement of Theorem 5.27 follows from Theorem 5.3, the Theorem of Banach–Steinhaus.   □

Interpolatory Quadratures

In the following $\mathbf{V}$ is a subspace of $C(D)$ with dimension $N \in \mathbb{N}$.

Definition 5.16 (Interpolatory Quadrature). *A quadrature on $\mathbf{V}$ is called interpolatory, if $M = N$ and if $T = \{t_1,\ldots,t_N\}$ is a fundamental system of $\mathbf{V}$.*

The existence of an interpolatory quadrature is guaranteed by the following theorem.

Theorem 5.28 (Interpolatory Quadratures). *Let I be an integral and let $\mathbf{V}$ be a subspace of $C(D)$ with dimension $N \in \mathbb{N}$. Then the following holds. For all fundamental systems T of $\mathbf{V}$ there is a uniquely determined interpolatory quadrature on $\mathbf{V}$ which is supported by T. In particular, an interpolatory quadrature exists.*

Proof. Assume T is a fundamental system, and let $L_1,\ldots,L_N \in \mathbf{V}$ be the Lagrange elements belonging to it. We define

$$\hat{I} = \sum_{j=1}^{N} A_j \hat{E}_j$$

by the choice of the coefficients

$$A_j = IL_j \tag{5.70}$$

for $j = 1, \ldots, N$. Since every $F \in \mathbf{V}$ can be represented in the form

$$F = \sum_{j=1}^{N} F(t_j)L_j,$$

it is obvious that $IF = \hat{I}F$ holds for all $F \in \mathbf{V}$. Hence $\hat{I}$ is an interpolatory quadrature on $\mathbf{V}$. Vice versa, if $\hat{I}$ is interpolatory, $IF = \hat{I}F$ holds in particular for $F = L_k$, and we get

$$IL_k = \hat{I}L_k = \sum_{j=1}^{N} A_j L_k(t_j) = A_k$$

for $k = 1, \ldots, N$, such that (5.70) is valid. So the coefficients of an interpolatory quadrature are uniquely determined by T, and given by (5.70). The existence of an interpolatory quadrature on $\mathbf{V}$ is guaranteed by the existence of a fundamental system, see Theorem 5.14. □

The Number of Nodes in a Quadrature

In what follows, we investigate the minimum number $m(I, \mathbf{V})$ of nodes in a quadrature which is exact on $\mathbf{V}$ with respect to the integral I. Because of Theorem 5.28 it satisfies

$$1 \leq m(I, \mathbf{V}) \leq N. \tag{5.71}$$

The lower bound in (5.71) seems to be unreasonable. But it occurs in quite non-trivial situations. We give two examples.

Let $D := S^{r-1}$ and let the integral be defined by

$$IF := \int_{S^{r-1}} F(x)d\omega(x) \text{ for } F \in C(S^{r-1}).$$

Assume that $\mathbf{V}$ is any finite-dimensional subspace of $C(S^{r-1})$ which consists of odd functions, only. Then the integral vanishes on $\mathbf{V}$, and so does the one-point quadrature $\hat{I} := 0 \cdot \hat{E}_1$. This is exact on $\mathbf{V}$, and we get $m(I, \mathbf{V}) = 1$.

By a similar reasoning we obtain, for $\mu \in \mathbb{N}$ and $r \in \mathbb{N} \setminus \{1\}$,

$$m\left(I, \overset{*}{\mathbb{H}}{}_{\mu}^{r}(S^{r-1})\right) = 1,$$

since the integral vanishes again on the subspace, but now because of Theorem 4.10.

In general a lower bound for $m(I, \mathbf{V})$ can be determined with the help of the following theorem.

Theorem 5.29 (Lower Bound for $m(I, \mathbf{V})$). *Let I be an integral on $C(D)$, and let $\mathbf{V}$ be a finite-dimensional subspace of $C(D)$. If $\mathbf{W}$ is a subspace of $\mathbf{V}$ such that $F \in \mathbf{W}$ implies $F^2 \in \mathbf{V}$, then*

$$m(I, \mathbf{V}) \geq \dim \mathbf{W}$$

is valid.

Proof. Let $\hat{I} = \sum_{j=1}^{M} A_j \hat{E}_j$ be an arbitrary quadrature which is exact on $\mathbf{V}$. Assume that $M < \dim \mathbf{W}$ holds. Then the linear system of equations

$$\hat{E}_j F = 0, \ j = 1, \dots, M,$$

has a nontrivial solution $F \in \mathbf{W}$. To be more explicit, $F(t_j) = 0$ holds for $j = 1, \dots, M$. Because of $F^2 \in \mathbf{V}$ this implies

$$0 < IF^2 = \hat{I}F^2 = \sum_{j=1}^{M} A_j \hat{E}_j F^2 = 0,$$

which is a contradiction. So $M \geq \dim \mathbf{W}$ must be valid. □

Note that a space of odd functions or of nonconstant harmonic functions does not contain a nonnegative function. So it does not contain the square of a function, except for the null-function. So the only possible choice of $\mathbf{W}$ is in their case $\mathbf{W} = [0]$. It furnishes the trivial lower estimate $m(I, \mathbf{V}) \geq 0$. But the situation changes if Theorem 5.29 is applied to a complete space of polynomials or of homogeneous polynomials of a certain degree.

Corollary 5.30 ($m(I, \mathbf{V})$ for Polynomial Spaces). *Assumptions as in Theorem 5.29. Then the following lower bounds hold for $\mu \in \mathbb{N}_0$:*

$$m\left(I, \mathbb{P}_\mu^r(D)\right) \geq \dim \mathbb{P}_{\lfloor \frac{\mu}{2} \rfloor}^r(D), \tag{5.72}$$

$$m\left(I, \overset{*}{\mathbb{P}}_\mu^r(D)\right) \geq \dim \overset{*}{\mathbb{P}}_{\lfloor \frac{\mu}{2} \rfloor}^r(D). \tag{5.73}$$

Proof. See Theorem 5.29 □

Remark. If D contains an interior point, then the result of Corollary 5.30 takes the form

$$m\left(I, \mathbb{P}_\mu^r(D)\right) \geq \binom{\lfloor \frac{\mu}{2} \rfloor + r}{r}, \tag{5.74}$$

$$m\left(I, \overset{*}{\mathbb{P}}_\mu^r(D)\right) \geq \binom{\lfloor \frac{\mu}{2} \rfloor + r - 1}{r - 1}, \tag{5.75}$$

see Theorem 3.14 together with (3.8) or (3.9), respectively. The bound of (5.74) is due to Stroud [70].

If no interior point is contained in D, the bounds need not be valid. To give an example, let $D := S^{r-1}$. (5.71) furnishes the inequality

$$m\left(I, \mathbb{P}_\mu^r(S^{r-1})\right) \leq \binom{\mu + r - 1}{r - 1} + \binom{\mu + r - 2}{r - 1}, \tag{5.76}$$

see (4.4). For $r = 2$ and $\mu \geq 12$, the upper bound is less than the right side of (5.74). Therefore (5.74) cannot be valid.

In what follows, space, integral and the subspace are arbitrary, again, but we consider positive quadratures, only. So let $p(I, \mathbf{V})$ denote the minimum number of nodes in a positive quadrature which is exact on $\mathbf{V}$ with respect to I. Because of Theorem 5.25 this is the minimum number of positive weights. If a positive quadrature does not exist, $p(I, \mathbf{V})$ is defined by $+\infty$.

Naturally, $p(I, \mathbf{V}) \geq m(I, \mathbf{V})$ holds, and if $\mathbf{W}$ satisfies the assumptions of Theorem 5.29, then we get

$$p(I, \mathbf{V}) \geq m(I, \mathbf{V}) \geq \dim \mathbf{W}.$$

A reasonable upper bound can be determined, if a positive quadrature actually exists, whatever the number of nodes may be.

Theorem 5.31 (Upper Bound for $p(I, \mathbf{V})$). *If a positive quadrature exists which is exact on $\mathbf{V}$ with respect to the integral I, then $p(I, \mathbf{V}) \leq N$ holds, where $N = \dim \mathbf{V}$.*

Proof. Assume that $\hat{I} = \sum_{j=1}^M A_j \hat{E}_j$, $M \in \mathbb{N}$, is a positive quadrature on $\mathbf{V}$ with respect to I. In view of Theorem 5.25 we may assume, without restriction of generality, that all weights are positive, i.e., that $A_j > 0$ holds for $j = 1, \dots, M$. $M \leq N$ implies $p(I, \mathbf{V}) \leq N$, as claimed. Next assume $M > N$. As $\hat{I}$ is exact on $\mathbf{V}$, we get

$$IF = \hat{I}F = \sum_{j=1}^M A_j F(t_j)$$

for all $F \in \mathbf{V}$. But, because of $M > N$, the restrictions $\hat{E}_j|_\mathbf{V}$ are linearly dependent (in $\mathbf{V}'$, the dual space of $\mathbf{V}$). So a vanishing, nontrivial linear combination of the restrictions exists, say

$$0 = \sum_{j=1}^M a_j F(t_j)$$

holds for all $F \in \mathbf{V}$, where some of the coefficients $a_j \in \mathbb{R}$ are different from zero, say, $a_k > 0$ holds for $k \in \{1, \dots, M\}$. It follows that the linear functional $\hat{K}_\lambda$, defined by

$$\hat{K}_\lambda F := \sum_{j=1}^M \left(A_j - \lambda a_j\right) F(t_j) \quad \text{for } F \in C(D),$$

is exact on $\mathbf{V}$ for arbitrary $\lambda \in \mathbb{R}$. Now let

$$\lambda^* := \sup\{\lambda : A_j - \lambda a_j \geq 0 \text{ for } j = 1, \dots, M\}.$$

Obviously, $\lambda^* > 0$ holds, and because of $a_k > 0$, λ^* is finite. Say, the supremum is attained for $j = l$, $l \in \{1, \ldots, M\}$, such that

$$A_l - \lambda^* a_l = 0, \quad A_j - \lambda^* a_j \geq 0$$

holds for $j = 1, \ldots, M$. $\hat{K}_{\lambda^*}$ is again a positive quadrature on $\mathbf{V}$, see Theorem 5.25, but with a number of nonvanishing weights now less than M.

Repeating this procedure, if necessary, we obtain positive quadratures on $\mathbf{V}$ with a decreasing number of nonvanishing, i.e., positive weights, until a positive quadrature with at most N positive weights is obtained. It follows that $p(I, \mathbf{V}) \leq N$ holds, as claimed.   □

Remark. The proof of Theorem 5.31 is constructive. The positive quadrature, which is finally obtained, is supported by at most N of the original nodes.

Positive Algebraic Quadratures

In what follows, the subspace $\mathbf{V}$ is one of the spaces

$$\mathbf{V}_\mu := \mathbb{P}^r_\mu(D), \ \mu \in \mathbb{N}_0.$$

Then Theorem 5.27 takes the following, handsome form.

Corollary 5.32 (Positive Algebraic Quadratures). *Let I be a given integral, and assume that the quadratures $\hat{I}_\mu$ are positive and exact of degree μ for $\mu \in \mathbb{N}_0$. Then*

$$\lim_{\mu \to \infty} \hat{I}_\mu F = IF$$

holds for all $F \in C(D)$.

Proof. All assumptions of Theorem 5.27 are satisfied, the density assumption because of the Theorem of Weierstrass, i.e., of Theorem 5.8.   □

Remark. The lower bounds occurring in Corollary 5.30 need not be attained, see Möller [36], [37], e.g., who made the minimal number of nodes subject to an intrinsic ideal theoretical investigation.

Gauß Quadrature

In the context of positive algebraic quadrature it is usual to introduce the following definition.

Definition 5.17 (Gauß Quadrature). *A quadrature $\hat{I}_\mu = \sum_{j=1}^{M} A_j \hat{E}_j$, $\mu \in \mathbb{N}_0$, based on the evaluation functionals $\hat{E}_j$, is called Gauß quadrature with respect to the integral I, if the following holds.*

    (i) $\hat{E}_1, \ldots, \hat{E}_M$ are linearly independent in the dual space of $\mathbb{P}^r_\mu(D)$.

    (ii) $\hat{I}_\mu F = IF$ holds for all $F \in \mathbb{P}^r_{2\mu}(D)$.

Gauß quadratures are well known in the cases $D = [-1, +1]$ and $D = S^1$. If a Gauß quadrature exists, then its nodes form a very particular geometric configuration.

Theorem 5.33 (Gauß Quadrature). *Let $r \in \mathbb{N}$, $\mu \in \mathbb{N}_0$. Assume I is an integral on $C(D)$, and let $G(\cdot, \cdot)$ be the reproducing kernel of $\mathbb{P}_\mu^r(D)$ belonging to the inner product defined by $\langle F_1, F_2 \rangle := I(F_1 F_2)$ for $F_1, F_2 \in C(D)$. Assume $\hat{I}_\mu = \sum_{j=1}^M A_j \hat{E}_j$ is a quadrature which is exact of degree μ and which is represented by means of linearly independent evaluation functionals in the sense of Definition 5.17, (i). Then $\hat{I}_\mu$ is a Gauß quadrature if and only if the following holds.*

(i)   *$\{t_1, \ldots, t_M\}$ is a fundamental system for $\mathbb{P}_\mu^r(D)$.*

        This implies $M = N := \dim \mathbb{P}_\mu^r(D)$.

(ii)   *The Lagrange elements satisfy $\langle L_j, L_k \rangle = A_j \delta_{j,k}$ for $j, k = 1, \ldots, N$.*

In particular, the weights of a Gauß quadrature are positive.

Proof. Assume $\hat{I}_\mu$ is a Gauß quadrature. $\hat{I}_\mu$ is a quadrature on $\mathbb{P}_{2\mu}^r(D)$, so Corollary 5.30 says that $M \geq \dim \mathbb{P}_\mu^r(D) = N$ is valid, but since the evaluation functionals are linearly independent, we get also $M \leq N$, i.e., $M = N$, and $t_1, \ldots, t_N$ form a fundamental system for $\mathbb{P}_\mu^r(D)$. Moreover, for fixed $j, k \in \{1, \ldots, N\}$ we obtain

$$\langle L_j, L_k \rangle = I(L_j L_k) = \hat{I}(L_j L_k) = \sum_{\nu=1}^N A_\nu (L_j L_k)(t_\nu) = A_j \delta_{j,k},$$

as claimed.

Vice versa, assume that (i) and (ii) hold. Every monomial $x^m \in \mathbb{P}_{2\mu}^r$ can be represented in the form

$$x^m = x^a \cdot x^b, \quad \text{where } x^a, \, x^b \in \mathbb{P}_\mu^r.$$

For $x \in D$, these monomials have the representations

$$x^a = \sum_{j=1}^N t_j^a \, L_j(x), \quad x^b = \sum_{k=1}^N t_k^b \, L_k(x),$$

and this implies

$$x^m = \sum_{j=1}^N \sum_{k=1}^N t_j^a t_k^b \, L_j(x) L_k(x).$$

So we get

$$\mathbb{P}_{2\mu}^r(D) \subset span\{L_j L_k \mid j, k = 1, \ldots, N\}.$$

Next let $F \in \mathbb{P}_{2\mu}^r(D)$ be arbitrary. F can be represented in the form

$$F = \sum_{j=1}^N \sum_{k=1}^N a_{jk} L_j L_k.$$

This implies
$$F(t_j) = a_{jj} \text{ for } j = 1, \ldots, N,$$

and we get

$$IF = \sum_{j=1}^{N} \sum_{k=1}^{N} a_{jk} I(L_j L_k) = \sum_{j=1}^{N} \sum_{k=1}^{N} a_{jk} \langle L_j, L_k \rangle = \sum_{j=1}^{N} \sum_{k=1}^{N} a_{jk} A_j \delta_{jk} = \sum_{j=1}^{N} A_j F(t_j)$$
$$= \hat{I}_\mu F.$$

$\hat{I}$ is a Gauß quadrature. In particular, $A_j = \langle L_j, L_j \rangle > 0$ holds for $j = 1, \ldots, N$, and the theorem is proved. $\square$

Corollary 5.34. *Assumptions as in Theorem 5.33. $\hat{I}_\mu$ is a Gauß quadrature if and only if $M = N = \dim \mathbb{P}_\mu^r(D)$, $t_1, \ldots, t_N$ form a fundamental system, and the fundamental matrix G takes the form*

$$G = diag\left(A_1^{-1}, \ldots, A_N^{-1}\right)$$

with positive diagonal elements.

Proof. If $\hat{I}_\mu$ is a Gauß quadrature, then the statement follows directly from Theorem 5.33, together with (5.30) and (5.34). Vice versa, assume $M = N$, and $t_1, \ldots, t_N$ form a fundamental system, where G has the diagonal form presented. From (5.30) and (5.34) we get

$$\left(\langle L_j, L_k \rangle\right) = G^{-1} = diag\left(A_1, \ldots, A_N\right),$$

and Theorem 5.33 says that $\hat{I}$ is a Gauß quadrature. $\square$

Remark. In the literature, a great variety of particular quadratures exists, which differ by domain, integral, dimension and degree. For an important collection we refer to Stroud [71].

In this section, D was an arbitrary, nonempty compact subset of $\mathbb{R}^r$. For particular subsets, more detailed results can be obtained. This holds, for instance, in the important cases $D = S^{r-1}$ and $D = B^r$, which are the subject of Chapter 6 and of Chapter 7, respectively.

5.7   Best Approximation in the Maximum Norm

In this section we continue our investigations on best appoximation problems, where we restrict ourselves to the case $X := C(D)$, D a nonempty compact subset of $\mathbb{R}^r$. $C(D)$ is provided again with the maximum norm,

$$\|F\| := \|F\|_\infty := \max\{|F(x)| : x \in D\} \text{ for } F \in C(D).$$

It is our aim to characterize the elements of best approximation for a function $F \in C(D)$ in a given subspace $\mathbf{V}$ of $C(D)$ of finite dimension $N \in \mathbb{N}$. Note that V^* best approximates F in $\mathbf{V}$ if and only if *zero* best approximates $F - V^*$. So it suffices to characterize the cases where *zero* is a best approximation. Recall that in our context a best approximation always exists. It can be characterized by the *Criterion of Kolmogoroff*, which is captured best from the theory of convex sets. This detour is due to Rivlin and Shapiro [59].

Theorem 5.35 (Carathéodory). *Let $A \subset \mathbb{R}^n$, $n \in \mathbb{N}$, and assume $x \in conv(A)$. Then a subset B of A with $card(B) \le n + 1$ exists such that $x \in conv(B)$.*

Proof. By assumption, x can be represented in the form

$$x = \sum_{j=0}^{m} \xi_j a_j, \quad where \quad a_j \in A, \ \xi_j > 0, \ \sum_{j=0}^{m} \xi_j = 1, \ m \in \mathbb{N}_0.$$

Among all representations of this kind there is one where m is minimal, and it satisfies to prove that $m \le n$ holds for this minimal m.

To this end let us assume, on the contrary, that $m > n$ is valid. Obviously, the elements $a_1 - a_0, \ldots, a_m - a_0$ are linearly dependent in $\mathbb{R}^n$. So there is a coefficient vector $(\eta_1, \ldots, \eta_m) \ne (0, \ldots, 0)$ such that

$$\sum_{j=1}^{m} \eta_j (a_j - a_0) = 0.$$

For some $k \in \{1, \ldots, m\}$, η_k does not vanish, where we even may assume $\eta_k > 0$. Now let us define, for $t \in \mathbb{R}$,

$$\xi_0(t) := \xi_0 + t \sum_{j=1}^{m} \eta_j,$$
$$\xi_j(t) := \xi_j - t\eta_j \text{ for } j = 1, \ldots, m.$$

The definitions are such that $\xi_j(0) > 0$ holds for $j = 0, 1, \ldots, m$. All of the inequalities remain valid, first, for increasing values of t, but the inequality for $j = k$ is violated for sufficiently large t. Together this yields that a positive number τ and a number $l \in \{1, \ldots, m\}$ exist such that

$$\xi_j(\tau) \ge 0 \text{ for } j = 0, 1, \ldots, m,$$

with equality holding for $j = l$. Besides we get

$$\sum_{j=0}^{m} \xi_j(\tau) a_j = \Big[\xi_0 + \tau \sum_{j=1}^{m} \eta_j\Big] a_0 + \sum_{j=1}^{m} (\xi_j - \tau \eta_j) a_j = x$$

and

$$\sum_{j=0}^{m} \xi_j(\tau) = 1 + \tau \sum_{j=1}^{m} \eta_j - \sum_{j=1}^{m} \tau\eta_j = 1.$$

In view of $\xi_l(\tau) = 0$ this means that x is represented as a convex combination from m elements of A, instead of $m+1$, which contradicts m being minimal. $\square$

Definition 5.18 (Separating Hyperplane). *The subsets $A, B \subset \mathbb{R}^n$, $n \in \mathbb{N}$, are separated by the affine hyperplane $H = \{x \in \mathbb{R}^n | \, ax - b = 0\}$, $0 \neq a \in \mathbb{R}^n$, $b \in \mathbb{R}$, if*

$$ax - b > 0 \quad \text{holds for all } x \in A, \quad ax - b < 0 \quad \text{holds for all } x \in B,$$

or vice versa.

Theorem 5.36 (Existence of a Separating Hyperplane). *Let A be a nonempty closed convex subset of $\mathbb{R}^n$ which does not contain the origin 0. Then an affine hyperplane exists which separates A and $\{0\}$.*

Proof. There is an element $y \neq 0$ in A. The set

$$\{x \in A : |x| \leq |y|\}$$

is compact and not empty, and $|x|$ attains its minimum value in this set at a point $a \in A$, where $|a| > 0$. Now let $x \in A$ be arbitrary. Since A is convex, we get

$$|\lambda x + (1 - \lambda)a| \geq |a|$$

for $0 \leq \lambda \leq 1$, and for $0 < \lambda \leq 1$ this is equivalent to

$$\lambda x^2 + 2(1 - \lambda)ax + (\lambda - 2)a^2 \geq 0.$$

This inequality remains valid for $\lambda = 0$, and we obtain

$$a(x - a) \geq 0 \quad \text{for } x \in A.$$

So the affine hyperplane

$$H = \{x \in \mathbb{R}^n | \, ax - \frac{a^2}{2} = 0\}$$

separates A and $\{0\}$, and the theorem is proved. $\square$

Now we recall that our original aim was to characterize the situation where *zero* is a best approximation. To this end we introduce the following definition.

Definition 5.19 (Extreme Points). *For $F \in C(D)$, $D \subset \mathbb{R}^n$ compact, the set of extreme points is defined by $\mathcal{E}(F) := \{x \in D : |F(x)| = \|F\|\}$.*

Note that $\mathcal{E}(F)$ is a compact set. Now let $V_1, \ldots, V_N$ form a basis in $\mathbf{V}$, and define

$$\Phi(F) := \left\{ \left(F(x)V_1(x), \ldots, F(x)V_N(x) \right)' \,\middle|\, x \in \mathcal{E}(F) \right\},$$

which is a subset of R^N. The definition enables us to formulate the following basic lemma.

Lemma 5.37. $F \in C(D)$ *is best approximated in* $\mathbf{V}$ *by zero if and only if the origin in* $\mathbb{R}^N$ *is the strict convex combination of at most* $N + 1$ *elements of* $\Phi(F)$.

Proof. In the case $F = 0$ we get $\Phi(F) = \{0\}$ and in view of the equation $0 = \frac{1}{2}(0 + 0)$, the statement is evident. So we may assume $F \neq 0$ in what follows. First assume that the origin in $\mathbb{R}^N$ is the strict convex combination of the points

$$\left(F(x_j)V_1(x_j), \ldots, F(x_j)V_N(x_j) \right)' \in \Phi(F), \quad j = 0, 1, \ldots, M,$$

where $M \leq N$. Say

$$0 = \sum_{j=0}^{M} \alpha_j F(x_j) V_k(x_j)$$

holds for $k = 1, \ldots, N$, where

$$\alpha_0 > 0, \ldots, \alpha_M > 0, \quad \sum_{j=0}^{M} \alpha_j = 1.$$

By a linear combination of the equations we obtain

$$0 = \sum_{j=0}^{M} \alpha_j F(x_j) G(x_j)$$

for all $G \in \mathbf{V}$. Now let $G \in \mathbf{V}$ be fixed. Then there exists a component, say an index $j \in \{0, \ldots, M\}$, such that

$$F(x_j)G(x_j) \leq 0.$$

It follows that

$$\|F - G\|^2 \geq |F(x_j) - G(x_j)|^2 \geq |F(x_j)|^2 = \|F\|^2,$$

where we used $x_j \in \mathcal{E}(F)$, finally. In other words, $\|F - G\| \geq \|F - 0\|$ holds for arbitrary $G \in \mathbf{V}$, and *zero* is a best approximation to F in $\mathbf{V}$.

Vice versa, assume that *zero* is a best approximation to F in $\mathbf{V}$, where $F \neq 0$. This implies

$$\|F - G\| \geq \|F\| > 0 \quad \text{for all } G \in \mathbf{V}. \tag{5.77}$$

Now let us assume, in addition, that *zero* is *not* the strict convex combination of at most $N+1$ elements of $\Phi(F)$. Then

$$0 \notin conv\big(\Phi(F)\big)$$

must hold in view of Theorem 5.35. We prove that this is a contradiction.

We begin with the remark that $\Phi(F)$ is the image of the nonempty and compact set $\mathcal{E}(F)$ under a continuous mapping, so it is a nonempty and compact set in $\mathbb{R}^N$, and so is $conv(\Phi(F))$. It follows from Theorem 5.36 that $\Phi(F)$ and $\{0\}$ are separated by a hyperplane, which means that $a \in \mathbb{R}^N$, $a \neq 0$, and $b \in \mathbb{R}$ exist such that the following inequalities hold,

$$\begin{aligned} aX - b &> 0 \quad \text{for } X \in \Phi(F), \\ -b &< 0 \,. \end{aligned}$$

These inequalities imply that

$$\sum_{k=1}^{N} a_k F(x) V_k(x) > b > 0$$

holds for arbitrary $x \in \mathcal{E}(F)$. With the definition

$$G := \sum_{k=1}^{N} a_k V_k \in \mathbf{V},$$

we write this in the form

$$F(x)G(x) > b > 0 \quad \text{for } x \in \mathcal{E}(F). \tag{5.78}$$

It follows that for every $z \in \mathcal{E}(F)$ there is an open neighbourhood U_z of z such that

$$F(x)G(x) > 0 \quad \text{for } x \in U_z$$

remains valid. Finally let

$$U := \bigcup \Big\{ U_z \mid z \in \mathcal{E}(F) \Big\}.$$

$D \setminus U$ is a compact set, since D is compact while U is open. Hence the following definition is admissable,

$$M := \left\{ \begin{array}{ll} \max\{|F(x)| : x \in D \setminus U\} & , \text{ if } D \setminus U \neq \emptyset, \\ 0 & , \text{ if } D \setminus U = \emptyset. \end{array} \right.$$

In both cases $M < \|F\|$ holds. Therefore an $\epsilon > 0$ exists with

$$M + \epsilon \|G\| < \|F\|.$$

This implies
$$|F(x) - \epsilon G(x)| < \|F\| \quad \text{for } x \in D \setminus U,$$

but using $F(x)G(x) > 0$, we get also

$$|F(x) - \epsilon G(x)| < \|F\| \quad \text{for } x \in U.$$

Both together yield
$$\|F - \epsilon G\| < \|F\|.$$

Because of $\epsilon G \in \mathbf{V}$ this contradicts (5.77). Hence *zero* must be the strict convex combination of at most $N+1$ elements of $\Phi(F)$. This finishes the proof. $\qquad\square$

Lemma 5.37 is the key to the following theorem.

Theorem 5.38 (Characterisation of Best Approximations). *Let* $\mathbf{V}$ *be a subspace of* $C(D)$ *of dimension* $N \in \mathbb{N}$. V^* *is a best approximation to* $F \in C(D)$ *in* $\mathbf{V}$ *if and only if there are* $M+1$ *extreme points*

$$x_0, x_1, \ldots, x_M \in \mathcal{E}(F - V^*),$$

where $M \leq N$, *and positive numbers* $\alpha_0, \alpha_1, \ldots, \alpha_M$, *such that*

$$\sum_{j=0}^{M} \alpha_j \Big(F(x_j) - V^*(x_j) \Big) G(x_j) = 0 \tag{5.79}$$

holds for all $G \in \mathbf{V}$.

Proof. According to our remark in the beginning of this section we may assume without restriction of generality that $V^* = 0$ best approximates F in $\mathbf{V}$. By Lemma 5.37 this holds if and only if

$$\sum_{j=0}^{M} \alpha_j F(x_j) V_k(x_j) = 0$$

is valid for $k = 1, 2, \ldots, N$ with extreme points $x_j \in \mathcal{E}(F)$ and positive coefficients α_j, as above, where $M \leq N$. This condition is equivalent to (5.79), and the theorem is proved. $\qquad\square$

Sometimes it is more convenient to use the following criterion.

Theorem 5.39 (Criterion of Kolmogoroff). *Assumptions as in Theorem 5.38.* V^* *best approximates* $F \in C(D)$ *in* $\mathbf{V}$ *if and only if*

$$\min \left\{ \Big(F(x) - V^*(x) \Big) G(x) \,\Big|\, x \in \mathcal{E}(F - V^*) \right\} \leq 0 \tag{5.80}$$

holds for all $G \in \mathbf{V}$.

Proof. Obviously, if V^* best approximates F in $\mathbf{V}$, then (5.79) holds for arbitrary $G \in \mathbf{V}$, and this implies (5.80). Vice versa, assume (5.80) is valid for some V^* and all $G \in \mathbf{V}$. Again it suffices to consider the case where $V^* = 0$. Now let $G \in \mathbf{V}$. Then an $x \in \mathcal{E}(F)$ exists such that

$$F(x)G(x) \leq 0$$

holds. This implies

$$\|F - G\|^2 \geq |F(x) - G(x)|^2 \geq F^2(x) = \|F\|^2.$$

Since this is valid for arbitrary $G \in \mathbf{V}$, $V^* = 0$ best approximates F in $\mathbf{V}$, as claimed. $\square$

Remark. In higher dimensional spaces only a few examples are known where a best approximation with respect to the maximum norm can be described by a formula. One important example is the elements L_j of an extremal Lagrange basis, as used in the proof of Theorem 5.21 in a rather general situation, see Problem 5.6. Two additional examples are presented in the following section, see (A) and (B).

5.8   Examples

(A) Best Approximation to Monomials on $\mathbf{B}^2$

Let $r = 2$ and $\mu \in \mathbb{N}$. As above, $\| \cdot \| = \| \cdot \|_{B^2}$ is the maximum norm on $C(B^2)$. In view of (4.13), the equation (4.48) takes the form

$$\sum_{|m|=\mu} R_m(x)t^m = \tfrac{1}{\pi}|t|^\mu T_\mu\left(\tfrac{tx}{|t|}\right). \tag{5.81}$$

Comparing this and (3.51) we see that

$$R_m = const \cdot V_m^{2,-1} \tag{5.82}$$

holds with some constant which depends on m. In other words, apart from a constant factor, the R_m are, in the present case, the Appell polynomials $V_m^{r,s}$, which belong to the degenerating index $s = -1$.

For every fixed $x \in B^2$, $R_m(x)$ occurs in (5.81) as the coefficient of a homogeneous polynomial which is bounded on S^1 by $\tfrac{1}{\pi}$. Hence the Kellogg bound

$$\|R_m(x)\| \leq \frac{1}{\pi}\binom{|m|}{m} \tag{5.83}$$

holds for $x \in B^2$ and $0 \neq m \in \mathbb{N}_0^2$, see Theorem 3.1. This enables us to prove the following theorem.

Theorem 5.40 (Best Approximation by Zero, Reimer). *For $0 \neq m \in \mathbb{N}_0^2$ the following holds.*

$$(i) \qquad \|R_m\|_{B^2} = \tfrac{1}{\pi}\binom{|m|}{m}, \tag{5.84}$$

$$(ii) \qquad R_m \text{ is best approximated in } \mathbb{P}^2_{|m|-1}(B^2) \text{ by zero.}$$

Proof. (i) Restricting (5.81) to the unit circle S^1 by inserting

$$x = \begin{pmatrix} \cos\phi \\ \sin\phi \end{pmatrix}, \quad t = \begin{pmatrix} \cos\psi \\ \sin\psi \end{pmatrix},$$

we obtain for $\mu \in \mathbb{N}$

$$
\begin{aligned}
\pi \sum_{k=0}^{\mu} R_{\mu-k,k}(\cos\phi,\sin\phi)\, t_1^{\mu-k} t_2^k &= \cos\mu(\phi-\psi) \\
&= \cos\mu\phi\cos\mu\psi + \sin\mu\phi\sin\mu\psi \\
&= \cos\mu\phi\cdot\left\{\binom{\mu}{0}t_1^\mu t_2^0 - \binom{\mu}{2}t_1^{\mu-2}t_2^2 \pm\cdots\right\} \\
&+ \sin\mu\phi\cdot\left\{\binom{\mu}{1}t_1^{\mu-1}t_2^1 - \binom{\mu}{3}t_1^{\mu-3}t_2^3 \pm\cdots\right\}.
\end{aligned}
$$

Comparing the coefficients which occur with $t_1^{\mu-k}t_2^k$ we get

$$
\begin{aligned}
R_{\mu-2\nu,2\nu}(\cos\phi,\sin\phi) &= \tfrac{(-1)^\nu}{\pi}\binom{\mu}{2\nu}\cos\mu\phi, \\
R_{\mu-2\nu-1,2\nu+1}(\cos\phi,\sin\phi) &= \tfrac{(-1)^\nu}{\pi}\binom{\mu}{2\nu+1}\sin\mu\phi,
\end{aligned}
\tag{5.85}
$$

for $\nu = 0,1,\ldots,\lfloor\frac{\mu}{2}\rfloor$ and $\nu = 0,1,\ldots,\lfloor\frac{\mu-1}{2}\rfloor$, respectively. In particular, the upper bound of (5.83) is attained on the unit circle, such that (i) is valid.

(ii) For $|m| = \mu \in \mathbb{N}$, it follows from (5.85) that there are 2μ equidistributed extreme points on the unit circle S^1, where R_m attains the values $\pm\|R_m\|_{B^2}$ with alternating signs. Now let G be an arbitrary element of $\mathbb{P}^2_{|m|-1}(B^2)$. Then $G(\cos\phi,\sin\phi)$ is a trigonometric polynomial of degree at most $\mu-1$, which cannot have the same sign distribution as $R_m(\cos\phi,\sin\phi)$ has at these 2μ extreme points. So we get

$$\min\left\{R_m(x)G(x)\mid x \in \mathcal{E}(R_m)\right\} \leq 0,$$

and (ii) follows from Theorem 5.39. This finishes the proof. $\qquad\square$

Note that $R_{0,0} = \frac{1}{2\pi}$ holds, such that the statements of Theorem 5.40 remain valid for $m = (0,0)$ with a slight modification, which concerns only the norm. The oscillating property of the polynomials R_m on the unit circle is demonstrated by Figure 5.2.

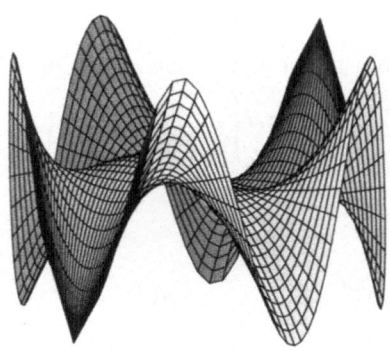

Figure 5.2. $R_{3,2}(x_1, x_2)$ on the Disk $x_1^2 + x_2^2 \le 1$.

There is an important application of Theorem 5.40, which concerns the best approximation to a monomial M_m on B^2 by polynomials of lower degree in the maximum norm. We present it after some introductory remarks.

Assume $\mu \in \mathbb{N}$, again. Identifying $Q_\mu = \frac{1}{\pi} T_\mu$ and $A_m = R_m$ we see that (5.81) is a realisation of the equation (3.41), while, in view of (5.84), equation (3.45) takes the form

$$R_m(x) = 2^{\mu-1} \|R_m\|_{B^2} \cdot x^m + TLD$$

for $|m| = \mu \in \mathbb{N}$. Now let R_m be normalized by the definition

$$\hat{R}_m := R_m / \|R_m\|_{B^2}, \tag{5.86}$$

which says that $\|\hat{R}_m\|_{B^2} = 1$ is valid. $\hat{R}_m$ has again the form

$$\hat{R}_m(x) := 2^{\mu-1} x^m + TLD, \tag{5.87}$$

for $|m| = \mu \in \mathbb{N}$, and the following theorem holds.

Theorem 5.41 (Best Approximation to a Monomial). Let $m \in \mathbb{N}_0^2$, $|m| = \mu \in \mathbb{N}$, $D_m := M_m - 2^{1-\mu} \hat{R}_m$. Then the following holds.

(i) $D_m \in \mathbb{P}_{\mu-1}^2$.

(ii) $E\left(M_m, \mathbb{P}_{\mu-1}(B^2)\right) = \|M_m - D_m\|_{B^2} = 2^{1-\mu}$.

(iii) $|c_m(P)| \le 2^{\mu-1} \|P\|_{B^2}$ is valid for all $P \in \mathbb{P}_\mu^2$ of the form
         $P(x) = c_m(P) x^m + TLD$.
         The bound is attained for $P = R_m$.

Proof. (i) follows immediately from (5.87).

(ii) In view of the definition of D_m we have

$$M_m - D_m = 2^{1-\mu}\hat{R}_m. \tag{5.88}$$

Because of Theorem 5.40, $M_m - D_m$ is best approximated in $\mathbb{P}^2_{\mu-1}$ by zero. Together with (i) it follows that D_m best approximates M_m in $\mathbb{P}^2_{\mu-1}$. This furnishes the first equality. The second equality follows from (5.88) because of $\|\hat{R}_m\|_{B^2} = 1$.

(iii) In the case $c_m(P) = 0$ the upper bound is evident. Next assume $c_m(P) \neq 0$. In view of (5.87) we get

$$2^{\mu-1}P(x) = c_m(P)\Big(\hat{R}_m(x) + TLD\Big),$$

and this implies

$$2^{\mu-1}|P\|_{B^2} = |c_m(P)| \cdot \|\hat{R}_m + TLD\|_{B^2} \geq |c_m(P)| \cdot \|\hat{R}_m\|_{B^2} = |c_m(P)|,$$

where we used again that $\hat{R}_m$ is best approximated in $\mathbb{P}^2_{\mu-1}(B^2)$ by *zero*. So the upper bound is valid in both cases. The last statement of (iii) follows immediately in view of (5.87), and the theorem is proved. $\square$

Corollary 5.42. *Let $m \in \mathbb{N}^2_0$. Then $|c_m(P)| \leq c_m(R_m)$ is valid for all $P \in \mathbb{P}^2_\mu$ which have the form $P(x) = c_m(P)x^m + TLD$ and satisfy $\|P\|_{B^2} \leq \|R_m\|_{B^2}$.*

Proof. For $m = 0$ the statement is evident. In the case $m \neq 0$ it suffices to prove that the statement is true if $\hat{R}_m$ takes the role of R_m. In this case the claim is that $|c_m(P)| \leq 2^{\mu-1}$ holds for all $P \in \mathbb{P}^2_\mu$ of the form $P(x) = c_m(P)x^m + TLD$ with $\|P\|_{B^2} \leq 1$. This is valid in view of Theorem 5.41, (iii). $\square$

Remark 1. It is worthwhile to compare Corollary 5.42 with Theorem 4.15 and with Theorem 4.16, which have the same structure, though space and norm are different in their case.

Remark 2. Theorem 5.41 says that D_m best approximates M_m in $\mathbb{P}^2_{\mu-1}(B^2)$ with respect to the maximum norm. In general, D_m is not uniquely determined by this property, see Gearhart [22].

Remark 3. In the case $r = 2$, $|m| = \mu \in \mathbb{N}$, the polynomials $A_m := \pi R_m$, which are also best approximated in $\mathbb{P}^r_{\mu-1}$ by zero, are generated by the *rational* function

$$|t|^\mu T_\mu\left(\tfrac{tx}{|t|}\right) = \sum_{|m|=\mu} A_m(x)\, t^m.$$

Moreover, their coefficients are integers, see (3.43). Of course, the question arises whether our results are transferrable to higher dimensions r by replacing T_μ by some properly chosen rational function Q_μ. Unfortunately, the answer is, in the general case, negative. For instance, in a cubic best approximation to the monomial $x_1^2 x_2 x_3$ on the unit ball in $\mathbb{R}^3$, an irrational coefficient occurs. See Problem 5.9.

Remark 4. Apart from a constant factor, R_m occurs also as the remainder in *Kergin interpolation* to the monomial M_m, see Bos [8].

(B) Best Approximation to Monomials on the Cube

Here we present an example where *zero* is a best approximation on the cube in the maximum norm. So let $r \in \mathbb{N}$,

$$D := C^r := [-1, +1]^r,$$

and the norm be given by $\| \cdot \| = \| \cdot \|_{C^r}$.

For $\nu \in \mathbb{N}$ we assume that

$$-1 = \xi_\nu^{(\nu)} < \xi_{\nu-1}^{(\nu)} < \cdots < \xi_0^{(\nu)} = +1$$

are the extreme points of the univariate Chebyshev polynomial T_ν in the interval $[-1, +1]$. We complete the definition by putting $\xi_0^{(0)} := 0$, such that

$$T_\nu(\xi_j^{(\nu)}) = (-1)^j$$

holds for arbitrary $\nu \in \mathbb{N}_0$ and $j = 0, 1, \ldots, \nu$. After that we introduce the node families

$$\Xi_\nu := \left\{ \xi_0^{(\nu)}, \xi_1^{(\nu)}, \ldots, \xi_\nu^{(\nu)} \right\},$$

and, for $m \in \mathbb{N}_0^r$, their Cartesian products

$$\Xi_m := \Xi_{m_1} \times \Xi_{m_2} \times \cdots \times \Xi_{m_r}.$$

The elements of Ξ_m are extreme points on C^r of the $r-$variate Chebyshev polynomial

$$T_m(x) := T_{m_1}(x_1) T_{m_2}(x_2) \cdots T_{m_r}(x_r),$$

whose maximum norm is given by

$$\|T_m\|_{C^r} = 1.$$

And with s denoting the number of vanishing components of m, T_m has the form

$$T_m(x) = 2^{|m|-r+s} x^m + TLD. \tag{5.89}$$

Now let us define the *multivariate divided difference* $[F; \Xi_m]$ for arbitrary polynomials $F \in \mathbb{P}^r$ as follows. For $j = 1, 2, \ldots, r$ we apply the univariate divided difference operator, which belongs to the nodes of Ξ_{m_j}, to F as a function of the $j - th$ argument. The result is independent of the order of these operations. So the definition is unambiguous, and $[F; \Xi_m]$ takes the form

$$[F; \Xi_m] = \sum_{\xi \in \Xi_m} \alpha(\xi) F(\xi),$$

where in particular

$$\alpha(\xi)T_m(\xi) > 0$$

holds for all $\xi \in \Xi_m$. This follows from the well-known representation formula for univariate divided differences. As a consequence we get the additional representation

$$[F; \Xi_m] = \sum_{\xi \in \Xi_m} |\alpha(\xi)| T_m(\xi) F(\xi). \tag{5.90}$$

Now let us introduce the polynomial spaces

$$\Pi_m^r := \{P \in \mathbb{P}_{|m|}^r \,|\, c_m(P) = 0\}$$

for $m \in \mathbb{N}_0^r$, and assume M_n is an element of Π_m^r. Because of $|n| \le |m|$ and $n \ne m$ there exists $j \in \{0, 1, \ldots, r\}$ such that $n_j < m_j$. This implies that the corresponding univariate divided difference of order m_j vanishes, and we get

$$[M_n; \Xi_m] = 0.$$

By a linear combination it follows that

$$[G; \Xi_m] = 0$$

holds for all $G \in \Pi_m^r$, and together with (5.90) we get

$$\sum_{\xi \in \Xi_m} |\alpha(\xi)| T_m(\xi) G(\xi) = 0$$

for arbitrary $G \in \Pi_m^r$. In view of $\Xi_m \subset \mathcal{E}(T_m)$ this says that the Criterion of Kolmogoroff is satisfied with $\mathbf{V} = \Pi_m^r$, $F = T_m$ and $\overset{*}{V} = 0$, see Theorem 5.39. Therefore, zero is a best approximation to T_m in Π_m^r for arbitrary $m \in \mathbb{N}_0^r$. This result can be used to prove the following theorem.

Theorem 5.43. *Let* $r \in \mathbb{N}$, $m \in \mathbb{N}_0^r$, $D_m := M_m - 2^{r-s-|m|}T_m$, *where* s *is the number of vanishing components of* m. *Then the following holds.*

(i) $D_m \in \Pi_m^r$.

(ii) $E\left(M_m, \Pi_m^r(C^r)\right) = \|M_m - D_m\|_{C^r} = 2^{r-s-|m|}$.

(iii) $|c_m(P)| \le 2^{|m|-r+s}\|P\|_{C^r}$ *for all* $P \in \mathbb{P}_{|m|}^r$,
where the bound is attained for $P = T_m$.

Proof. (i) follows immediately from (5.89). (ii) By the definition of D_m we have

$$M_m - D_m = 2^{r-s-|m|}T_m, \tag{5.91}$$

so $M_m - D_m$ is best approximated in $\Pi_m^r(C^r)$ by zero, since the right side is. Together with (i) it follows that D_m best approximates M_m in Π_m^r. This furnishes the first equality. Because of $\|T_m\|_{C^r} = 1$, the second equality follows from (5.91).

(iii) Let $P \in \mathbb{P}_{|m|}^r$. In the case $c_m(P) = 0$ the upper bound is evident. Next assume $c_m(P) \neq 0$. In view of (5.89) we get

$$2^{|m|-r+s} P(x) = c_m(P)\Big(T_m(x) + G(x)\Big)$$

with some $G \in \Pi_m^r$. It follows that

$$2^{|m|-r+s}\|P\|_{C^r} = |c_m(P)| \cdot \|T_m + G\|_{C^r} \geq |c_m(P)| \cdot \|T_m\|_{C^r} = |c_m(P)|,$$

where we used again that T_m is best approximated in Π_m^r by *zero*. So the upper bound is valid in both cases. The final statement of (iii) follows immediately from (5.89), and the theorem is proved.   □

Remark. Theorem 5.43 generalizes results of Sloss [68] and of Ehlich and Zeller [15]. Zero best approximates T_m even in the larger space $\tilde{\Pi}_m^r$ which is spanned by the monomials M_n where $n_\nu < m_\nu$ holds for some component, see [46]. For the use of more general product lattices in best approximation see Ehlich and Haußmann [16]. Note also that Theorem 5.41 and Theorem 5.43 agree in their structure. We could also add a corollary to Theorem 5.43 which is comparable with Corollary 5.42.

(C) Interpolation in $\mathbb{P}_\mu^r(\Sigma^{r-1})$ on the Equidistant Grid

Here we give an example where a fundamental system and the corresponding Lagrange elements are known in formula. The domain is now the simplex $\Sigma^{r-1} = \{x \in \mathbb{R}^r \mid x \geq 0, x_1 + \ldots + x_r = 1\}$.

Theorem 5.44 (Equidistant Grid on Σ^{r-1}). *Let $r \in \mathbb{N} \setminus \{1\}$. Then the following is valid.*

(i) *For $\mu \in \mathbb{N}_0$,* $\dim \mathbb{P}_\mu^r(\Sigma^{r-1}) = \dbinom{\mu + r - 1}{r - 1}$ *holds.*   (5.92)

(ii) *For $\mu \in \mathbb{N}$ the equidistant grid* $\Delta_\mu^{r-1} = \left\{\frac{n}{\mu} \mid n \in \mathbb{N}_0^r, |n| = \mu\right\}$

        is a fundamental system for $\mathbb{P}_\mu^r(\Sigma^{r-1})$.

(iii) *For $\mu \in \mathbb{N}$ the Lagrange elements in $\mathbb{P}_\mu^r(\Sigma^{r-1})$ which belong to Δ_μ^r*

        are represented by the polynomials

$$L_m(x) := \prod_{\nu=1}^{r} \prod_{\kappa=0}^{m_\nu - 1} \frac{\mu x_\nu - \kappa}{m_\nu - \kappa}, \qquad m \in \mathbb{N}_0^r, |m| = \mu,$$ (5.93)

        i.e., $L_m \in \mathbb{P}_\mu^r$ holds together with $L_m\left(\frac{n}{\mu}\right) = \delta_{m,n}$ for $m, n \in \mathbb{N}_0^r$,

        $|m| = \mu = |n|$.

Proof. Σ^{r-1} is a subset of the hyperplane $H = \{x \in \mathbb{R}^r \mid x_1 + \cdots + x_r = 1\}$ and contains an interior point relatively to H. By an adequate application of Theorem 3.14 this yields

$$\dim \mathbb{P}_\mu^r(\Sigma^{r-1}) = \dim \mathbb{P}_\mu^r(H) = \dim \mathbb{P}_\mu^r(\mathbb{R}^{r-1}),$$

where

$$\dim \mathbb{P}_\mu^r(\mathbb{R}^{r-1}) = \dim \mathbb{P}_\mu^{r-1} = \binom{\mu + r - 1}{r - 1},$$

see (3.8). So (i) is valid.

(ii) and (iii). It is obvious that the $L_m(x)$ are polynomials of degree $\mu = |m|$ which satisfy $L_m(\frac{n}{\mu}) = \delta_{m,n}$ for all $m, n \in \mathbb{N}_0^r$ with $|m| = \mu = |n|$, where the occuring arguments $\frac{n}{\mu}$ are exactly the nodes of the grid Δ_μ^{r-1}. It follows that the restrictions $L_m|_{\Sigma^{r-1}}$ are linearly independent elements of $\mathbb{P}_\mu^r(\Sigma^{r-1})$. Their number is given by

$$card\left(\Delta_\mu^{r-1}\right) = \binom{\mu + r - 1}{r - 1} = \dim \mathbb{P}_\mu^r(\Sigma^{r-1}),$$

where we used the result of (i). Because of $det\left(L_m(\frac{n}{\mu})\right) = 1$, the grid is a fundamental system, see Theorem 5.14. Hence (ii) is valid, and of course, the L_m are the corresponding Lagrange elements, which finishes the proof. $\qquad\square$

(D) Interpolation in the Space $\overset{*}{\mathbb{H}}{}_\mu^3(S^2)$

The unit sphere S^2 has the parameter representation

$$x(\phi, \psi) = (e_1 \cos \phi + e_2 \sin \phi) \sin \psi + e_3 \cos \psi, \tag{5.94}$$

where the parameter domain is $0 \le \phi < 2\pi$, $0 \le \psi \le \pi$ (polar coordinates). Now let $\mu \in \mathbb{N}$ be fixed. The reproducing kernel of $\overset{*}{\mathbb{H}}{}_\mu^3(S^2)$ has the form

$$G_\mu^3 = const \cdot P_\mu,$$

where $F_\mu = C_\mu^{\frac{1}{2}}$ is the Legendre polynomial of degree μ, see (4.13). From Corollary 3.11 we get

$$\dim \overset{*}{\mathbb{H}}{}_\mu^3(S^2) = 2\mu + 1,$$

which agrees with the dimension of the space of the trigonometrical polynomials of degree at most μ. So it is natural to try to interpolate at $2\mu + 1$ points which are equidistantly distributed on a parallel circle, say at the points

$$t_j := x(\phi_j, \psi), \quad \phi_j := \frac{2\pi j}{2\mu+1}, \tag{5.95}$$

$j = -\mu, \ldots, 0, \ldots, \mu$, where $\psi \in (0, \pi)$ is fixed. For $j, k \in \{-\mu, \ldots, 0, \ldots, \mu\}$ we get

$$t_j t_k = \cos^2 \psi + \sin^2 \psi \cdot \cos(\phi_j - \phi_k),$$

and the fundamental matrix takes the form

$$G = \Big[g(\phi_j - \phi_k, \psi) \Big]_{j,k \in \{0, \pm 1, \ldots, \pm \mu\}}, \tag{5.96}$$

where g is defined by

$$g(\phi, \psi) := G_\mu^3 \Big(\cos^2 \psi + \sin^2 \psi \cdot \cos \phi \Big). \tag{5.97}$$

G is circulant, so we can write it in the form

$$G = \sum_{\nu = -\mu}^{\mu} g(\phi_\nu, \psi) \cdot Z^\nu,$$

where

$$Z := \begin{pmatrix} 0 & 1 & & \\ & & \ddots & \\ & & & 1 \\ 1 & & & 0 \end{pmatrix}$$

is the $(2\mu + 1) \times (2\mu + 1)$ circulant basic matrix. Note that $Z^{2\mu+1} = I$ holds. It is our aim to calculate the determinant of G from the eigenvalues.

Obviously, $g(\phi, \psi)$ can be expanded with respect to the variable ϕ in the form

$$g(\phi, \psi) = \frac{a_0}{2} + \sum_{\nu=1}^{\mu} a_\nu \cos \nu \phi = \frac{1}{2} \sum_{\nu=-\mu}^{\mu} a_\nu \cos \nu \phi, \tag{5.98}$$

where the $a_\nu = a_\nu(\psi)$ depend on the latitude, and where we put $a_{-\nu} := a_\nu$ for $\nu = 1, \ldots, \mu$. Because of

$$spec(Z) = \Big\{ e^{\frac{2\pi \kappa i}{2\mu+1}} \mid \kappa = -\mu, \ldots, 0, \ldots, \mu \Big\},$$

the eigenvalues λ_κ of G are known and given by

$$\lambda_\kappa = \sum_{\nu=-\mu}^{\mu} g\Big(\phi_\nu, \psi \Big) \cdot e^{\frac{2\pi \nu \kappa i}{2\mu+1}}, \quad \kappa = -\mu, \ldots, 0, \ldots, \mu.$$

So we obtain, using (5.98) and the discrete orthogonality of the trigonometric polynomials,

$$spec(G) = \frac{2\mu+1}{2} \Big\{ a_{-\mu}, \ldots, a_0, \ldots, a_\mu \Big\},$$

and it follows that

$$\det G = \Big(\frac{2\mu+1}{2} \Big)^{2\mu+1} \cdot a_0 \cdot a_1^2 \cdots a_\mu^2. \tag{5.99}$$

Therefore, $T = \{t_{-\mu}, \ldots, t_{\mu}\}$ is a fundamental system if and only if none of the a_ν vanishes, which depends on the choice of ψ. So it is necessary to evaluate the $a_\nu = a_\nu(\psi)$ as a function of ψ, i.e., of the latitude of the parallel circle. This takes some steps.

As in Section 4.1, we use the polynomials

$$\begin{aligned} A_\nu(x_1, x_2) &= \Re(x_1 + ix_2)^\nu, \\ B_\nu(x_1, x_2) &= \Im(x_1 + ix_2)^\nu, \end{aligned}$$

again for $\nu \in \mathbb{N}_0$, and recall that

$$\{A_0\} \qquad \text{is a basis in } \overset{*}{\mathbb{H}}{}^2_0,$$

$$\{A_\nu, B_\nu\} \quad \text{is a basis in } \overset{*}{\mathbb{H}}{}^2_\nu \text{ for } \nu \in \mathbb{N}.$$

The bases can be used to produce a basis for $\overset{*}{\mathbb{H}}{}^3_\mu$, which consists of the polynomials

$$\begin{aligned} C_{\mu\nu}(x) &:= A_\nu(x_1, x_2) \cdot |x|^{\mu-\nu} P_\mu^{(\nu)}\left(\tfrac{x_3}{|x|}\right), & \nu &= 0, 1, \cdots, \mu, \\ S_{\mu\nu}(x) &:= B_\nu(x_1, x_2) \cdot |x|^{\mu-\nu} P_\mu^{(\nu)}\left(\tfrac{x_3}{|x|}\right), & \nu &= 1, \cdots, \mu, \end{aligned} \tag{5.100}$$

see Appendix (A), Theorem A.1. We restrict these polynomials onto the unit sphere by inserting $x = x(\phi, \psi)$, and we obtain, in polar coordinates, a basis for $\overset{*}{\mathbb{H}}{}^3_\mu(S^2)$, which consists of the functions

$$\begin{aligned} C_{\mu\nu}(x(\phi, \psi)) &= \cos\nu\phi \cdot P_\mu^\nu(\cos\psi), & \nu &= 0, 1, \cdots, \mu, \\ S_{\mu\nu}(x(\phi, \psi)) &= \sin\nu\phi \cdot P_\mu^\nu(\cos\psi), & \nu &= 1, \cdots, \mu, \end{aligned}$$

where the *Legendre functions* P_μ^ν are defined by

$$P_\mu^\nu(\xi) := (1 - \xi^2)^{\frac{\nu}{2}} \cdot P_\mu^{(\nu)}(\xi) \tag{5.101}$$

for $-1 \le \xi \le +1$ and $\nu = 0, 1, \ldots, \mu$.

With the help of the formula

$$\int_{S^2} F(x) d\omega(x) = \int_0^\pi \int_0^{2\pi} F(x(\phi, \psi)) \sin\psi \, d\phi d\psi,$$

it is now easy to prove that the polynomials (5.100) are pairwise orthogonal with respect to the inner product $\langle \cdot, \cdot \rangle$ induced by the surface integral on S^2. So it follows from (1.5) that the reproducing kernel of $\overset{*}{\mathbb{H}}{}^3_\mu(S^2)$ can be represented for $x, y \in S^2$ in the form

$$G^3_\mu(xy) = \frac{C_{\mu 0}(x)C_{\mu 0}(y)}{\|C_{\mu 0}\|_2^2} + \sum_{\nu=1}^{\mu} \left[\frac{C_{\mu\nu}(x)C_{\mu\nu}(y)}{\|C_{\mu\nu}\|_2^2} + \frac{S_{\mu\nu}(x)S_{\mu\nu}(y)}{\|S_{\mu\nu}\|_2^2} \right].$$

Besides we get

$$\|C_{\mu 0}\|_2^2 = 2A_{\mu 0}^{-1}, \quad \|C_{\mu\nu}\|_2^2 = \|S_{\mu\nu}\|_2^2 = A_{\mu\nu}^{-1} \text{ for } \nu = 1, \ldots, \mu,$$

with the positive numbers $A_{\mu\nu}$ defined by

$$A_{\mu\nu}^{-1} = \pi \int\limits_0^\pi (\sin\psi)^{2\nu+1} \left[P_\mu^{(\nu)}(\cos\psi) \right]^2 d\psi$$

for $\mu, \nu \in \mathbb{N}_0$. Now let us insert above two points from the same parallel, say the points $x = x(\phi, \psi)$ and $y = x(\bar\phi, \psi)$. In view of (5.97) this furnishes the identity

$$g(\phi - \bar\phi, \psi) = \tfrac{1}{2} A_{\mu 0} \left[P_\mu^{(0)}(\cos\psi) \right]^2 + \sum_{\nu=1}^{\mu} A_{\mu\nu} \left[(\sin\psi)^\nu P_\mu^{(\nu)}(\cos\psi) \right]^2 \cos\nu(\phi - \bar\phi).$$

Here we set $\bar\phi$ to *zero* and get by a comparison of the result with (5.98)

$$a_\nu(\psi) = A_{\mu\nu} \left[(\sin\psi)^\nu P_{\mu 0}^{(\nu)}(\cos\psi) \right]^2$$

for $\nu = 0, 1, \ldots, \mu$. So the eigenvalues are determined, and (5.99) yields

$$\det G = c_\mu (1 - \xi^2)^{\mu(\mu+1)} \left[P_\mu^{(0)}(\xi) \right]^2 \prod_{\nu=1}^{\mu} \left[P_\mu^{(\nu)}(\xi) \right]^4, \tag{5.102}$$

where $c_\mu = A_{\mu 0} \cdot \prod_{\nu=1}^{\mu} A_{\mu\nu}^2$ is a positive constant, on which we have no influence, but where $\xi = \cos\psi$ is determined by the latitude of the parallel circle in consideration.

Let us discuss our result for $\mu \in \mathbb{N}$. At the poles, i.e., for $\xi \in \{+1, -1\}$, the fundamental determinant vanishes at a very high order. This makes interpolation impossible, and useless even on neighbouring parallel circles. At the equator, i.e., for $\xi = 0$, it vanishes also, though at a lower order, caused by the fact that $P_\mu^{(\nu)}(0)$ vanishes for odd $\mu - \nu$. However the number of latitudes where the determinant vanishes is finite, so it is possible to interpolate in $2\mu + 1$ equidistributed nodes on almost every parallel circle, where the best choice is the parallel where the right side of (5.102) attains its maximum value. This parallel circle is located near the

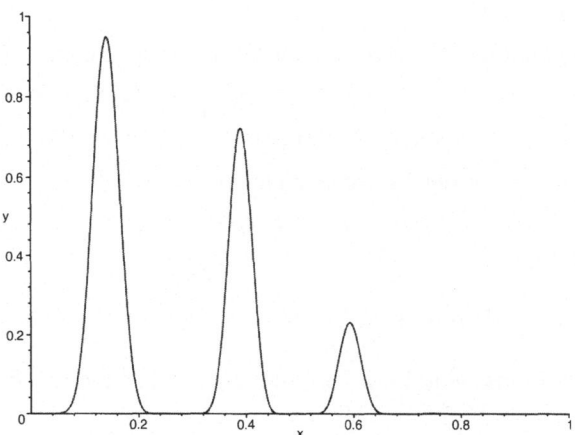

Figure 5.3.   det G in Dependence of $\xi = \cos\psi$.
$\xi = 0$: Equator, $\xi = 1$: Northpole.

equator, see Figure 5.3. Actually, Sündermann [72] used a parallel near the equator to obtain Lebesgue constants of the remarkably low order $\mathcal{O}(\mu^{\frac{3}{4}})$. This result is attainable also from Corollary 5.20 by maximizing the minimum eigenvalue as a function of the latitude, see Reimer and Sündermann [57]. The norms $\|C_{\mu\nu}\|_2^2$ are known in formula, see Problem 5.10. Note that an extremal fundamental system guarantees only the order $\mathcal{O}(\mu)$.

The Lagrange Elements

By the arguments from above, and again in view of the Appendix (A), Theorem A.1, the general element of $\overset{*}{\mathbb{H}}{}^3_\mu(S^2)$ can be written by means of polar coordinates in the form

$$F(x(\phi,\psi)) = \frac{a_0}{2}\cdot P^0_\mu(\cos\psi) + \sum_{\nu=1}^{\mu}\left[a_\nu\cos\nu\phi + b_\nu\sin\nu\phi\right]\cdot P^\nu_\mu(\cos\psi),$$

where it is convenient to use the Legendre functions (5.101), again. This enables us to prove the following theorem.

Theorem 5.45 (Interpolation in $\overset{*}{\mathbb{H}}{}^3_\mu(S^2)$ on a Parallel Circle). Let $\mu \in \mathbb{N}_0$, $\psi_0 \in (0,\pi)$, and define the node system $T = (t_0,\ldots,t_{2\mu})$ by

$$\phi_j := \tfrac{2\pi j}{2\mu+1}, \quad t_j := x(\phi_j,\psi_0),$$

for $j = 0, 1, \ldots, 2\mu$. Then the following is valid.

(i) T is a fundamental system for $\overset{*}{\mathbb{H}}{}^3_\mu(S^2)$ if and only if $P_\mu^{(\nu)}(\cos\psi_0) \neq 0$
      holds for $\nu = 0, 1, \ldots, \mu$.

(ii) If T is a fundamental system, then the Lagrange element L_0 is given by

$$L_0(x(\phi,\psi)) = \frac{1}{2\mu + 1} \cdot \left(\frac{P_\mu^0(\cos\psi)}{P_\mu^0(\cos\psi_0)} + 2 \cdot \sum_{\nu=1}^{\mu} \frac{P_\mu^\nu(\cos\psi)}{P_\mu^\nu(\cos\psi_0)} \cos\nu\phi \right).$$

(iii) L_0 and the remaining Lagrange elements L_j satisfy

$$L_j(x(\phi,\psi)) = L_0(x(\phi - \phi_j, \psi))$$

for $j = 0, 1, \ldots, 2\mu$.

Proof. (i) follows immediately from (5.102).

(ii) Assume T is a fundamental system, and define the function $F \in \overset{*}{\mathbb{H}}{}^3_\mu(S^2)$ by

$$F(x(\phi,\psi)) := \frac{1}{2\mu + 1} \cdot \left(\frac{P_\mu^0(\cos\psi)}{P_\mu^0(\cos\psi_0)} + 2 \cdot \sum_{\nu=1}^{\mu} \frac{P_\mu^\nu(\cos\psi)}{P_\mu^\nu(\cos\psi_0)} \cos\nu\phi \right).$$

Actually,

$$F(x(\phi,\psi_0)) = \frac{\sin(2\mu + 1)\frac{\phi}{2}}{(2\mu + 1)\sin\frac{\phi}{2}}$$

holds where, apart from a constant factor, the right side is the Dirichlet kernel. It
follows that

$$F(t_k) = F(x(\phi_k, \psi_0)) = \delta_{0,k}$$

for $k \in \{0, 1, \ldots, 2\mu\}$, and hence $F = L_0$. In particular, L_0 has the representation,
claimed.

(iii) In view of (ii) it is clear that $L_0(x(\phi - \phi_j))$ is an element of $\overset{*}{\mathbb{H}}{}^3_\mu(S^2)$, written
in polar coordinates. Besides, in view of the particular node distribution we get

$$L_0(x(\phi_k - \phi_j, \psi_0)) = \delta_{j,k},$$

now for arbitrary $j, k \in \{0, 1, \ldots, 2\mu\}$, which implies

$$L_0(x(\phi - \phi_j)) = L_j(x(\phi, \psi))$$

for $j = 0, 1, \ldots, 2\mu$, as claimed.   □

5.9   Problems

Problem 5.1. Let X, Y, Z be normed linear spaces, and let $A \in \mathcal{L}(X,Y)$, $B \in \mathcal{L}(Y,Z)$. Prove $\|B \circ A\| \leq \|B\| \cdot \|A\|$.

Problem 5.2. Let **V** be an inner product space of real functions on D, where $\dim \mathbf{V} = N \in \mathbb{N}$. Assume that $A_1, \ldots, A_N \in \mathbf{V}$ and $B_1, \ldots, B_N \in \mathbf{V}$ are biorthogonal, i.e., assume that $\langle A_j, B_k \rangle = \delta_{j,k}$ holds for $j, k \in \{1, \ldots, N\}$. Prove that the reproducing kernel of **V** has the representation

$$G(x, y) = \sum_{j=1}^{N} A_j(x) B_j(y) \ \text{ for } x, y \in D.$$

Problem 5.3. Let $t_1, \ldots, t_N$ form a fundamental system for the rotation-invariant subspace **V** of $\mathbb{P}^r(S^{r-1})$, $r \in \mathbb{N} \setminus \{1\}$, whose reproducing kernel is $K(xy)$. Denote the corresponding Lagrange elements by $L_1, \ldots, L_N$, and define the kernel functions K_j for $j = 1, \ldots, N$ by $K_j(x) := K(t_j x)$ for $x \in S^{r-1}$. Prove the formula

$$K(xy) = \sum_{j=1}^{N} L_j(x) K_j(y) \ \text{ for } x, y \in S^{r-1}.$$

Problem 5.4. Let $t_1, \ldots, t_N$ form a fundamental system for the subspace **V** of $C(D)$. For $F, G \in \mathbf{V}$ define

$$[F, G] := \sum_{j=1}^{N} F(t_j) G(t_j).$$

Then $[\cdot, \cdot]$ is an inner product on **V**, and the Lagrange elements $L_1, \ldots, L_N$ belonging to the fundamental system are orthonormal with respect to $[\cdot, \cdot]$.

Problem 5.5. Let $\mu \in \mathbb{N}_0$, $r \in \mathbb{N} \setminus \{1\}$, $N := \dim \overset{*}{\mathbb{P}}{}^r_\mu(S^{r-1})$. The matrix

$$\begin{pmatrix} (t_1 t_1)^\mu & , \ldots, & (t_1 t_\mu)^\mu \\ \vdots & & \vdots \\ (t_\mu t_1)^\mu & , \ldots, & (t_\mu t_\mu)^\mu \end{pmatrix},$$

$t_1, \ldots, t_N \in S^{r-1}$, is positive semidefinite. It is positive definite if and only if $t_1, \ldots, t_N$ form a fundamental system for $\overset{*}{\mathbb{P}}{}^r_\mu(S^{r-1})$.

Problem 5.6. Assume $L_1, \ldots, L_N$ form an extremal basis in the subspace X of $C(D)$, such that every L_j is best approximated in $X_j := \mathrm{span}\{L_k \mid k \neq j\}$ by zero, see proof of Theorem 5.21. Show that the Kolmogoroff criterion is satisfied. How many extreme points are necessary to decide this?

Problem 5.7. Consider the space $\overset{*}{\mathbb{P}}{}^r_\mu(B^r)$, and prove: The monomial $M_{\mu e_1}$ is best approximated in $\mathrm{span}\{M_m : |m| = \mu, \ m \neq \mu e_1\}$ by zero.

Problem 5.8. Assume $D \subset \mathbb{R}^r$ is symmetric with respect to the hyperplane $x_j = 0$, i.e.,

$$(\ldots, x_j, \ldots)' \in D \quad implies \quad (\ldots, -x_j, \ldots)' \in D.$$

And assume that the subspace $\mathbf{V}$ of $C(D)$ has the property that

$$\mathbf{V} \ni F(\ldots, x_j, \ldots) \quad implies \quad F(\ldots, -x_j, \ldots) \in \mathbf{V}.$$

Then the following holds. If $F \in C(D)$ is an even or odd function with respect to the variable x_j, then a best approximation to F in $\mathbf{V}$ exists which is in this variable also even or odd, respectively.

Problem 5.9. Use the result of Problem 5.8 in order to prove

$$E\big(M_{2,1,1}; \mathbb{P}_3^3(B^3)\big) \; = \; \big\|M_{2,1,1} - (3 - 2\sqrt{2})\, M_{0,1,1}\big\|_{B^3} \; = \; \tfrac{1}{2}(3 - 2\sqrt{2}).$$

Problem 5.10. Calculate the values of $\|C_{\mu\nu}\|_2^2$ and $\|S_{\mu\nu}\|_2^2$.

Hint: Apart from a constant factor, $P_\mu^{(\nu)}$ is the kernel function of a certain space of spherical harmonics.

Chapter 6

Approximation on the Sphere

This chapter is devoted to the particular, but important case where the domain is the unit sphere S^{r-1}, where $r \in \mathbb{N} \setminus \{1\}$. This is the simplest rotation-invariant domain, so the theory is basic, and taking a plain and valid form. Several results can be transferred to the ball, see Section 7, with an important application to tomography in Section 8.

6.1 Orthogonal Projections and Laplace Series

In this section the basic space X is the space $C(S^{r-1})$ or $L^2(S^{r-1})$, respectively. It is provided with the norm $\| \cdot \|_2$ which is induced by the inner product $\langle \cdot, \cdot \rangle = \langle \cdot, \cdot \rangle_{S^{r-1}}$ which belongs to the surface integral, see (1.6).

Our main concern is the orthogonal projections

$$\Omega_\nu : \quad X \to \overset{*}{\mathbb{H}}{}_\nu^r(S^{r-1})$$

and

$$\Pi_\mu : \quad X \to \mathbb{P}_\mu^r(S^{r-1}),$$

for $\nu, \mu \in \mathbb{N}_0$, respectively. They are related by

$$\Pi_\mu = \sum_{\nu=0}^{\mu} \Omega_\nu, \tag{6.1}$$

see (5.23) and Theorem 4.11. Note that in $\| \cdot \|_2$ the orthogonal projection is always a minimal projection, see Theorem 5.13. So the following theorem is fundamental.

Theorem 6.1 (Convergence of the Orthogonal Projections). *In $\| \cdot \|_2$, the orthogonal projections Π_μ, $\mu = 0, 1, \ldots$, are pointwise convergent to the identity on X.*

Proof. We want to apply Theorem 5.3, the Theorem of Banach and Steinhaus. So we check the assumptions.

(i) It follows from Theorem 5.12 that $\|\Pi_\mu\|_2 = 1$ holds for arbitrary $\mu \in \mathbb{N}_0$. So assumption (i) is satisfied.

(ii) Let $E := \mathbb{P}^r(S^{r-1}) = \bigcup_{\mu=0}^\infty \mathbb{P}_\mu^r(S^{r-1})$. By Theorem 5.8, the Theorem of Weierstrass, E is dense in $C(S^{r-1})$ with respect to $\| \cdot \|_\infty$, but in view of $\| \cdot \|_2 \leq \sqrt{\omega_{r-1}} \cdot \| \cdot \|_\infty$ also in $\| \cdot \|_2$. This implies that E is dense in X.

Now let P be an element of E, say $P \in \mathbb{P}_\kappa^r(S^{r-1})$ where $\kappa \in \mathbb{N}_0$. It follows that $P \in \mathbb{P}_\mu^r(S^{r-1})$ holds for all $\mu \geq \kappa$. But Π_μ is a projection, so we get $\Pi_\mu P = P$ for $\mu \geq \kappa$, and this implies

$$\lim_{\mu \to \infty} \Pi_\mu P = P$$

in $\| \cdot \|_2$. Since P was an arbitrary element of E, assumption (ii) is also valid, and the Π_μ are pointwise convergent to the identity in $\| \cdot \|_2$, as claimed. $\qquad\square$

In view of (6.1) the result of Theorem 6.1 can also be written in the following form,

$$\sum_{\nu=0}^\infty \Omega_\nu F = F \tag{6.2}$$

holds for all $F \in X$ in the sense of $\| \cdot \|_2$–convergence of the infinite series.

Definition 6.1 (Laplace Series). *The* Laplace series operator $L : X \to X$ *is defined by*

$$L := \sum_{\nu=0}^\infty \Omega_\nu$$

in the sense of pointwise convergence in $\| \cdot \|_2$. *For* $F \in X$

$$LF = \sum_{\nu=0}^\infty \Omega_\nu F$$

is called the Laplace series *of* F.

In view of this definition, $\Pi_\mu F$ is the μ-th partial sum of the Laplace series of F, and Theorem 6.1 takes the following form.

Corollary 6.2 (Convergence of a Laplace Series).

$$LF = F \quad \text{holds for all} \quad F \in X.$$

Corollary 6.2 says that

$$\int_{S^{r-1}} \left[F(x) - \sum_{\nu=0}^\mu (\Omega_\nu F)(x) \right]^2 d\omega(x)$$

converges to zero. We express this also by saying that the partial sums converge
in the quadratic mean, but have to add that convergence need not take place in
every single point $x \in S^{r-1}$. Nor does Corollary 6.2 give us an idea of the rate
of convergence, which actually depends on the *smoothness* of the function. We
explain this in what follows.

Let $F \in L^2(S^{r-1})$ and put $F_\nu := \Omega_\nu F$ for $\nu \in \mathbb{N}_0$. Then the Laplace series takes
the form

$$F = \sum_{\nu=0}^{\infty} F_\nu,$$

while $\Pi_\mu F$ is the μ-th partial sum. The F_ν are pairwise orthogonal, and Theorem
6.1 yields

$$\|F\|_2^2 = \sum_{\nu=0}^{\infty} \|F_\nu\|_2^2.$$

Now recall Theorem 4.13, by which the spaces $\overset{*}{\mathbb{H}}{}_\nu^r(S^{r-1})$ are eigenspaces of the
Laplace–Beltrami operator $\tilde{\Delta}$ with respect to the eigenvalue $-\nu(\nu + r - 2)$. So we
get, in particular,

$$\tilde{\Delta}^k F_\nu = \left[-\nu(\nu + r - 2) \right]^k F_\nu \tag{6.3}$$

for $\nu \in \mathbb{N}_0$ and $k \in \mathbb{N}_0$. But, provided $\tilde{\Delta}^k F$ exists at all, this need not say that

$$\tilde{\Delta}^k F = \sum_{\nu=0}^{\infty} \left[-\nu(\nu + r - 2) \right]^k F_\nu$$

is valid, i.e., $\tilde{\Delta}^k$ and the summation operator need not commute. This difficulty
can be mastered with the help of the following definition.

Definition 6.2 (The spaces $L^{k,2}(S^{r-1})$). *For $k \in \mathbb{N}_0$ the space $L^{k,2}(S^{r-1})$ consists
of all $F \in L^2(S^{r-1})$ with the following properties.*

$$(i) \qquad \tilde{\Delta}^k F \text{ exists almost everywhere },$$
$$(ii) \qquad \tilde{\Delta}^k F \in L^2(S^{r-1}),$$
$$(iii) \qquad \tilde{\Delta}^k F = \sum_{\nu=0}^{\infty} \tilde{\Delta}^k F_\nu.$$

Theorem 6.3 (Order of Convergence in $L^{k,2}(S^{r-1})$). *Let $r \in \mathbb{N} \setminus \{1\}$, $k \in \mathbb{N}_0$ and
$F \in L^{k,2}(S^{r-1})$. Then*

$$\|F - \Pi_\mu F\|_2 \leq (\mu + 1)^{-2k} \|\tilde{\Delta} F\|_2$$

holds for arbitrary $\mu \in \mathbb{N}_0$.

Proof. Let $F \in L^{k,2}(S^{r-1})$, where $k \in \mathbb{N}_0$. By the assumption,

$$\tilde{\Delta}^k F = \sum_{\nu=0}^{\infty} \tilde{\Delta}^k F_\nu$$

is the Laplace series of $\tilde{\Delta}^k F$, and in view of (6.3) we get

$$\sum_{\nu=0}^{\infty} [\nu(\nu+r-2)]^{2k} \|F_\nu\|_2^2 = \sum_{\nu=0}^{\infty} \|\tilde{\Delta}_\nu^k F_\nu\|_2^2 = \|\tilde{\Delta}^k F\|_2^2 < \infty.$$

For arbitrary $\mu \in \mathbb{N}_0$ this implies

$$\begin{aligned}
\|F - \Pi_\mu F\|_2^2 &= \sum_{\nu=\mu+1}^{\infty} \|F_\nu\|_2^2 \\
&\leq \sum_{\nu=\mu+1}^{\infty} \left[\frac{\nu(\nu+r-2)}{(\mu+1)(\mu+r-1)} \right]^{2k} \|F_\nu\|_2^2 \\
&\leq [(\mu+1)(\mu+r-1)]^{-2k} \|\tilde{\Delta}^k F\|_2^2,
\end{aligned}$$

and because of $r \geq 2$ the statement is true. □

6.2 Minimal Projection in the Uniform Norm

In this and in the following sections we consider the space $C(S^{r-1})$, only. In view of (5.23), the orthogonal projections Ω_ν and Π_μ can be represented with the help of the corresponding reproducing kernel, i.e., with the abbreviations $G_\nu = G_\nu^r$ and $\Gamma_\mu = \Gamma_\mu^r$ they can be represented in the form

$$(\Omega_\nu F)(x) = \int_{S^{r-1}} F(t) G_\nu(tx) d\omega(t), \qquad (6.4)$$

$$(\Pi_\mu F)(x) = \int_{S^{r-1}} F(t) \Gamma_\mu(tx) d\omega(t), \qquad (6.5)$$

for $F \in C(D)$, $x \in S^{r-1}$, and $\nu, \mu \in \mathbb{N}_0$, where the kernel functions are explicitly known, see (4.13) and (4.30) or(4.31), respectively.

Apart from the maximum norm $\| \cdot \|_\infty$, $C(S^{r-1})$ is provided with the norm $\| \cdot \|_2$ induced by the inner product $\langle \cdot, \cdot \rangle$. For linear operators $L : C(S^{r-1}) \to C(S^{r-1})$, which are bounded in the uniform norm, we use the operator norm

$$\|L\|_\infty := \|L\|_{\infty,\infty} = \max\left\{ \|LF\|_\infty : F \in C(S^{r-1}), \|F\|_\infty \leq 1 \right\}.$$

Now let us recall that, in the $\| \cdot \|_2$–setting, the orthogonal projection is always the uniquely determined minimal projection. In the uniform norm, this need not hold. But, surprisingly, if the image space is rotation-invariant, it does. This is the subject of what follows.

Minimal Projection in the Uniform Norm

We are going to prove the following theorem.

Theorem 6.4 (Berman, Daugavet). *Let $r \in \mathbb{N} \setminus \{1\}$, and assume $\mathbf{V}$ is a rotation-invariant subspace of $\mathbb{P}^r(S^{r-1})$ with finite dimension. Then the orthogonal projection with respect to the inner product $\langle \cdot, \cdot \rangle$,*

$$\Pi : C(S^{r-1}) \to \mathbf{V},$$

is a minimal projection in $\| \cdot \|_\infty$.

Proof. The proof takes several steps.

1) Because of Theorem 4.11 we can write $\mathbf{V}$ in the form

$$\mathbf{V} = \bigoplus_{\nu=0}^{\infty} \epsilon_\nu \overset{*}{\mathbb{H}}{}^r_\nu(S^{r-1}), \tag{6.6}$$

$\epsilon_\nu \in \{0,1\}$, where only finite many ϵ_ν are different from zero. Besides, the reproducing kernel of $\mathbf{V}$ is given by

$$G(xy) = \sum_{\nu=0}^{\infty} \epsilon_\nu G_\nu(xy). \tag{6.7}$$

For $F \in C(S^{r-1})$ and $x \in S^{r-1}$ the equation (5.23) takes the form

$$(\Pi F)(x) = \int\limits_{S^{r-1}} F(t)G(tx)d\omega(t),$$

and with the help of the inequality of Schwarz we obtain

$$\left| (\Pi F)(x) \right|^2 \leq \int_{S^{r-1}} F^2(t)d\omega(t) \cdot \int_{S^{r-1}} G^2(tx)d\omega(t).$$

This implies

$$\|\Pi F\|_\infty^2 \leq \|F\|_\infty^2 \cdot \omega_{r-1} \cdot G(1),$$

where we used finally the reproducing property of G. Therefore Π is a bounded linear operator which satisfies

$$\|\Pi\|_\infty^2 \leq G(1) \cdot \omega_{r-1},$$

and Π is a projection even in the $\| \cdot \|_\infty$–setting.

2) Now recall Definition 1.2, i.e., for $F \in C(S^{r-1})$ and rotations $A \in \mathbf{A}^r$ let $F_A \in C(S^{r-1})$ be defined by $F_A(\cdot) = F(A \cdot)$. By $\mathrm{E}_A F := F_A$ this defines a map

$$\mathrm{E}_A \; : \; C(S^{r-1}) \to C(S^{r-1}).$$

As the ranges of $\mathrm{E}_A F$ and of F are the same, E_A is an *isometry* and

$$\|\mathrm{E}_A\|_\infty = 1$$

holds.

3) Next assume that $\Omega \; : \; C(S^{r-1}) \to \mathbf{V}$ is any projection, bounded in the uniform norm. We compare its norm with the norm of the orthogonal projection, which takes again some steps.

For fixed $F \in C(S^{r-1})$ and $x \in S^{r-1}$ we define the continuous real function Φ on $\mathbf{A}^r$ by

$$\Phi(A) := \Big(\mathrm{E}_A \Omega \mathrm{E}_{A'} F\Big)(x).$$

The expression

$$(BF)(x) := \int_{\mathbf{A}^r} \Big(\mathrm{E}_A \Omega \mathrm{E}_{A'} F\Big)(x) dA \; / \int_{\mathbf{A}^r} dA \tag{6.8}$$

is well defined, if we agree, for convenience, that the integral is the right-invariant integral on the group $\mathbf{A}^r$, as explained in the Remark to Theorem 1.7. Actually, this defines a linear operator

$$\mathrm{B} \; : \; C(S^{r-1}) \to C(S^{r-1}),$$

which we write symbolically in the form

$$\mathrm{B} = \int_{\mathbf{A}^r} \mathrm{E}_A \Omega \mathrm{E}_{A'} dA \; / \int_{\mathbf{A}^r} dA.$$

Note that $F \in C(S^{r-1})$ implies

$$\mathrm{E}_A \Omega \mathrm{E}_{A'} F \in \mathbf{V},$$

where we use, one after another, the rotation invariance of $C(S^{r-1})$, the projection property of Ω, and the rotation invariance of $\mathbf{V}$. Moreover, as the integral is a bounded linear operator and as $\mathbf{V}$ is closed, we obtain $BF \in \mathbf{V}$, while $F \in \mathbf{V}$ implies even $BF = F$. Together this means that B is a *projection* $C(S^{r-1}) \to \mathbf{V}$, whose uniform norm can be estimated as follows.

For arbitrary $F \in C(S^{r-1})$ and $x \in S^{r-1}$ we get

$$
\begin{aligned}
|(BF)(x)| & \leq \int_{\mathbf{A}^r} |(\mathrm{E}_A \Omega \mathrm{E}_{A'} F)(x)| dA \; / \int_{\mathbf{A}^r} dA \\
& \leq \int_{\mathbf{A}^r} \|\mathrm{E}_A\|_\infty \|\Omega\|_\infty \|\mathrm{E}_{A'}\|_\infty \|F\|_\infty dA \; / \int_{\mathbf{A}^r} dA \\
& = \|\Omega\|_\infty \cdot \|F\|_\infty.
\end{aligned}
$$

It follows that

$$\|B\|_\infty \le \|\Omega\|_\infty$$

holds for arbitrary projections $\Omega : C(S^{r-1}) \to \mathbf{V}$. In other words, B is a minimal projection in $\| \cdot \|_\infty$.

4) We want to identify the minimal projection B with the orthogonal projection Π. Both are bounded linear operators, and hence continuous. So it suffices to compare their actions on a dense subset of $C(S^{r-1})$, say on the space E of all spherical polynomials, which is dense on $C(S^{r-1})$ in view of the Theorem of Weierstrass. Recalling (4.22), we write E in the form

$$E = \bigoplus_{\nu=0}^\infty \overset{*}{\mathbb{H}}{}^r_\nu(S^{r-1}).$$

Moreover, every $\overset{*}{\mathbb{H}}{}^r_\nu(S^{r-1})$ is spanned by the kernel functions $G_\nu(t \cdot)$, $t \in S^{r-1}$, see Theorem 5.15. Together this yields

$$E = span\Big\{G_\nu(t \cdot) \,\big|\, t \in S^{r-1},\ \nu \in \mathbb{N}_0\Big\}.$$

Therefore it suffices to identify the actions of B and of Π onto the kernel functions

$$G_\nu(t \cdot),\ t \in S^{r-1},\ \nu \in \mathbb{N}_0,$$

where, in view of (6.6), we have to distinguish the cases $\epsilon_\nu = 1$ and $\epsilon_\nu = 0$.

4a) First we assume $\epsilon_\nu = 1$. In this case we get $G_\nu(t \cdot) \in \mathbf{V}$, and since B and Π are projections onto $\mathbf{V}$, we obtain, for all $t \in S^{r-1}$,

$$BG_\nu(t \cdot) = G_\nu(t \cdot) = \Pi G_\nu(t \cdot).$$

4b) Next we assume $\epsilon = 0$. In this case we get $G_\nu(t \cdot) \in \mathbf{V}^\perp$, where $\mathbf{V}^\perp$ is the orthogonal complement of $\mathbf{V}$ in E. It follows immediately that $\Pi G_\nu(t \cdot) = 0$ holds, and it is left to prove that $BG_\nu(t \cdot)$ vanishes also.

To this end let $\big\{S_{\nu,1}, \ldots, S_{\nu,N}\big\}$ be an orthonormal basis in $\overset{*}{\mathbb{H}}{}^r_\nu(S^{r-1})$, such that

$$G_\nu(t \cdot) = \sum_{j=1}^N S_{\nu,j}(t)S_{\nu,j}(\cdot)$$

holds, see (1.5). Then we get, for $x \in S^{r-1}$,

$$BG_\nu(t \cdot)\big|_x = \sum_{j=1}^N S_{\nu,j}(t)\big(BS_{\nu,j}\big)(x). \tag{6.9}$$

This says that, for fixed x and as a function of t, $BG_\nu(t \cdot)|_x$ is an element of $\mathbf{V}^\perp$. We are going to prove that it belongs also to $\mathbf{V}$. This requires a detailed investigation of the action of B, where we write, temporarily, the euclidean inner product in the form $x'y$, instead of xy.

So let $A \in \mathbf{A}^r$ be an arbitrary rotation. For $x \in S^{r-1}$, we get, step by step,

$$E_{A'}G_\nu(t' \cdot)\big|_x \;=\; G_\nu(t'A'x) \;=\; \sum_{j=1}^N S_{\nu,j}(At)S_{\nu,j}(x),$$

$$\Omega E_{A'}G_\nu(t' \cdot)\big|_x \;=\; \sum_{j=1}^N S_{\nu,j}(At)\big(\Omega S_{\nu,j}\big)(x),$$

$$E_A\Omega E_{A'}G_\nu(t' \cdot)\big|_x \;=\; \sum_{j=1}^N S_{\nu,j}(At)\big(\Omega S_{\nu,j}\big)(Ax),$$

and hence

$$BG_\nu(t' \cdot)\big|_x \;=\; \sum_{j=1}^N \int_{\mathbf{A}^r} S_{\nu,j}(At)\big(\Omega S_{\nu,j}\big)(Ax)\, dA \Big/ \int_{\mathbf{A}^r} dA.$$

Now let $C \in \mathbf{A}^r$, and put $u := C't$. Then we get

$$\begin{aligned}
BG_\nu(t' \cdot)\big|_{Cx} &\;=\; \sum_{j=1}^N \int_{\mathbf{A}^r} S_{\nu,j}(ACu)\big(\Omega S_{\nu,j}\big)(ACx)\, dA \Big/ \int_{\mathbf{A}^r} dA \\
&\;=\; \sum_{j=1}^N \int_{\mathbf{A}^r} S_{\nu,j}(Au)\big(\Omega S_{\nu,j}\big)(Ax)\, dA \Big/ \int_{\mathbf{A}^r} dA \\
&\;=\; BG_\nu(u' \cdot)\big|_x,
\end{aligned} \tag{6.10}$$

where, on the right side, C is cancelled by the right-invariance of the integral. In particular, for $C \in \mathbf{A}_t^r$ we get $u = t$ and hence

$$BG_\nu(t' \cdot)\big|_{Cx} \;=\; BG_\nu(t' \cdot)\big|_x.$$

Now we arrive at a point where we have to distinguish the cases $r = 2$ and $r \geq 3$. First let $r \geq 3$. From Corollary 1.6 it follows that

$$BG_\nu(t' \cdot)\big|_x \;=\; g_t(t'x) \tag{6.11}$$

holds with some function $g_t \in C[-1,1]$, which possibly depends on t. However, let $u \in S^{r-1}$ be another point on the sphere. Then a rotation $C \in \mathbf{A}^r$ exists such that $u = C't$, and using (6.10), (6.11), and again (6.10), we get

$$\begin{aligned}
g_u(u'x) &\;=\; BG_\nu(u' \cdot)\big|_x \;=\; BG_\nu(t' \cdot)\big|_{Cx} \\
&\;=\; g_t(t'Cx) \;=\; g_t(u'x).
\end{aligned}$$

The values of $u'x$ cover the interval $[-1,1]$, while x varies in S^{r-1}, so it follows that

$$g_u = g_t = \cdots = g$$

is a fixed function of $C[-1,1]$, and (6.11) takes the form

$$BG_\nu(t' \cdot)\big|_x = g(t'x),$$

which indicates that

$$BG_\nu(t' \cdot)\big|_x = BG_\nu(x' \cdot)\big|_t$$

holds for arbitrary $x, t \in S^{r-1}$. Because of this symmetry we can conclude that, as a function of t, $BG_\nu(t' \cdot)\big|_x$ is an element of $\mathbf{V}$, whilst we know already from (6.9) that it belongs to $\mathbf{V}^\perp$. So it is contained in $\mathbf{V} \cap \mathbf{V}^\perp = [0]$. In other words, it vanishes, and we obtain

$$BG_\nu(t \cdot) = 0 = \Pi G_\nu(t \cdot)$$

for all $t \in S^{r-1}$.

Next let $r = 2$, and recall that we are still considering the case $\epsilon_\nu = 0$. To prove (6.11), Corollary 1.6 is no longer available. But in view of (4.22), E has now the simple representation

$$E = span\{A_0, A_1, B_1, \cdots\},$$

with the functions

$$\begin{aligned} A_\kappa(x_1, x_2) &= \Re(x_1 + ix_2)^\kappa, \\ B_\kappa(x_1, x_2) &= \Im(x_1 + ix_2)^\kappa, \end{aligned}$$

for $\kappa \in \mathbb{N}_0$, compare Section 4.1. We write the elements x, t of S^{r-1} in polar coordinates,

$$x = \begin{pmatrix} \cos\phi \\ \sin\phi \end{pmatrix}, \quad t = \begin{pmatrix} \cos\tau \\ \sin\tau \end{pmatrix},$$

and the rotations in the form

$$A = \begin{pmatrix} \cos\alpha, \sin\alpha \\ -\sin\alpha, \cos\alpha \end{pmatrix}.$$

Then, with $f(\alpha) := \Phi(A)$, the integral (1.12) takes the simple form

$$I^2\Phi = \int_0^{2\pi} f(\alpha)d\alpha,$$

and for $F \in S^1$ we get

$$(E_{A'}F)(x) = F\Big(A'x\Big) = F\Big(\cos(\phi + \alpha), \sin(\phi + \alpha)\Big).$$

In particular we obtain, for arbitrary $\kappa \in \mathbb{N}_0$,

$$
\begin{aligned}
(\mathrm{E}_{A'}A_\kappa)(x) &= \cos(\kappa(\phi + \alpha)) &= A_\kappa(x) \cdot \cos \kappa\alpha - B_\kappa(x) \cdot \sin \kappa\alpha, \\
(\mathrm{E}_{A'}B_\kappa)(x) &= \sin(\kappa(\phi + \alpha)) &= A_\kappa(x) \cdot \sin \kappa\alpha + B_\kappa(x) \cdot \cos \kappa\alpha.
\end{aligned}
\tag{6.12}
$$

Now we return to our original problem. $G_\nu(t \cdot)$ is a linear combination of A_ν and B_ν, so it suffices to prove $\mathrm{B}A_\nu = 0 = \mathrm{B}B_\nu$. The action of $\mathrm{E}_{A'}$, and hence of E_A to these functions is ruled by (6.12). For instance we get

$$
\Omega \mathrm{E}_{A'} A_\nu = \Omega A_\nu \cdot \cos \nu\alpha - \Omega B_\nu \cdot \sin \nu\alpha,
$$

and a similar equation holds for B_ν instead of A_ν. But ΩA_ν and ΩB_ν are linear combinations of the functions A_κ and B_κ which belong to $\mathbf{V}$, where $\epsilon_\kappa = 1$ and hence $\kappa \neq \nu$ holds. Using (6.12) once more, we find that $\mathrm{E}_A\Omega \mathrm{E}_{A'} A_\nu$ is a linear combination of these functions with coefficients which contain $\cos \kappa\alpha$ or $\sin \kappa\alpha$ as an additional factor to $\cos \nu\alpha$ or $\sin \nu\alpha$, respectively. By the orthogonality of the trigonometric functions this yields

$$
\begin{aligned}
\mathrm{B}A_\nu &= \tfrac{1}{2\pi} \int_0^{2\pi} \mathrm{E}_A\Omega \mathrm{E}_{A'} A_\nu \, d\alpha = 0, \\
\mathrm{B}B_\nu &= \tfrac{1}{2\pi} \int_0^{2\pi} \mathrm{E}_A\Omega \mathrm{E}_{A'} B_\nu \, d\alpha = 0,
\end{aligned}
$$

and therefore we get

$$
\mathrm{B}G_\nu(t \cdot) = 0 = \Pi G_\nu(t \cdot)
$$

for all $t \in S^{r-1}$, as in the case $r = 3$.

4c) Summarizing the results of 4a) and of 4b) we may state that

$$
\mathrm{B}G_\nu = \Pi G_\nu
$$

holds for all $\nu \in \mathbb{N}_0$ and $t \in S^{r-1}$. It follows that B and Π agree in their action on the whole space E, and hence even on $C(S^{r-1})$, such that $\Pi = \mathrm{B}$ holds. B is a minimal projection in $\| \cdot \|_\infty$, so the theorem is proved. $\square$

Remark. Theorem 6.4 is based on the identity

$$
\Pi = \int_{\mathbf{A}^r} \mathrm{E}_A\Omega \mathrm{E}_{A'} dA \Big/ \int_{\mathbf{A}^r} dA,
$$

which holds for arbitrary projections

$$
\Omega : C(S^{r-1}) \to \mathbf{V}.
$$

In this generality it seems to be due to Daugavet 1974, [12]. In the particular case $r = 2$ and $\mathbf{V} := \mathbb{P}^2_\mu(S^1)$, where, by introducing polar coordinates, $C(S^1)$ and $\mathbf{V}$ can be identified with $C_{2\pi}$, the space of 2π–periodic continuous real functions, and with the subspace of all trigonometric polynomials of degree at most μ, respectively, the identity takes the form

$$\Pi = \tfrac{1}{2\pi} \int_0^{2\pi} E_\alpha \Omega E_{-\alpha} d\alpha,$$

where E_α is the shift operator defined by $E_\alpha f(\cdot) = f(\cdot + \alpha)$. In this case, Πf is the μ–th trigonometric Fourier partial sum of f, the reproducing kernel function can be identified with the Dirichlet kernel, and the identity is due to Berman 1952, [7], and others. It is called the *Berman–Marcinkiewicz identity*.

The Minimal Uniform Projection Norm

Theorem 6.4 is of great importance, as it descibes one of the minimal projections onto $\mathbb{P}^r_\mu(S^{r-1})$ explicitly, namely to agree with the orthogonal projection Π_μ. Naturally, in view of the Theorem of Banach and Steinhaus, or of Theorem 5.11, respectively, our main interest is now directed to the growth order of the corresponding Lebesgue constants $\|\Pi_\mu\|_\infty$. They are the crucial point if we want to determine the convergence or the approximation order of any sequence of bounded linear operators. We determine the minimal Lebesgue constants in what follows.

Theorem 6.5 (Minimal Uniform Projection Norm). *Let* $r \in \mathbb{N} \setminus \{1\}$. *The minimal uniform norm of a projection* $C(S^{r-1}) \to \mathbb{P}^r_\mu(S^{r-1})$ *is given by*

$$\|\Pi_\mu\|_\infty = \omega_{r-2} \int_{-1}^{1} |\Gamma_\mu(\xi)| (1 - \xi^2)^{\frac{r-3}{2}} d\xi.$$

Proof. Let $F \in C(S^{r-1})$ satisfy $\|F\|_\infty \leq 1$. From (6.5) we get, for arbitrary $x \in S^{r-1}$

$$|(\Pi_\mu F)(x)| \leq \int_{S^{r-1}} |\Gamma_\mu(tx)| d\omega(t).$$

The integral on the right side does not depend on the choice of x. Moreover, by the help of (1.27) we get

$$|(\Pi_\mu F)(x))| \leq \omega_{r-2} \int_{-1}^{1} |\Gamma_\mu(\xi)| (1 - \xi^2)^{\frac{r-3}{2}} d\xi.$$

Obviously, this yields

$$\|\Pi_\mu\|_\infty \leq \omega_{r-2} \int_{-1}^{1} |\Gamma_\mu(\xi)| (1 - \xi^2)^{\frac{r-3}{2}} d\xi. \tag{6.13}$$

We want to show that equality holds. To this end we approximate the staircase function $sgn\,\Gamma_\mu$ on the interval $[-1, +1]$ by a continuous function $E \in C[-1, +1]$, as follows.

Γ_μ is, apart from a constant factor, the Jacobi polynomial $P_\mu^{\frac{r-1}{2},\frac{r-3}{2}}$, see (4.31). So Γ_μ has exactly μ zeros $\xi_1 < \cdots < \xi_\mu$ in the open interval $(-1,+1)$. Moreover, let $\xi_0 := -1$, $\xi_{\mu+1} := +1$, and let $a := \min\{\xi_{\nu+1} - \xi_\nu : \nu = 0, 1, \ldots, \mu\}$ be the minimum distance between successive nodes.

Now let $\epsilon > 0$ be a given number. For $\delta \in (0, \frac{a}{3}]$ we define the 1-spline function $E \in C[-1,1]$ by the following properties.

   (i)   The spline nodes are ξ_0, $\xi_1 - \delta$, $\xi_1 + \delta$, ..., $\xi_\mu - \delta$, $\xi_\mu + \delta$, $\xi_{\mu+1}$.

   (ii) $E(\xi) = sgn\,\Gamma_\mu(\xi)$ holds for $\xi \in [-1,1] \setminus \bigcup_{\nu=0}^{\mu} [\xi_\nu - \delta, \xi_\nu + \delta]$.

Moreover, we assume that δ is so small that

$$\omega_{r-2}\Gamma_\mu(1) \int_{-1}^{1} |E(\xi) - sgn\Gamma_\mu(\xi)|(1 - \xi^2)^{\frac{r-3}{2}} d\xi < \epsilon \tag{6.14}$$

holds. After that, let us define $F \in C(S^{r-1})$ by

$$F(x) := E(x_1) \quad for \quad x \in S^{r-1}.$$

Obviously, $\|F\|_\infty = 1$ is valid, such that the following holds,

$$\|\Pi_\mu\|_\infty \geq \|\Pi_\mu F\|_\infty \geq |(\Pi_\mu F)(e_1)|$$

$$= \left| \int_{S^{r-1}} F(t)\Gamma_\mu(te_1)\,d\omega(t) \right| = \left| \int_{S^{r-1}} E(t_1)\Gamma_\mu(t_1)\,d\omega(t) \right|$$

$$= \omega_{r-2} \left| \int_{-1}^{1} E(\xi)\Gamma_\mu(\xi)(1 - \xi^2)^{\frac{r-3}{2}} d\xi \right|$$

$$\geq \omega_{r-2} \int_{-1}^{1} \left(sgn\Gamma_\mu(\xi) + E(\xi) - sgn\Gamma_\mu(\xi) \right) \Gamma_\mu(\xi)(1 - \xi^2)^{\frac{r-3}{2}} d\xi$$

$$\geq \omega_{r-2} \int_{-1}^{1} |\Gamma_\mu(\xi)|(1 - \xi^2)^{\frac{r-3}{2}} d\xi$$

$$- \omega_{r-2}\,\Gamma_\mu(1) \int_{-1}^{1} |E(\xi) - sgn\Gamma_\mu(\xi)|(1 - \xi^2)^{\frac{r-3}{2}} d\xi,$$

where we used in the last step that

$$\max\{|\Gamma_\mu(\xi)| : -1 \leq \xi \leq 1\} = \Gamma_\mu(1) \tag{6.15}$$

holds, which follows from (4.30) and (2.16). Now we use (6.14) and get

$$\|\Pi_\mu\|_\infty \geq \omega_{r-2} \int_{-1}^{1} |\Gamma_\mu(\xi)|(1-\xi^2)^{\frac{r-3}{2}} d\xi - \epsilon.$$

This holds for arbitrary $\epsilon > 0$, so we obtain

$$\|\Pi_\mu\|_\infty \geq \omega_{r-2} \int_{-1}^{1} |\Gamma_\mu(\xi)|(1-\xi^2)^{\frac{r-3}{2}} d\xi.$$

In view of (6.13), equality must hold, as claimed.   □

The explicit expression for $\|\Pi_\mu\|_\infty$, given in Theorem 6.5, can be used to evaluate the exact growth order of the minimal projection constants. However, the calculation is lengthy, so we report on the results, only.

In the case $r = 2$, where Π_μ can be identified with the Dirichlet operator, it is well known that

$$\|\Pi_\mu\|_\infty \sim \tfrac{4}{\pi^2} \log \mu \tag{6.16}$$

holds for $\mu \to \infty$. See, e.g., Davis [13], p.358. In the case of an arbitrary $r \geq 3$, the corresponding asymptotics are given by

$$\|\Pi_\mu\|_\infty \sim \frac{2}{\pi^{\frac{3}{2}}} \frac{\Gamma(\frac{r-2}{4})\Gamma(\frac{r}{4})}{\Gamma(\frac{r-1}{2})} \cdot \mu^{\frac{r-2}{2}} \tag{6.17}$$

for $\mu \to \infty$, see Ragozin [43] for the order, and Daugavet [12] for the asymptotics. Note that the minimal projection constants are unbounded in both cases.

Theorem 6.6 (Divergence of Projections in the Uniform Norm). *For $r \in \mathbb{N} \setminus \{1\}$, no sequence of projections $L_\mu : C(S^{r-1}) \to \mathbb{P}_\mu^r(S^{r-1})$, $\mu \in \mathbb{N}_0$, converges in the uniform norm pointwise to the identity on $C(S^{r-1})$.*

Proof. Let any sequence of projections be given. Every L_μ is a bounded linear map $C(S^{r-1}) \to C(S^{r-1})$. In view of $\|L_\mu\|_\infty \geq \|\Pi_\mu\|_\infty$ and of (6.16) and (6.17), respectively, the assumption (i) of Theorem 5.3 is violated. So the statement is true.   □

Remark. Theorem 6.6 says, in other words, that to every sequence of projections L_μ a function $F \in C(S^{r-1})$ exists such that

$$\|F - L_\mu F\|_\infty$$

does not converge to zero. In this sense, Theorem 6.6 belongs to the class of so-called *negative results*. However, we recall Theorem 5.11, which promises convergence if the algebraic minimal deviation $E_\mu(F)$, see (5.43), converges to zero rapidly enough.

6.3   Interpolation on the Sphere

In this section we consider *interpolatory* projections $C(S^{r-1}) \rightarrow \mathbb{P}^r_\mu(S^{r-1})$, where $r \geq 2$. Of course, all that is said about interpolation in Section 5.4 in general or with respect to rotation-invariant subspaces of $C(S^{r-1})$ remains valid, but in the present case there are particulars arising from the algebraic stucture, which we investigate separately in what follows.

Fundamental Systems on Hypercircles

We construct fundamental systems $T^{r,\mu}$ for $\mathbb{P}^r_\mu(S^{r-1})$, $\mu \in \mathbb{N}$, whose nodes are distributed on $\mu + 1$ hypercircles. A hypercircle is the nonempty intersection of S^{r-1} and a hyperplane, and can degenerate to a single point, the pole. The method is recursive, this means for $r \geq 3$ it constructs a fundamental system $T^{r,\mu}$ from fundamental systems $T^{r-1,0}, T^{r-1,1}, \ldots, T^{r-1,\mu}$, which are already known. The method can be started with $r = 3$ because of the following theorem.

Theorem 6.7 ($T^{2,\mu}$–**Configurations**). *For $\mu \in \mathbb{N}_0$, every family of $2\mu + 1$ pairwise different points of S^1 is a fundamental system for $\mathbb{P}^2_\mu(S^1)$.*

Proof. By the substitution $x_1 = \cos\phi$, $x_2 = \sin\phi$, the space $\mathbb{P}^2_\mu(S^1)$ can be identified with the space of the trigonometric polynomials of degree μ, and it is well known that for this space every $2\mu + 1$ pairwise different points of the interval $[0, 2\pi)$ form a fundamental system. $\qquad \square$

For arbitrary $r \in \mathbb{N} \setminus \{1\}$ and $\mu \in \mathbb{N}_0$ let us define

$$N^r_\mu := \dim \mathbb{P}^r_\mu(S^{r-1}).$$

But in what follows, we assume $r \geq 3$. The construction method is based on the identity

$$N^r_\mu = \sum_{\nu=0}^{\mu} N^{r-1}_\nu, \qquad (6.18)$$

which follows from (4.4). For $\nu = 0, 1, \ldots, \mu$ it distributes N^{r-1}_ν of the nodes at a time on a hypercircle

$$H_\nu = \left\{ x \in S^{r-1} \mid e_r x = \xi_\nu \right\},$$

where the ξ_ν, $-1 < \xi_\nu \leq +1$, are pairwise different and arbitrary, except for the condition $\xi_0 := 1$. Note that $H_0 = \{e_r\}$ is the only hypercircle which degenerates to a single point, in consistency with the exceptional value $N^{r-1}_0 = 1$. We need the following lemma.

Lemma 6.8. *Let $r \geq 3$, $\nu \in \mathbb{N}_0$, and assume $T^{r-1,\nu} = \{\bar{t}_j \mid j = 1, \ldots, N^{r-1}_\nu\}$ is a fundamental system for the space $\mathbb{P}^{r-1}_\nu(S^{r-2})$. The points $t_j \in S^{r-1}$ are defined by*

$$t_j = \begin{pmatrix} \sqrt{1 - \xi_\nu^2} \cdot \bar{t}_j \\ \xi_\nu \end{pmatrix}$$

for $j = 1, \ldots, N_\nu^{r-1}$. Then the system of equations

$$F(t_j) = z_j, \quad j = 1, \ldots, N_\nu^{r-1}$$

has a solution $F \in \mathbb{P}_\nu^r$ for arbitrary real numbers z_j on the right side.

Proof. For every fixed right side the system

$$\bar{F}(\bar{t}_j) = z_j, \quad j = 1, \ldots, N_\nu^{r-1},$$

has a solution $\bar{F} \in \mathbb{P}_\nu^{r-1}(S^{r-2})$, and hence a solution $\bar{F}$ in $\mathbb{P}_\nu^{r-1}$ itself. This is a polynomial in the variables $x_1, \ldots, x_{r-1}$, which we arrange in $\bar{x} = (x_1, \ldots, x_{r-1})' \in \mathbb{R}^{r-1}$. Now we define $F \in \mathbb{P}_\nu^r$ by

$$F(x) := \bar{F}\left(\frac{1}{\sqrt{1-\xi_\nu^2}} \cdot \bar{x}\right),$$

if $\nu > 0$, and by the constant z_1, if $\nu = 0$ and hence $N_\nu^{r-1} = 1$. Obviously, in both cases F is a polynomial as wanted. $\qquad\square$

In what follows we assume that the fundamental systems $T^{r-1,\nu}$, for $\mathbb{P}_\nu^{r-1}(S^{r-2})$, $\nu = 0, 1, \ldots, \mu$, are given in the form

$$T^{r-1,\nu} = \left\{\bar{t}_j^{r-1,\nu} \mid j = 1, \ldots, N_\nu^{r-1}\right\}$$

with $\bar{t}_j^{r-1,\nu} \in S^{r-2}$. We define the points $t_j^{r,\nu} \in H_\nu$ for $j = 0, 1, \ldots, N_\nu^{r-1}$ and $\nu = 0, \ldots, \mu$ by

$$t_j^{r,\nu} := \begin{pmatrix} \sqrt{1-\xi_\nu^2} \cdot \bar{t}_j^{r-1,\nu} \\ \xi_\nu \end{pmatrix}. \tag{6.19}$$

Then the following holds.

Theorem 6.9 (Recursive Construction of Fundamental Systems, Sündermann).
Let $r \geq 3$ and assume that the nodes occurring in

$$T^{r,\mu} = \left\{t_j^{r,\nu} \mid j = 1, \ldots, N_\nu^{r-1}, \; \nu = 0, 1, \ldots, \mu\right\}$$

are constructed from the fundamental systems $T^{r-1,\nu}$ with the help of (6.19). Then $T^{r,\mu}$ is a fundamental system for $\mathbb{P}_\mu^r(S^{r-1})$.

Proof. In view of (6.18) the number of elements contained in $T^{r,\mu}$ is equal to $\dim \mathbb{P}_\mu^r(S^{r-1})$, and in view of Definition 5.8, $T^{r,\mu}$ is a fundamental system if the corresponding evaluation functionals are linearly independent. An equivalent condition is that the interpolation problem

$$F(t_j^{r,\nu}) = y_j^\nu, \quad j = 1, \ldots, N_\nu^{r-1}, \; \nu = 0, 1, \ldots, \mu, \tag{6.20}$$

has a solution $F \in \mathbb{P}_\mu^r(S^{r-1})$ for arbitrary real values y_j^ν on the right side. Actually, this can be proved as follows.

In view of Lemma 6.8 we can solve, in the first step, the problem

$$Q_\mu(t_j^{r,\mu}) = y_j^\mu, \quad j = 1, \ldots, N_\mu^{r-1},$$

by some polynomial $Q_\mu \in \mathbb{P}_\mu^r$. After that we put

$$P_\mu := Q_\mu.$$

$F = P_\mu$ satisfies the equations (6.20) for $\nu = \mu$, but not necessarily the remaining ones. Therefore we solve, in the second step, the problem

$$Q_{\mu-1}(t_j^{r,\mu-1}) = y_j^{\mu-1} - P_\mu(t_j^{r,\mu-1}), \quad j = 1, \ldots, N_{\mu-1}^{r-1},$$

by some polynomial $Q_{\mu-1} \in \mathbb{P}_{\mu-1}^r$, which is possible again because of Lemma 6.8. After that we put

$$P_{\mu-1} := P_\mu + \frac{x_r - \xi_\mu}{\xi_{\mu-1} - \xi_\mu} \cdot Q_{\mu-1}.$$

It is obvious that $P_{\mu-1} \in \mathbb{P}_\mu^r$ holds again, but in addition, $F = P_{\mu-1}$ satisfies the equations (6.20) both for $\nu = \mu$ and for $\nu = \mu - 1$. We continue this procedure by solving, in the general case, i.e., for $\kappa = \mu - 1, \ldots, 0$, the problem

$$Q_\kappa(t_j^{r,\kappa}) = y_j^\kappa - P_{\kappa+1}(t_j^{r,\kappa}), \quad j = 1, \ldots, N_\kappa^{r-1},$$

by some $Q_\kappa \in \mathbb{P}_\kappa^r$, and by putting

$$P_\kappa := P_{\kappa+1} + \prod_{\lambda=\kappa+1}^{\mu} \frac{x_r - \xi_\lambda}{\xi_\kappa - \xi_\lambda} \cdot Q_\kappa.$$

The construction is such that $F = P_\kappa \in \mathbb{P}_\mu^r$ solves the equations (6.20) for $\nu \in \{\mu, \mu - 1, \ldots, \kappa\}$. In particular, $F = P_0$ is an element of $\mathbb{P}_\mu^r$ which solves all of the equations (6.20). Of course, the restriction of F onto S^{r-1} solves the same problem. Therefore (6.20) has a solution in $\mathbb{P}_\mu^r(S^{r-1})$ for an arbitrary right side, and $T^{r,\mu}$ is a fundamental system for this space. $\qquad \square$

The basic idea of the proof of Theorem 6.9 is due to Sündermann, [72], who used it in the construction of particular fundamental systems $T^{3,\mu}$. His fundamental systems are based on fundamental systems $T^{2,\nu}$ whose $N_\nu^2 = 2\nu + 1$ nodes are distributed equidistantly on S^1, while the choice of the $\xi_\nu = \xi_\nu^\mu$, as a function of μ, is more sophisticated, as these points are given by

$$\xi_\nu^\mu := (-1)^\nu \cos \frac{\nu\pi}{2\mu+1}, \quad \nu = 0, 1, \ldots, \mu. \tag{6.21}$$

In some sense, this choice provides the sphere S^2 with a best possibly balanced node distribution. It is promissory also in the general case.

Node Distance and Covering Radius

We can judge the geometric quality of a node system $T = \{t_1, \ldots, t_N\}$ with the help of two measures, which are introduced in what follows.

First let us introduce the trigonometric, or geodetic distance $\arccos(xy)$ as the natural measure for the distance of two points $x, y \in S^{r-1}$. Because of $|x - y|^2 = 2(1 - xy)$ it is, in general, different from the euclidean distance in $\mathbb{R}^r$, but asymptotically the distances are equal if y tends to x.

Definition 6.3 (Node Distance, Covering Radius). *Let* $T = \{t_1, \ldots, t_N\}$ *be an arbitrary system of nodes* $t_j \in S^{r-1}$, $r \geq 2$. *The node distance* $\delta(T)$ *and the covering radius* $\rho(T)$ *of* T *are introduced by the definitions*

$$\delta(T) := \min_{j \in \{1, \ldots, N\}} \min_{j \neq k \in \{1, \ldots, N\}} \arccos(t_j t_k), \qquad (6.22)$$

$$\rho(T) := \max_{x \in S^{r-1}} \min_{j \in \{1, \ldots, N\}} \arccos(xt_j). \qquad (6.23)$$

Obviously, if T is intended to be a fundamental system of high quality, it is not desirable that nodes appear in clusters, and the fundamental determinant even vanishes if two nodes are equal. So $\delta(T)$ is a rough quality measure, which should attain a value as large as possible.

Besides, every point $x \in S^{r-1}$ should be close by some node. This is measured by the value of $\rho(T)$, which should be as small as possible. It is called the covering radius of T for the following reason.

A subset of S^{r-1},

$$C(t, \alpha) := \{x \in S^{r-1} \mid xt \geq \cos\alpha\}, \quad t \in S^{r-1}, \ \alpha \geq 0,$$

is called a *(spherical) cap* with *center* t and *radius* α. It degenerates to the center for $\alpha = 0$. Actually, $\rho(T)$ is the minimal value of α satisfying

$$S^{r-1} \subset \bigcup_{j=1}^{N} C(t_j, \alpha). \qquad (6.24)$$

It it easy to see that a fundamental system $T^{2,\mu}$, consisting of $2\mu + 1$ equidistantly distributed nodes on S^1, satisfies

$$\delta(T^{2,\mu}) = \frac{2\pi}{2\mu + 1}, \quad \rho(T^{2,\mu}) = \frac{\pi}{2\mu + 1}. \qquad (6.25)$$

Moreover, the following holds.

Theorem 6.10 (Covering Radius of the Constructed Fundamental Systems). *Assumptions as in Theorem 6.9. Moreover assume that a constant* c_{r-1} *exists such that*

$$\rho(T^{r-1,\nu}) \leq \frac{c_{r-1}}{\nu + 1}$$

holds for all $\nu \in \mathbb{N}_0$. For $\mu \in \mathbb{N}_0$ let $\xi_\nu = \xi_\nu^\mu$ be defined for $\nu \in \{0, \ldots, \mu\}$ by (6.21). Then a constant c_r exists such that

$$\rho(T^{r,\mu}) \leq \frac{c_r}{\mu + 1}$$

is valid for all $\mu \in \mathbb{N}_0$.

Before we begin the proof, we introduce the following lemma.

Lemma 6.11. *Let a and b be positive numbers, and define*

$$A_{\mu,\nu} := 1 - 2 \sin^2 \frac{a\nu}{2\mu + 1} \cdot \sin^2 \frac{b}{\nu + 1}$$

for $\mu \in \mathbb{N}_0$ and $\nu = 0, 1, \ldots, \mu$. Then a constant $c > 0$ exists such that

$$\arccos A_{\mu,\nu} \leq \frac{c}{\mu + 1}$$

holds for $\mu \in \mathbb{N}_0$ and $\nu \in \{0, 1, \ldots, \mu\}$.

Proof. Let $\nu \in \{0, 1, \ldots, \mu\}$. Because of $-1 \leq A_{\mu,\nu} \leq +1$, a number $\gamma_{\mu,\nu}$ exists such that

$$A_{\mu,\nu} = \cos \frac{\gamma_{\mu,\nu}}{\mu + 1}, \quad 0 \leq \frac{\gamma_{\mu,\nu}}{\mu + 1} \leq \pi,$$

holds. It follows that

$$\sin^2 \frac{\gamma_{\mu,\nu}}{2(\mu + 1)} = \frac{1}{2}(1 - A_{\mu,\nu}) = \sin^2 \frac{a\nu}{2\mu + 1} \sin^2 \frac{b}{\nu + 1}.$$

This implies

$$\frac{2}{\pi} \cdot \frac{\gamma_{\mu,\nu}}{2(\mu + 1)} \leq \frac{a\nu}{2\mu + 1} \cdot \frac{b}{\nu + 1} < \frac{ab}{2(\mu + 1)},$$

and hence , with $c := \frac{ab\pi}{2}$,

$$\arccos A_{\mu,\nu} \leq \frac{c}{\mu + 1},$$

as claimed.   □

Proof of Theorem 6.10. Let $x \in S^{r-1}$ be given. We write it in the form

$$x = \begin{pmatrix} \sqrt{1 - \xi^2} \cdot \bar{t} \\ \xi \end{pmatrix},$$

where $\bar{t} \in S^{r-2}$, $\xi = \cos\phi$, $0 \leq \phi \leq \pi$. The $\xi_\nu = \xi_\nu^\mu$, defined by (6.21), are the abscissae of an equidistant grid on S^1 with trigonometric mesh size $\frac{2\pi}{2\mu+1}$. To be more precise,

$$\{\xi_0, \xi_1, \ldots, \xi_\mu\} = \left\{ \cos \frac{2k\pi}{2\mu + 1} : k = 0, 1, \ldots, \mu \right\}$$

is valid. It follows that either

$$0 \leq \phi \leq \frac{\pi}{2\mu + 1}$$

holds, or an index $\nu \in \{1, \ldots, \mu\}$ exists such that

$$-\frac{\pi}{2\mu + 1} \leq \phi - \arccos \xi_\nu \leq \frac{\pi}{2\mu + 1}$$

is valid. In the first case we get, in view of $t_0^{r,0} = e_r$, which follows from (6.19) and (6.21),

$$\arccos x t_0^{r,0} \leq \frac{\pi}{2\mu + 1}. \tag{6.26}$$

In the second case, where $\nu \in \{1, \ldots, \mu\}$, we get

$$\cos(\phi - \arccos \xi_\nu) \geq \cos \frac{\pi}{2\mu + 1},$$

and hence

$$\xi \cdot \xi_\nu + \sqrt{1 - \xi^2} \cdot \sqrt{1 - \xi_\nu^2} \geq \cos \frac{\pi}{2\mu + 1}.$$

Now let us introduce the intermediate point

$$y := \begin{pmatrix} \sqrt{1 - \xi_\nu^2} \cdot \bar{t} \\ \xi_\nu \end{pmatrix}.$$

It satisfies

$$xy = \xi \cdot \xi_\nu + \sqrt{1 - \xi^2} \cdot \sqrt{1 - \xi_\nu^2} \geq \cos \frac{\pi}{2\mu + 1},$$

and hence

$$\arccos xy \leq \frac{\pi}{2\mu + 1}.$$

Moreover by the assumption on $T^{r-1,\nu}$, there is a node $\bar{t}_k^{r-1,\nu} \in T^{r-1,\nu}$ such that

$$\arccos(\bar{t}\,\bar{t}_k^{r-1,\nu}) \leq \rho(T^{r-1,\nu}) \leq \frac{c_{r-1}}{\nu + 1}$$

holds. In view of (6.19) this yields

$$
\begin{aligned}
y t_k^{r,\nu} &= \xi_\nu^2 + (1 - \xi_\nu^2) \cdot \bar{t}\,\bar{t}_k^{r-1,\nu} \\
&\geq \xi_\nu^2 + (1 - \xi_\nu^2) \cdot \begin{cases} \cos \frac{c_{r-1}}{\nu+1} &, \text{if } \nu + 1 \geq \frac{c_{r-1}}{\pi}, \\ -1 &, \text{if } 2 \leq \nu + 1 < \frac{c_{r-1}}{\pi}, \end{cases}
\end{aligned}
$$

and hence

$$
y t_k^{r,\nu} \geq \begin{cases} 1 - 2 \sin^2 \frac{\nu\pi}{2\mu+1} \cdot \sin^2 \frac{c_{r-1}}{2(\nu+1)} &, \text{if } \nu + 1 \geq \frac{c_{r-1}}{\pi}, \\ \cos \frac{2\nu\pi}{2\mu+1} &, \text{if } 2 \leq \nu + 1 < \frac{c_{r-1}}{\pi}. \end{cases}
$$

Now it follows from Lemma 6.11 that a constant $c > 0$ exists such that

$$\arccos yt_k^{r,\nu} \leq \begin{cases} \frac{c}{\mu+1} & , \text{ if } \nu + 1 \geq \frac{c_{r-1}}{\pi}, \\ \frac{2\nu\pi}{2\mu+1} & , \text{ if } 2 \leq \nu + 1 < \frac{c_{r-1}}{\pi}. \end{cases}$$

Therefore, a constant c' exists such that

$$\arccos yt_k^{r,\nu} \leq \frac{c'}{\mu+1} \tag{6.27}$$

holds, independently of the location of ν in the set $\{1, \ldots, \mu\}$.

Now we use the triangular inequality of the geodetic distance, and get the estimate

$$\begin{aligned} \arccos xt_k^{r,\nu} &\leq \arccos xy + \arccos yt_k^{r,\nu} \\ &\leq \frac{\pi}{2\mu+1} + \frac{c'}{\mu+1}. \end{aligned}$$

This inequality remains valid for $\nu = k = 0$, see (6.26).

So let us define the constant $c_r := \frac{\pi}{2} + c'$. Then we may summarize our results as follows. To every $x \in S^{r-1}$ and $\mu \in \mathbb{N}_0$ a node $t_k^{r,\nu} \in T^{r,\mu}$ exists such that

$$\arccos xt_k^{r,\nu} \leq \frac{c_r}{\mu+1}$$

holds. Since the bound does not depend on x, we obtain

$$\rho(T^{r,\mu}) \leq \frac{c_r}{\mu+1}.$$

This finishes the proof.   □

Remark. In view of (6.25) we can satisfy the assumption of Theorem 6.10 for $r = 3$ by defining the $T^{2,\nu}$ by the help of $2\nu + 1$ equidistributed nodes on S^1. Then we get, recursively for $r = 2, 3, \ldots$, positive constants $c_2, c_3, \ldots$, and sequences of fundamental systems $T^{r,\mu}$, $\mu \in \mathbb{N}_0$, whose covering radii satisfy

$$\rho(T^{r,\mu}) \leq \frac{c_r}{\mu+1}.$$

Extremal Fundamental Systems

If the Exchange Algorithm of Section 5.5 is to be started, a fundamental system $T^{r,\mu}$, constructed recursively along Theorem 6.9, may be expected to be a better choice for $T^{(0)}$, than a random one. If, finally, the extremal fundamental system $T = \{t_1, \ldots, t_N\}$ is calculated, then the nodes must have a certain minimal distance. We prove this with the help of the following lemma.

Lemma 6.12 (Riesz). *Let g denote a real trigonometric polynomial of degree $\mu \in \mathbb{N}$, and assume that*

$$1 = g(0) = \|g\|_\infty = \max\{|g(\phi)| : \phi \in \mathbb{R}\}$$

is valid. Then $g(\phi) > 0$ holds for $-\frac{\pi}{2\mu} < \phi < \frac{\pi}{2\mu}$.

Proof. Assume the statement is false. Then g has a zero τ satisfying $0 < |\tau| < \frac{\pi}{2\mu}$. Without loss of generality we may assume $\tau > 0$, since $g(-\phi)$ satisfies the same assumptions, as $g(\phi)$ does. Now let us define

$$f(\phi) := \cos \mu\phi - g(\phi),$$

which is a trigonometric polynomial of degree μ satisfying $f(0) = 0$. Note that $f(\tau) > 0$ holds, while $f(\frac{j\pi}{\mu})$ has the sign of $(-1)^j$ for $j = 1, 2, \ldots, 2\mu$. So f has 2μ roots in the interval $0 < \phi < 2\pi$, if multiplicities are counted, and one additional at $\phi = 0$. Together this implies $f = 0$, $g(\phi) = \cos \mu\phi$ and hence $g(\tau) > 0$, which is a contradiction. So the statement is true. □

Theorem 6.13 (Node Distance in an Extremal Fundamental System). *Let $T = \{t_1, \ldots, t_N\}$ be an extremal fundamental system for a subspace $\mathbf{V}$ of $\mathbb{P}_\mu^r(S^{r-1})$, where $r \in \mathbb{N} \setminus \{1\}$, $\mu \in \mathbb{N}$, and $N > 1$. Then the node distance of T satisfies*

$$\delta(T) \geq \frac{\pi}{2\mu}.$$

Proof. Let $t_j, t_k \in T$ be different nodes, i.e., assume $j, k \in \{1, \ldots, N\}$, $j \neq k$. Then some $u_j \in S^{r-1} \cap span\{t_j, t_k\}$ exists such that $ut_j = 0$ is valid. Now let $L_j \in \mathbf{V}$ be the Lagrange element belonging to t_j, note that $\|L_j\|_\infty = 1$ holds, and define

$$g(\phi) := L_j(t_j \cos \phi + u_j \sin \phi).$$

This is a trigonometrical polynomial of degree μ satisfying $1 = g(0) = \|g\|_\infty$, and obviously, there exists some $\phi \in [-\pi, +\pi)$ such that $t_k = t_j \cos \phi + u_j \sin \phi$, which implies $g(\phi) = 0$ and

$$t_j t_k = \cos \phi.$$

From Lemma 6.12 it follows that ϕ is not contained in the interval $\left(-\frac{\pi}{2\mu}, +\frac{\pi}{2\mu}\right)$. This implies $t_j t_k \leq \cos \frac{\pi}{2\mu}$, and hence $\arccos t_j t_k \geq \frac{\pi}{2\mu}$. Since this is valid for every pair of nodes $t_j \neq t_k$, we obtain $\delta(T) \geq \frac{\pi}{2\mu}$, as claimed. □

In the case $r = 2$ we get complete information about the extremal fundamental systems for $\mathbb{P}^2(S^1)$ by the following theorem.

Theorem 6.14 (The Extremal Fundamental Systems $T^{2,\mu}$). *$T = \{t_0, t_1, \ldots, t_{2\mu}\}$ is an extremal fundamental system for $\mathbb{P}_\mu^2(S^1)$ if and only if the nodes are equidistantly distributed on S^1.*

Proof. Again we identify the space $\mathbb{P}^2_\mu(S^1)$ and the space of trigonometric polynomials of degree μ. It suffices to prove that a system of nodes $0 = \phi_0 < \phi_1 < \cdots < \phi_{2\mu} < 2\pi$ is an extremal fundamental system for this space if and only if $\phi_j = \frac{2\pi j}{2\mu+1}$ holds for $j = 0, 1, \ldots, 2\mu$.

First let us define the nodes by $\phi_j := \frac{2\pi j}{2\mu+1}$ for $j = 0, 1, \ldots, 2\mu$, and let us introduce the trigonometric polynomial l_0 of degree μ by

$$l_0(\phi) := \frac{\sin(2\mu+1)\frac{\phi}{2}}{(2\mu+1)\sin\frac{\phi}{2}} = \frac{1}{2\mu+1}(1 + 2\cos\phi + \cdots + 2\cos\mu\phi),$$

which is, apart from a constant factor, the *Dirichlet kernel*. Then the translates

$$l_j(\phi) = l_0(\phi - \phi_j), \quad j = 0, 1, \ldots, 2\mu,$$

are the trigonometric Lagrange elements belonging to our nodes, and $\|l_j\|_\infty = 1$ holds, obviously, for $j = 0, 1, \ldots, 2\mu$. In other words, the system $\{\phi_0, \phi_1, \ldots, \phi_{2\mu}\}$ is an extremal fundamental system.

Vice versa, assume, without restriction of generality, that the nodes $0 = \phi_0 < \phi_1 < \cdots < \phi_{2\mu} < 2\pi$ form an extremal fundamental system. We have to show that they are equidistantly distributed in the period.

It is well known that the Lagrange elements can be represented in the form

$$l_j(\phi) = \lambda_j(\phi)/\lambda_j(\phi_j)$$

for $j = 0, 1, \ldots, 2\mu$, where λ_j is defined by

$$\lambda_j(\phi) := \prod_{\substack{k=0 \\ k \neq j}}^{2\mu} \sin\frac{1}{2}(\phi - \phi_k).$$

Moreover, let

$$\omega(\phi) := \prod_{k=0}^{2\mu} \sin\frac{1}{2}(\phi - \phi_k).$$

For every fixed $j \in \{0, 1, \ldots, 2\mu\}$ we get

$$\omega(\phi) = \sin\frac{1}{2}(\phi - \phi_j) \cdot \lambda_j(\phi),$$

which implies

$$\omega''(\phi_j) = \lambda'_j(\phi_j).$$

But in view of

$$\|\lambda_j\|_\infty = |\lambda_j(\phi_j)| \cdot \|l_j\|_\infty = |\lambda_j(\phi_j)|,$$

$\lambda'_j(\phi_j) = 0$ must hold, and we get

$$\omega(\phi_k) = 0 = \omega''(\phi_k)$$

for $k = 0, 1, \ldots, 2\mu$.

Now let $\Omega(\phi) := \omega(2\phi)$. Ω is a nontrivial trigonometric polynomial of degree $2\mu+1$. It satisfies

$$\Omega(\phi + \pi) = \omega(2\phi + 2\pi) = -\sin\frac{1}{2}(2\phi - \phi_j) \cdot \lambda_j(2\phi) = -\omega(2\phi) = -\Omega(\phi)$$

and

$$\Omega\left(\frac{\phi_k}{2}\right) = 0 = \Omega''\left(\frac{\phi_k}{2}\right)$$

for $j, k = 0, 1, \ldots, 2\mu$. It follows that

$$\Omega'' - (2\mu + 1)^2\Omega,$$

which is a trigonometric polynomial of degree 2μ, vanishes at the $4\mu + 2$ pairwise different points

$$\frac{\phi_k}{2} \in [0, \pi) \quad and \quad \frac{\phi_k}{2} + \pi \in [\pi, 2\pi),$$

$k = 0, 1, \ldots, 2\mu$. So it must vanish identically, i.e., Ω satisfies the differential equation

$$\Omega'' - (2\mu + 1)^2\Omega = 0,$$

together with the intial value $\Omega(0) = \Omega(\frac{\phi_0}{2}) = 0$. It follows that Ω has the form

$$\Omega(\phi) = const \cdot \sin\frac{\phi}{2\mu + 1}$$

with some nonvanishing constant. Because of $\Omega(\frac{\phi_k}{2}) = \omega(\phi_k) = 0$ it follows that

$$\phi_k = \frac{2\pi k}{2\mu + 1}$$

for $k = 0, 1, \ldots, 2\mu$. The nodes are equidistantly distributed in the period, as claimed. □

For $r \geq 3$ there is no comparable result known. Numerically, extremal fundamental systems on S^2 have been investigated by Reimer and Sündermann [56] for $\mu \leq 36$, and by Sloan and Womersley [67] for $\mu \leq 128$.

Minimal Interpolatory Projections in the Uniform Norm

In analogy to Definition 5.7 (minimal projection), we introduce now the following one.

Definition 6.4 (Minimal Interpolatory Projection). *Let* $\emptyset \neq D \subset \mathbb{R}^r$, $r \in \mathbb{N}$. *An interpolatory projection* $\Lambda^* : C(D) \to Y$ *onto a finite-dimensional subspace* Y *of* $C(D)$ *is called* minimal *if*

$$\|\Lambda^*\|_\infty \leq \|\Lambda\|_\infty$$

holds for all interpolatory projections Λ.

Interpolatory projections $\Lambda_\mu : C(S^{r-1}) \to \mathbb{P}_\mu^r(S^{r-1})$ satisfy

$$\|\Pi_\mu\|_\infty \leq \|\Lambda_\mu\|_\infty,$$

even if they are minimal, and in view of Theorem 6.6, no sequence of interpolatory projections converges in the uniform norm pointwise to the identity. Though this result occurs as a corollary to Theorem 6.6, we formulate it, for historical reasons, as a separate theorem.

Theorem 6.15 (Faber). *Let* $r \in \mathbb{N} \setminus \{1\}$. *To every sequence of interpolatory projections* $\Lambda_\mu : C(S^{r-1}) \to \mathbb{P}_\mu^r(S^{r-1})$, $\mu = 0, 1, \ldots$, *there exists some* $F \in C(S^{r-1})$ *such that* $\|F - \Lambda_\mu F\|_\infty$ *does not converge to zero.*

In the case $r = 2$, where the space $\mathbb{P}_\mu^r(S^{r-1})$ can be identified with the space of all real trigonometric polynomials of degree at most μ, the result of Theorem 6.15 is due to G. Faber 1914, [17]. In historical review, it has to be counted among the most stimulating theorems — in spite of its *negative character*.

Interpolatory projections are useful in quadrature (Section 6.4), and if they are applied to sufficiently smooth functions. Actually, Theorem 5.11 yields, for $F \in C(S^{r-1})$,

$$\|F - \Lambda_\mu F\|_\infty \leq \left(1 + \|\Lambda_\mu\|_\infty\right) E_\mu(F),$$

where $E_\mu(F)$ is the minimal deviation of F in $\mathbb{P}_\mu^r(S^{r-1})$, see (5.43). So the growth of $\|\Lambda_\mu\|_\infty$, though it is at least the growth of the minimal projection given by (6.16) or by (6.17), respectively, can be matched by a quick enough decreasing $E_\mu(F)$.

In the case $r \geq 3$, a best possible result would be to find a sequence of *interpolatory* projections of the order

$$\|\Lambda_\mu\|_\infty = \mathcal{O}\left(\mu^{\frac{r-2}{2}}\right) \quad (?) \tag{6.28}$$

for $\mu \to \infty$, see (6.17) again. But it is one of the most important open questions in multivariate interpolation, whether this growth order can be realized.

$\|\Lambda_\mu\|_\infty$ depends on the choice of the supporting fundamental system T_μ. In the following, we give a short report on *numerical experiments* of Sloan and Womersley [66], which concern the case $r = 3$ and the following criteria.

(1)   Minimize $\|\Lambda_\mu\|_\infty$, directly.
        This is the most interesting case, but difficult to handle.
(2)   Maximize $\lambda_{\min}$ (in view of Corollary 5.20).
(3)   Maximize Δ (extremal fundamental system).
(4)   Minimize some potential energy function.

Experiments have been performed for the degrees $0 < \mu \le 30$ and show that the Lebesgue constants increase on the order of the criteria, where those obtained by (2) or (3) are not far above the minimal interpolation norm obtained by (1), while criterion (4) furnishes unreasonably large norms. Besides, the minimal interpolation norm seems to behave like

$$\|\Lambda_\mu^*\|_\infty \approx 0.7\,\mu + 1.8, \tag{6.29}$$

which could reject the possibility of (6.28), which would have to be replaced by the more pessimistic conjecture

$$\|\Lambda_\mu^*\|_\infty = \mathcal{O}\left(\mu^{\frac{r-1}{2}}\right) \quad (?) \tag{6.30}$$

We finish with the remark, that, in view of Corollary 5.20, the order of (6.30) holds if an inequality of the form

$$\lambda_{\min} \ge const \cdot \frac{\lambda_1 + \cdots + \lambda_N}{N} \quad (?) \tag{}$$

is realized with some positive constant.

6.4 Quadrature on the Sphere

In this section, $r \in \mathbb{N} \setminus \{1\}$ and $\mu \in \mathbb{N}_0$ are fixed. We assume again that the inner product $\langle \cdot , \cdot \rangle$ is induced by the surface integral, i.e., by the integral defined by

$$IF := \int_{S^{r-1}} F(x)d\omega(x)$$

for $F \in C(S^{r-1})$. The reproducing kernel of $\mathbb{P}_\mu^r(S^{r-1})$ is given again by $K(xy)$, $x, y \in S^{r-1}$, where

$$K = \Gamma_\mu^r.$$

Now let $T = \{t_1, \ldots, t_N\}$, $N = \dim \mathbb{P}_\mu^r(S^{r-1})$, be a fundamental system for $\mathbb{P}_\mu^r(S^{r-1})$. The corresponding Lagrange elements L_j and kernel functions

$$K_j = K(t_j \cdot), \quad j = 1, \ldots, N,$$

play the main role in what follows. For instance, the fundamental matrix takes the form

$$G = \left(K(t_j t_k)\right) = \begin{pmatrix} K(1) & & & \\ & K(1) & & * \\ & & \ddots & \\ & * & & K(1) \end{pmatrix}. \tag{6.31}$$

We are interested in the interpolatory quadrature

$$\hat{I}F = \sum_{j=1}^{N} A_j F(t_j), \tag{6.32}$$

which belongs to the integral I, the space $\mathbb{P}_\mu^r(S^{r-1})$, and the fundamental system T. The weights of such a quadrature are given by (5.70). In view of Theorem, 5.27 our particular interest is directed to positively weighted quadratures. If they are to be used in the evaluation of inner products $\langle F_1, F_2 \rangle$, where $F_1, F_2 \in \mathbb{P}_\mu^r(S^{r-1})$, then it is desirable that their degree of exactness be 2μ, at least. For this reason, Gauß quadratures fit our concept, precisely, see Definition 5.17 and Theorem 5.33.

Gauß Quadrature on the Sphere

(a) Characterisation

In view of Corollary 5.34, a Gauß quadrature is always interpolatory and characterized by a fundamental matrix of diagonal form. In the actual case, this condition is equivalent to G taking the diagonal form

$$G = diag\Big(K(1), \ldots, K(1)\Big). \tag{6.33}$$

In particular, a Gauß quadrature on $\mathbb{P}_\mu^r(S^{r-1})$ must have the constant weights

$$A_j = \frac{1}{K(1)} = \frac{\omega_{r-1}}{N} \tag{6.34}$$

for $j = 1, \ldots, N$, see (4.6), and so it must have the form

$$\hat{I}F = \frac{1}{K(1)} \sum_{j=1}^{N} F(t_j). \tag{6.35}$$

The exact characterisation (6.33) imposes a strong condition on the *geometric configuration* of the nodes. Corresponding questions occured in *discrete geometry* within the setting of *spherical designs*.

Definition 6.5 (Spherical Design). *Let* $r \in \mathbb{N} \setminus \{1\}$. *For* $\mu \in \mathbb{N}_0$ *a system* $T = \{t_1, \ldots, t_M\}$ *of points* $t_j \in S^{r-1}$ *is called a* spherical μ-design, *if*

$$\sum_{j=1}^{M} H(t_j) = 0 \;\; holds \; for \; all \;\; H \in \bigoplus_{\nu=1}^{\mu} \overset{*}{\mathbb{H}}_\nu^r. \tag{6.36}$$

A spherical 2μ-design is called tight, *if* $M = N = \dim \mathbb{P}_\mu^r(S^{r-1})$.

Among others, the following theorem says that the concept of Gauß quadratures and of tight spherical 2μ-designs are essentially the same.

Theorem 6.16 (Characterisation of Gauß Quadratures). *Let $r \in \mathbb{N} \setminus \{1\}$, $\mu \in \mathbb{N}$, $T = \{t_1, \dots, t_N\}$ a fundamental system for $\mathbb{P}_\mu^r(S^{r-1})$, and let the L_j and the K_j be the corresponding Lagrange elements and kernel functions, respectively. Then the following statements are equivalent.*

(i)	T supports a Gauß quadrature.
(ii)	$\sum_{j=1}^{N} L_j^2(x) = 1$ for $x \in S^{r-1}$.
(iii)	$\sum_{j=1}^{N} K_j^2(x) = K^2(1)$ for $x \in S^{r-1}$.
(iv)	$\langle K_j, K_k \rangle = K(1)\delta_{j,k}$ for $j, k = 1, \dots, N$.
(v)	$L_j = K_j/K(1)$ for $j = 1, \dots, N$.
(vi)	T is a tight spherical 2μ-design.

Proof. First we give a circulant proof for the equivalence of the statements $(ii)-(v)$. It uses the identity

$$K(xy) = \sum_{j=1}^{N} K_j(x)L_j(y), \qquad (6.37)$$

which holds for arbitrary $x, y \in S^{r-1}$. Identifying x and y we get the identity

$$K(1) = \sum_{j=1}^{N} K_j(x)L_j(x), \qquad (6.38)$$

and using the Schwarz inequality we obtain

$$K^2(1) \leq \sum_{j=1}^{N} K_j^2(x) \sum_{j=1}^{N} L_j^2(x), \qquad (6.39)$$

for all $x \in S^{r-1}$.

Now let us assume that (ii) holds. Then we get

$$K^2(1) \leq \sum_{j=1}^{N} K_j^2(x),$$

again for all $x \in S^{r-1}$. If equality does not hold for some $x \in S^{r-1}$, then integration over S^{r-1} yields, by the reproducing property of K,

$$K^2(1) \cdot \omega_{r-1} < N \cdot K(1),$$

which is a contradiction in view of (6.34). Therefore (iii) is valid. Next assume (iii) holds. Inserting $x = t_k$ we obtain

$$\sum_{j \neq k} K^2(t_j t_k) = 0.$$

In view of $\langle K_j, K_k \rangle = K(t_j t_k)$ this implies (iv).

(iv) says that $K(t_j \cdot)/K(1)$ satisfies the defining interpolation conditions of L_j. This yields (v).

(v) implies (ii), see (6.38), such that $(ii) - (v)$ are all equivalent, as claimed.

In particular, (iv) is equivalent to (6.33), which turned out already to be equivalent to (i). It follows that $(i) - (v)$ are equivalent, and it is left to prove, for instance, the equivalence of (i) and (vi).

So assume (i) is valid. A Gauß quadrature which is supported by T has the form (6.35). It follows that

$$\hat{I}F = \frac{1}{K(1)} \sum_{j=1}^{N} F(t_j)$$

holds for arbitrary $F \in \mathbb{P}^r_{2\mu}(S^{r-1})$. However, for

$$F = H \in \bigoplus_{\nu=1}^{2\mu} \overset{*}{\mathbb{H}}{}^r_\nu$$

the right side vanishes since the constant 1, which belongs to $\overset{*}{\mathbb{H}}{}^r_0$, is orthogonal to H, see Theorem 4.10. It follows that T is a tight spherical 2μ-design.

Vice versa, assume T is a tight spherical 2μ-design, define the functional $\hat{I}$ by (6.35), and let $F \in \mathbb{P}^r_{2\mu}(S^{r-1})$ be arbitrary. In view of the diagram (4.26), with 2μ instead of μ, we can write F in the form $F = H_0 + H$ where H_0 is some constant and where

$$H \in \bigoplus_{\nu=1}^{2\mu} \overset{*}{\mathbb{H}}{}^r_\nu.$$

This implies

$$\hat{I}F = \frac{\omega_{r-1}}{N} \sum_{j=1}^{N} (H_0 + H) = \omega_{r-1} H_0 = I H_0 = I(H_0 + H) = IF.$$

Therefore, $\hat{I}$ is a Gauß quadrature, which is supported by T. This finishes the proof.   □

(b) Existence

For $r = 2$, the kernel takes the form $K = \frac{1}{2\pi}(U_\mu + U_{\mu-1})$, see (4.30) and Section 2.1. It follows that

$$K(\cos \phi) = \frac{\sin(2\mu + 1)\frac{\phi}{2}}{2\pi \sin \frac{\phi}{2}}$$

is the *Dirichlet kernel*. Using this fact, it is easy to see that all $2\mu+1$ equidistantly distributed nodes on S^1 satisfy condition (*iv*) of Theorem 6.16. So they support a Gauß quadrature, whose weights are given by (6.34), and take now the form

$$A_j = \frac{2\pi}{2\mu + 1}, \quad j = 1, \ldots, 2\mu + 1.$$

For a direct existence proof see Problem 6.1.

Next let $r \geq 3$. We begin with some preliminaries. First let $\xi_1, \ldots, \xi_\mu$ denote the zeros of the Jacobi polynomial $P_\mu^{(\frac{r-1}{2}, \frac{r-3}{2})}$, and hence of K, see (4.31). Moreover, put $\xi_0 := 1$. We may assume that these points are in the order

$$-1 < \xi_\mu < \cdots < \xi_1 < \xi_0 = 1. \tag{6.40}$$

The following lemma compares these points with the zeros

$$-1 < \zeta_\mu < \cdots < \zeta_1 < 1$$

of the Gegenbauer polynomial $C_\mu^{\frac{r}{2}}$.

Lemma 6.17. *Let $r \in \mathbb{N} \setminus \{1\}$, $\mu \in \mathbb{N}$. The zeros ξ_ν of $K = \Gamma_\mu^r$ and the zeros ζ_κ of $C_\mu^{\frac{r}{2}}$ interlace as follows,*

$$-1 < \xi_\mu < \zeta_\mu < \cdots < \xi_1 < \zeta_1 < 1.$$

In particular, $|\xi_\nu| < -\xi_\mu$ holds for $\nu = 1, \ldots, \mu - 1$.

Proof. Put $\zeta_{\mu+1} := -1$. With the help of the symmetry $C_\mu^{\frac{r}{2}}(-\xi) = (-1)^\mu C_\mu^{\frac{r}{2}}(\xi)$, and of (2.13), we get

$$(-1)^\mu \{C_\mu^{\frac{r}{2}}(-1) + C_{\mu-1}^{\frac{r}{2}}(-1)\} = C_\mu^{\frac{r}{2}}(1) - C_{\mu-1}^{\frac{r}{2}}(1) > 0.$$

Together with the well-known interlacing property of the roots of orthogonal polynomials this yields, in view of (4.30),

$$(-1)^\nu K(\zeta_\nu) < 0$$

for $\nu = 1, \ldots, \mu+1$, and the first statement is true. Moreover, since $C_\mu^{\frac{r}{2}}$ is an even or odd function, its roots satisfy $\zeta_{\mu+1-\nu} = -\zeta_\nu$ for $\nu = 1, \ldots, \mu$. For $\nu = 1, \ldots, \mu - 1$ we get, in particular, $-\xi_\mu > -\zeta_{\mu+1-\nu} = \zeta_\nu > \xi_\nu > \xi_\mu$, which finishes the proof. $\square$

In the following, we denote by $l_\mu, \ldots, l_1, l_0$ the Lagrange elements in $\mathbb{P}_\mu^1$, which belong to the nodes $\xi_\mu < \cdots < \xi_1 < \xi_0$. After these preliminaries, we assume that a Gauß quadrature of degree μ exists, i.e., that a fundamental system $T = \{t_1, \ldots, t_N\}$ exists such that

$$\frac{1}{K(1)} \sum_{k=1}^{N} F(t_k) = \int_{S^{r-1}} F(t) d\omega(t) \tag{6.41}$$

holds for all $F \in \mathbb{P}^r_{2\mu}(S^{r-1})$. We investigate some implications arising from this assumption.

In view of Theorem 6.16, (iv), we get

$$K(t_j t_k) = K(1)\delta_{j,k} \tag{6.42}$$

for $j, k = 1, \ldots, N$. In particular, $t_j t_k \in \{\xi_\mu, \ldots, \xi_1\}$ holds for $j \neq k$. By means of the *hypercircles*

$$H^j_\kappa := \{x \in S^{r-1} \mid x t_j = \xi_\kappa\},$$

$\kappa = 0, 1, \ldots, \mu$ and $j = 1, \ldots, N$, we can formulate (6.42) in a more geometric form. First we note that

$$T \subset \bigcup_{\kappa=0}^{\mu} H^j_\kappa$$

holds for all $j \in \{1, \ldots, N\}$, and that H^j_0 degenerates to the one-point set $\{t_j\}$. Next let $t_j \in T$ be a fixed node, and define

$$n_\kappa := card(T \cap H^j_\kappa)$$

for $\kappa = 0, \ldots, \mu$, where $n_0 = 1$ holds by our remark from above.

For $f \in \mathbb{P}^1_{2\mu}$ we define $F \in \mathbb{P}^r_{2\mu}$ by $F(\,\cdot\,) := f(t_j \,\cdot\,)$, and (6.41) takes the form

$$\frac{1}{K(1)} \sum_{k=1}^{N} f(t_j t_k) = \int_{S^{r-1}} f(t_j t)d\omega(t).$$

The integral on the right side is independent of the value of t_j. Moreover, using (1.27) we get

$$\int_{S^{r-1}} f(t_j t)d\omega(t) = \omega_{r-2} \int_{-1}^{1} f(\xi)(1 - \xi^2)^{\frac{r-3}{2}} d\xi.$$

Inserting this above, and ordering the left side by the values of ξ_κ, which $t_j t_k$ attains n_κ–times, we get

$$\int_{-1}^{1} f(\xi)(1 - \xi^2)^{\frac{r-3}{2}} d\xi = \frac{\omega_{r-1}}{\omega_{r-2}} \cdot \frac{1}{N} \cdot \sum_{\kappa=0}^{\mu} n_\kappa f(\xi_\kappa) \tag{6.43}$$

for all $f \in \mathbb{P}^1_{2\mu}$, where we used (6.34). In particular, inserting $f = l^2_\kappa$ we find that the n_κ are positive and not dependent on the choice of t_j, i.e., on j. Note that (6.43) is a so-called *Radau quadrature*.

We summarize our result by the formula

$$card(T \cap H^j_\kappa) = n_\kappa \in \mathbb{N}, \tag{6.44}$$

which holds for $\kappa = 0, 1, \ldots, \mu$ and arbitrary $j = 1, \ldots, N$, thus characterizing the geometry of the Gauß nodes.

Next let $t_j \in T$ be fixed again. Without loss of generality we assume $t_j = e_r$. If $t_k \neq t_j$ is another node, then $t_j t_k = \xi_\kappa$ holds for some $\kappa \in \{1, \ldots, \mu\}$, and t_k can be written in the form

$$t_k = \xi_\kappa e_r + \sqrt{1 - \xi_\kappa^2} \begin{pmatrix} \bar{u}_k \\ 0 \end{pmatrix}, \tag{6.45}$$

where $\bar{u}_k \in S^{r-2}$ is uniquely determined by t_k. For $k = j$, (6.45) remains valid with $\kappa = 0$ and an arbitrary $\bar{u}_k \in S^{r-2}$.

Now let us apply the quadrature (6.41) to a polynomial of the particular form

$$F(x) = l_\kappa(x_r) G(x_1, \ldots, x_{r-1}),$$

where $G \in \mathbb{P}_\mu^{r-1}$ is arbitrary. Then we obtain

$$\frac{\omega_{r-1}}{N} \sum_{k=1}^{N} F(t_k) = \int_{S^{r-1}} l_\kappa(x_r) G(\bar{x}) d\omega(x)$$

with $\bar{x}$ defined by $\bar{x} := (x_1, \ldots, x_{r-1})' \in \mathbb{R}^{r-1}$ for $x = (x_1, \ldots, x_r)' \in \mathbb{R}^r$. Because of (6.45), we can bring this by some changes on the right side to the form

$$\frac{\omega_{r-1}}{N} \sum_{k}^{(\kappa)} G\left(\sqrt{1 - \xi_\kappa^2}\, \bar{u}_k\right) = \int_{-1}^{1} \left(\int_{S^{r-2}} l_\kappa(\xi) G\left(\sqrt{1 - \xi^2}\, \bar{u}\right) d\bar{\omega}(\bar{u}) \right) (1 - \xi^2)^{\frac{r-3}{2}} d\xi, \tag{6.46}$$

where $d\bar{\omega}(\bar{u})$ is the surface element of S^{r-2}, and where the symbol $\sum_k^{(\kappa)}$ indicates that the sum is extended over all k satisfying $t_j t_k = e_r t_k = \xi_\kappa$, which is the same as $t_k \in H_\kappa^j$. Their number is n_κ.

For $G \in \overset{*}{\mathbb{P}}{}_\iota^{r-1}$, $\iota \in \{\mu, \mu - 1\}$, (6.46) takes the form

$$\sum_k^{(\kappa)} G(\bar{u}_k) = c_{\kappa,\iota} \cdot \int_{S^{r-2}} G(\bar{u}) d\bar{\omega}(\bar{u})$$

with some constant $c_{\kappa,\iota}$. For odd ι, the integral on the right side vanishes, and so does the sum on the left side. So let us define

$$d_{\kappa,\mu} := \begin{cases} c_{\kappa,\mu}, & \text{if } \mu \text{ is even,} \\ c_{\kappa,\mu-1}, & \text{if } \mu \text{ is odd.} \end{cases}$$

Then

$$\sum_k^{(\kappa)} G(\bar{u}_k) = d_{\kappa,\mu} \cdot \int_{S^{r-2}} G(\bar{u}) d\bar{\omega}(\bar{u})$$

holds both for $G \in \overset{*}{\mathbb{P}}{}_{\mu}^{r-1}$ and for $G \in \overset{*}{\mathbb{P}}{}_{\mu-1}^{r-1}$, and hence for all $G \in \mathbb{P}_{\mu}^{r-1}(S^{r-2})$, see (4.3). The constant can be determined by inserting $G = 1$. This yields $n_\kappa = d_{\kappa,\mu} \cdot \omega_{r-2}$, and we obtain finally the quadratures

$$\int_{S^{r-2}} G(\bar{u}) d\bar{\omega}(\bar{u}) = \frac{\omega_{r-2}}{n_\kappa} \sum_k^{(\kappa)} G(\bar{u}_k) \tag{6.47}$$

for $\kappa = 1, \ldots, \mu$, which all hold for arbitrary $G \in \mathbb{P}_{\mu}^{r-1}(S^{r-2})$.

We are now going to prove that, in general, not all of these quadratures are as exact as stated. It suffices to consider the case $\kappa = \mu$.

So let $\kappa := \mu$. Because of $n_\mu \in \mathbb{N}$ there is some $t_k \in H_\mu^j$, and if $t_l \in H_\mu^j$, $t_l \neq t_k$, is another node on this hypercircle, then we get, in view of (6.45),

$$\xi_\lambda = t_k t_l = \xi_\mu^2 + (1 - \xi_\mu^2)\bar{u}_k \bar{u}_l$$

with some $\lambda \in \{1, \ldots, \mu\}$. This implies

$$\bar{u}_k \bar{u}_l = \frac{\xi_\lambda - \xi_\mu^2}{1 - \xi_\mu^2} \leq \frac{\xi_1 - \xi_\mu^2}{1 - \xi_\mu^2}. \tag{6.48}$$

First we assume $r = 3$ and $\mu \geq 3$. In Appendix B it is proved that in this case $\xi_1 < \xi_\mu^2$ holds. This implies $\bar{u}_k \bar{u}_l < 0$, and hence $\arccos \bar{u}_k \bar{u}_l > \frac{\pi}{2}$. But it is impossible to distribute more than three points with such a distance on S^1, such that $n_\mu \leq 3$ must hold. Now define $G \in \mathbb{P}_3^2(S^1)$ by

$$G(\bar{u}) := \prod_p^{(\mu)} (1 - \bar{u}\bar{u}_p)$$

for $\bar{u} \in S^1$, where the product is extended over all p satisfying $t_p \in H_\mu^j$. G is positive on S^1 almost everywhere. So we get, by inserting G in the corresponding equation (6.47),

$$0 < \int_{S^1} G(\bar{u}) d\bar{\omega}(\bar{u}) = \frac{\omega_1}{n_\mu} \sum_q^{(\mu)} G(\bar{u}_q) = 0,$$

which is a contradiction. Therefore the assumption that a Gauß quadrature exists was false. A Gauß quadrature does not exist in the case $r = 3$, $\mu \geq 3$.

Next we assume $r \geq 4$. With the help of Lemma 6.17 we obtain $\xi_\lambda \leq -\xi_\mu$, and (6.48) implies

$$\bar{u}_k \bar{u}_l \leq -\frac{\xi_\mu}{1 - \xi_\mu} = \frac{|\xi_\mu|}{1 + |\xi_\mu|} < \frac{1}{2}.$$

It follows that

$$|\bar{u}_k - \bar{u}_l|^2 = 2(1 - \bar{u}_k \bar{u}_l) > 1.$$

But the number of points with such a distance, which can be distributed on S^{r-2}, is bounded above by some constant $a(r)$. This implies $n_\mu \leq a(r)$. But by Corollary 5.30, the number of nodes n_κ in the positive quadrature (6.47) is also bounded below, and we get

$$\dim \mathbb{P}^{r-1}_{\lfloor \frac{\mu}{2} \rfloor}(S^{r-2}) \leq n_\kappa \leq a(r).$$

This is a contradiction for large μ, i.e., to every $r \geq 4$ a number $\mu(r)$ exists such that for $\mu \geq \mu(r)$ a Gauß quadrature does not exist.

This result is the essential part of the following theorem.

Theorem 6.18 (Nonexistence of Gauß Quadratures, Bannai–Damerell, Bos). *For* $(r, \mu) \geq (3, 3)$, *a Gauß quadrature on* $\mathbb{P}^r_\mu(S^{r-1})$ *does not exist.*

Proofs. The first proof is due to Bannai and Damerell [3]. They proved by number theoretical arguments that a tight spherical design does not exist, see Theorem 6.16, (*vi*).

A number theoretical proof in analysis looks strange. Actually, a quite different proof was given later by Bos [9], who uses probability measures to prove that the identity $\sum_{j=1}^N L_j^2(x) = 1$ is impossible, see Theorem 6.16, (*ii*).

In our introducing discussion we gave a proof, restricted to the cases $r = 3$, $\mu \geq 3$, and $r \geq 4$, $\mu \geq \mu(r)$, which uses interpolatory arguments, only. □

Remark. Let $\mu = 1$, $r \in \mathbb{N} \setminus \{1\}$, and assume all vertices of a *regular simplex* are located on S^{r-1}. Then they support a Gauß quadrature. In the language of geometry, they form a *tight spherical 2-design*. The proof is left to the reader, see Problem 6.3.

The case $\mu = 2$ is more sophisticated. For instance, if a Gauß quadrature exists, then the space dimension must have the form $r = p^2 - 3$, where p is an odd integer number, see, e.g., [49].

Product Gauß Quadratures

A Gauß quadrature, if it exists, is exact on $\mathbb{P}^r_{2\mu}(S^{r-1})$ and acting at the minimum number of nodes which is possible in view of Corollary 5.30, and which is given by $N = \dim \mathbb{P}^r_\mu(S^{r-1})$. Vice versa, a quadrature on $\mathbb{P}^r_{2\mu}(S^{r-1})$ which acts on exactly N nodes is interpolatory and hence a Gauß quadrature. For $(r, \mu) \geq (3, 3)$ such a quadrature does not exist, such that the number M of nodes in every quadrature on $\mathbb{P}^r_{2\mu}(S^{r-1})$ satisfies $M > N$, i.e., the lower bound of Corollary 5.30 is never attained. For this reason we have to admit a larger number of nodes in what follows.

Recall that we are interested in positive quadratures which are exact on $\mathbb{P}^r_{2\mu}(S^{r-1})$. A good recommendation is *product Gauß quadratures*. They are derived from univariate Gauß quadratures based on $\mu + 1$ nodes, which are exact of degree $2\mu + 1$. So it is not surprising that a product Gauß quadrature will prove to be more precise than required, namely to be exact on $\mathbb{P}^r_{2\mu+1}(S^{r-1})$.

Preliminaries

Before we define the announced multivariate quadratures recursively, we present some well-known facts on particular univariate Gauß–Jacobi quadratures. For details we refer to Davis [13].

Let $\lambda > -\frac{1}{2}$. The Gegenbauer polynomials of index λ are orthogonal polynomials with respect to the weight function $(1 - \xi^2)^{\lambda - \frac{1}{2}}$, see Theorem 2.3. Now let

$$-1 < \eta_{\mu+1}^\lambda < \cdots < \eta_1^\lambda < 1$$

be the zeros of $C_{\mu+1}^\lambda$, where $\mu \in \mathbb{N}_0$. Then the corresponding interpolatory quadrature,

$$\int_{-1}^{1} f(\xi)(1 - \xi^2)^{\lambda - \frac{1}{2}} d\xi \ = \ \sum_{\nu=1}^{\mu+1} c_\nu^\lambda f(\eta_\nu^\lambda), \tag{6.49}$$

is uniquely determined and valid for all $f \in \mathbb{P}_{2\mu+1}^1$. It is called the *Gauß–Gegenbauer quadrature* of *index* λ.

In this case, the weight function is an even function. This implies

$$\eta_\nu^\lambda \ = \ -\eta_{\mu+1-\nu}^\lambda, \tag{6.50}$$
$$c_\nu^\lambda \ = \ c_{\mu+1-\nu}^\lambda, \tag{6.51}$$

both for $\nu = 1, \ldots, \mu + 1$. We write the nodes also in trigonometric form

$$\eta_\nu^\lambda \ = \ \cos \psi_\nu^\lambda, \ \ where \ 0 < \psi_\nu^\lambda < \pi.$$

Then the ψ_ν^λ are uniquely determined, and satisfy

$$\psi_{\mu+1-\nu}^\lambda = \pi - \psi_\nu^\lambda, \tag{6.52}$$

again for $\nu = 1, \ldots, \mu + 1$. Moreover, for $f \in \mathbb{P}_{2\mu+1}^1$, the expression

$$g(\phi) \ := \ f(\cos \phi)$$

is a cosine polynomial of degree $2\mu + 1$, and, vice versa, every cosine polynomial of this degree can be written in this form, and from (6.49) we obtain

$$\int_{0}^{\pi} g(\phi)(\sin \phi)^{2\lambda} d\phi \ = \ \sum_{\nu=1}^{\mu+1} c_\nu^\lambda \, g(\psi_\nu^\lambda). \tag{6.53}$$

Note that the cosine polynomials are exactly the even trigonometric polynomials. We call (6.53) a *trigonometric*, or more precisely, a *cosine Gauß quadrature*.

After these preliminaries, we return to our original problem.

Recursive Construction

First let $r = 2$, and put $\lambda := 0$. Then the quadrature (6.49) is based on the zeros of $T_{\mu+1}$, which are given by

$$\eta_\nu^0 = \cos \psi_\nu^0, \qquad \psi_\nu^0 = \frac{2\nu - 1}{2\mu + 2} \cdot \pi,$$

for $\nu = 1, \ldots, \mu + 1$. It is not difficult to prove that the corresponding weights are the constants

$$c_\nu^0 = \frac{\pi}{\mu + 1},$$

$\nu = 1, \ldots, \mu + 1$, see Problem 6.2.

Now let $F \in \mathbb{P}_{2\mu+1}^2$. Then we get

$$\int_{S^1} F(x) d\omega(x) = \int_{-1}^{1} \left\{ F(\xi, \sqrt{1 - \xi^2}) + F(\xi, -\sqrt{1 - \xi^2}) \right\} \frac{d\xi}{\sqrt{1 - \xi^2}}$$

$$= \int_{0}^{\pi} \left\{ F(\cos \phi, \sin \phi) + F(\cos \phi, -\sin \phi) \right\} d\phi.$$

The integrand is an even trigonometric polynomial of degree $2\mu+1$. So the integral can be evaluated exactly by means of the trigonometric Gauß quadrature (6.53), where λ is put to zero. So we obtain

$$\int_{S^1} F(x) d\omega(x) = \frac{\pi}{\mu + 1} \sum_{\nu=1}^{\mu+1} \left\{ F(\cos \psi_\nu^0, \sin \psi_\nu^0) + F(\cos \psi_\nu^0, -\sin \psi_\nu^0) \right\}.$$

This quadrature is exact on $\mathbb{P}_{2\mu+1}^2(S^1)$, has positive weights, and is based on $2\mu + 2$ well-known nodes. By a change in the notation we bring it to the form

$$\int_{S^1} F(x) d\omega(x) = \frac{\pi}{\mu + 1} \sum_{j=1}^{M_\mu^2} F(x_j^2), \tag{6.54}$$

where $x_1^2, \ldots, x_{2\mu+2}^2 \in S^{r-1}$ and $M_\mu^2 = 2(\mu + 1)$.

Next let $r \geq 3$, and assume that a positive quadrature on $\mathbb{P}_{2\mu+1}^{r-1}(S^{r-2})$ is known, say

$$\int_{S^{r-2}} G(\bar{x}) d\bar{\omega}(\bar{x}) = \sum_{j=1}^{M_\mu^{r-1}} A_j^{r-1} G(\bar{x}_j^{r-1}) \tag{6.55}$$

holds for all $G \in \mathbb{P}_{2\mu+1}^{r-1}(S^{r-2})$ with weights $A_j^{r-1} > 0$. In the case $r = 3$ this assumption can be realized by identifying (6.55) with (6.54). We want to construct a similar quadrature for the integral

$$
\int_{S^{r-1}} F(x)d\omega(x) \;=\; \int_{B^{r-1}} \left\{ F(\bar{x}, \sqrt{1-\bar{x}^2}) + F(\bar{x}, -\sqrt{1-\bar{x}^2}) \right\} \frac{d\bar{x}}{\sqrt{1-\bar{x}^2}},
$$

where $F \in \mathbb{P}^r_{2\mu+1}(S^{r-1})$. Substituting $\bar{x} = \rho\,\bar{u}$, $\bar{u} \in S^{r-2}$, $0 \le \rho \le 1$, we obtain

$$
\int_{S^{r-1}} F(x)d\omega(x)
$$

$$
= \int_0^1 \int_{S^{r-2}} \left\{ F(\rho\,\bar{u}, \sqrt{1-\rho^2}) + F(\rho\,\bar{u}, -\sqrt{1-\rho^2}) \right\} \rho^{r-2} \cdot \frac{d\bar{\omega}(\bar{u})d\rho}{\sqrt{1-\rho^2}}
$$

$$
= \int_0^{\frac{\pi}{2}} \int_{S^{r-2}} \left\{ F(\sin\phi \cdot \bar{u}, \cos\phi) + F(\sin\phi \cdot \bar{u}, -\cos\phi) \right\} (\sin\phi)^{r-2}\, d\bar{\omega}(\bar{u})d\phi
$$

$$
= \int_0^{\pi} \left(\int_{S^{r-2}} F(\sin\phi \cdot \bar{u}, \cos\phi)\, d\bar{\omega}(\bar{u}) \right) (\sin\phi)^{r-2} d\phi.
$$

Here the inner integral does not change its value if $\bar{u}$ is replaced by $-\bar{u}$. So it is an even trigonometric polynomial of degree $2\mu+1$ with respect to ϕ, and we may evaluate the outer integral by means of the trigonometric Gauß quadrature (6.53) with $\lambda := \frac{r-2}{2}$. This yields

$$
\int_{S^{r-1}} F(x)d\omega(x) \;=\; \sum_{\nu=1}^{\mu+1} c_\nu^{\frac{r-2}{2}} \int_{S^{r-2}} F\!\left(\sin\psi_\nu^{\frac{r-2}{2}} \cdot \bar{u}, \cos\psi_\nu^{\frac{r-2}{2}} \right) d\bar{\omega}(\bar{u}).
$$

For every fixed ν, the integrand is now, with respect to $\bar{u}$, a spherical polynomial on S^{r-2} of degree $2\mu + 1$, which we may evaluate by means of the quadrature (6.55). So we get

$$
\int_{S^{r-1}} F(x)d\omega(x) \;=\; \sum_{\nu=1}^{\mu+1} \sum_{j=1}^{M_\mu^{r-1}} c_\nu^{\frac{r-2}{2}} A_j^{r-1} F(x_{j,\nu}^r), \tag{6.56}
$$

with the nodes defined by

$$
x_{j,\nu}^r := \begin{pmatrix} \sin\psi_\nu^{\frac{r-2}{2}} \cdot \bar{x}_j^{r-1} \\[4pt] \cos\psi_\nu^{\frac{r-2}{2}} \end{pmatrix} \in S^{r-1}
$$

for $j = 1, \ldots, M_\mu^{r-1}$ and $\nu = 1, \ldots, \mu+1$. If $\sin\psi_\nu^{\frac{r-2}{2}}$ vanishes, then exactly M_μ^{r-1} of the nodes are equal and coincide with e_r. This case occurs if and only if $C_{\mu+1}^{\frac{r-2}{2}}$ is an odd polynomial, i.e., if μ is even, and then exactly once, namely for $\nu = \frac{\mu+2}{2}$.

Summarizing we state that, by a change of the notation, we can bring (6.56) to the form

$$\int_{S^{r-1}} F(x)d\omega(x) = \sum_{j=0}^{M_\mu^r} A_j^r F(x_j^r),$$
(6.57)

with nodes $x_j^r \in S^{r-1}$ and weights $A_j^r > 0$ for $j = 1, \ldots, M_\mu^r$, where

$$M_\mu^r \leq (\mu+1)M_\mu^{r-1}.$$
(6.58)

Actually, (6.57) is the desired quadrature on $\mathbb{P}_{2\mu+1}^r(S^{r-1})$. In view of its construction it is called a *product Gauß quadrature*.

By the method presented we are now able to construct a product Gauß quadrature recursively for $r = 3, 4, \ldots$, and if we extend our definition of a product Gauß quadrature to the initial quadrature (6.54), then such a quadrature exists for all dimensions $r \geq 2$.

Theorem 6.19 (Product Gauß Quadratures, Stroud). *For $r \geq 2$ and arbitrary $\mu \in \mathbb{N}_0$, a product Gauß quadrature exists with the following properties.*

> (i)   *The quadrature is exact on $\mathbb{P}_{2\mu+1}^r(S^{r-1})$.*
> (ii) *The number of nodes is $M_\mu^r \leq 2(\mu+1)^{r-1}$.*
> (iii) *The weights are positive.*

Proof. It is left to prove (ii), but this inequality follows from (6.58), which holds for $r \geq 3$, together with the initial value $M_\mu^2 = 2(\mu+1)$. □

Remark. A product Gauß quadrature is always positive and exact on $\mathbb{P}_{2\mu}^r(S^{r-1})$, such that our original problem is solved. The minimum number of nodes, in the sense of Corollary 5.30, which occur necessarily in every quadrature which is exact on $\mathbb{P}_{2\mu+1}^r(S^{r-1})$, is given by

$$N_\mu^r = \dim \mathbb{P}_\mu^r(S^{r-1}).$$

It is worthwile to compare M_μ^r and N_μ^r. Actually we get

$$\begin{aligned} M_\mu^2 &\leq N_\mu^2 + 1, \\ M_\mu^3 &\leq 2N_\mu^3, \\ M_\mu^r &\lesssim (r-1)!\, N_\mu^r, \end{aligned}$$
(6.59)

for $r \geq 2$ and $\mu \to \infty$, see (4.2). In general the number of nodes is far less than the number promised by Theorem 5.31.

Reduction to Simplices

Let I denote the surface integral on S^{r-1}, again. IF vanishes, if $F(x_1, \ldots, x_r)$ is odd with respect to any of the variables x_ν. We say $F \in C(S^{r-1})$ is *totally even*, if it is even with respect to all of the variables, i.e., if

$$F(\ldots, -x_\nu, \ldots) = F(\ldots, x_\nu, \ldots)$$

holds for $\nu = 1, \ldots, r$. A function F is always the sum of a uniquely determined totally even part $G \in C(S^{r-1})$, and another part which is odd with respect to at least one of the variables. G can be represented in the form

$$G(x_1, x_2, \ldots, x_r) = 2^{-r} \sum_{\pm} F(\pm x_1, \pm x_2, \ldots, \pm x_r).$$

Here the sum is extended over all of the 2^r possible combinations of the signs $\pm$. By our remark from above it follows that $IF = IG$ holds.

For even degree, say for μ replaced by 2μ, i.e., for $F \in \mathbb{P}_{2\mu}^{r-1}(S^{r-1})$, we can write G in the form

$$G(x_1, \ldots, x_r) = H(x_1^2, \ldots, x_r^2),$$

where $H \in \mathbb{P}_\mu^r(\Sigma^{r-1})$. The simplex Σ^{r-1} is defined in Section 1.1. Moreover, let us introduce the positive sector of S^{r-1} by the definition

$$S_{ps}^{r-1} := \{x \in S^{r-1} \mid x \geq 0\}.$$

Then we get, using the symmetries of S^{r-1} and the substitution $u_\nu := x_\nu^2$ for $\nu = 1, \ldots, r$, which maps S_{ps}^{r-1} bijectively onto the simplex Σ^{r-1},

$$IF = \int_{S^{r-1}} F(x)d\omega(x) = 2^r \int_{S_{ps}^{r-1}} G(x)d\omega(x) = \int_{\Sigma^{r-1}} H(u)\frac{d\sigma(u)}{\sqrt{u_1}\cdots\sqrt{u_r}} =: JH.$$

So we obtain a quadrature on $\mathbb{P}_{2\mu}^r(S^{r-1})$ for the integral I, if we evaluate the integral J with the help of a quadrature on $\mathbb{P}_\mu^r(\Sigma^{r-1})$. If this is interpolatory, for instance acting on the equidistant grid, which is a fundamental system by Theorem 5.44, then the number of nodes which are used in the evaluation of JH is given by

$$\dim \mathbb{P}_\mu^r(\Sigma^{r-1}) = \binom{\mu + r - 1}{r - 1},$$

see (5.92). However, each single evaluation of H makes 2^r evaluations of F necessary. On the other hand, some of these nodes coincide. We formulate this result as a corollary to Theorem 6.19 and to Theorem 5.31.

Corollary 6.20 (Number of Nodes in a Quadrature on $\mathbb{P}_{2\mu}^r(S^{r-1})$). *For $r \geq 2$ and $\mu \in \mathbb{N}_0$ a quadrature exists on $\mathbb{P}_{2\mu}^r(S^{r-1})$, which uses at most*

$$\tilde{M}_\mu^r := 2^r \cdot \binom{\mu + r - 1}{r - 1} \sim 2^{r-1} \cdot N_\mu^r$$

nodes, where $N_\mu^r = \dim \mathbb{P}_\mu^r(S^{r-1})$.

Remark 1. For $r > 3$ the number of nodes occurring in Corollary 6.20 is asymptotically far less than in a product Gauß quadrature, see (6.59). However, the weights need not be positive.

Remark 2. The number of nodes can be reduced further by writing JH as an integral on the domain E^{r-1}, instead of Σ^{r-1}. For the definition of E^{r-1} see Section 1.1, again. By an affine linear transform, E^{r-1} can be mapped to a regular simplex in $\mathbb{R}^{r-1}$, where additional symmetries are available to be used in the construction of a quadrature. See Grundmann and Möller [23].

Remark 3. In order to describe the exact minimum number of nodes, which is necessary in a quadrature, it is favourable to use the theory of polynomial ideals, see Möller [36], [37].

Quadrature on Extremal Fundamental Systems

Interpolatory quadratures for $\mathbb{P}_\mu^r(S^{r-1})$ which are supported by an extremal fundamental system have been investigated numerically by Reimer and Sündermann [56] and by Sloan and Womersley [67]. Up to the degree $\mu = 128$, the calculated weights were positive, but a general proof has not yet been found, except for the case $\mu = 2$, but arbitrary r, see Reimer [50]. In practice these quadratures are very precise. However, in comparison with a product Gauß quadrature, the calculation of their nodes is expensive, such that their use is limited to the case where the nodes are accessible in tabular form.

Quadratures on Subspaces of $\mathbb{P}_\mu^r(S^{r-1})$

The surface integral of S^{r-1} vanishes for all odd functions. So quadratures which are exact on $\overset{*}{\mathbb{P}}_\mu^r(S^{r-1})$ make sense, only, if μ is even. If μ is even, such a quadrature can be identified with a quadrature on a space $\overset{*}{\mathbb{P}}_\mu^r(S_+^{r-1})$, where S_+^{r-1} is a *hemisphere*, defined by $S_+^{r-1} := \{x \in S^{r-1} \mid ex \geq 0\}$ for some $e \in S^{r-1}$. We do not pursue this idea in detail, but refer to Bannai and Damerell [4], who proved the nonexistence of so-called *tight antipodal spherical designs*, which corresponds again to the nonexistence of a corresponding Gauß quadrature.

For quadratures based on extremal fundamental systems we refer again to Reimer and Sündermann [56].

Quadratures for the subspaces $\overset{*}{\mathbb{H}}_\mu^r(S^{r-1})$, $\mu \in \mathbb{N}$, are useless, again since the integral vanishes on them — similarly for the spaces $\mathbb{P}_m^r(S^{r-1})$, $m \in \mathbb{N}_0^r$, except all components of m are even.

6.5 Geometry of Nodes and Weights in a Positive Quadrature

In the following, I is again the surface integral of S^{r-1}, while

$$\hat{I} = \sum_{j=1}^{M} A_j \hat{E}_j \tag{6.60}$$

is a quadrature on $\mathbb{P}_{2\mu}^r(S^{r-1})$, $r \in \mathbb{N} \setminus \{1\}$, $\mu \in \mathbb{N}$, with positive weights, and supported by a node system $T = \{t_1, \ldots, t_M\}$, $t_j \in S^{r-1}$. In principle we are allowed to assume, without loss of generality, that the restrictions of the evaluation functionals

$$\hat{E}_j F = F(t_j),$$

$j = 1, \ldots, M$, onto $\mathbb{P}_{2\mu}^r(S^{r-1})$ are linearly independent, where we recall the reduction method used in the proof of Theorem 5.31. But in what follows we consider problems only where such an assumption is not necessary. Actually, our aim is to show that a positive quadrature is always accompanied by a covering of the sphere by M small caps, and by a very regular distribution of the weights. Recall that a cap is defined by

$$C(t, \alpha) := \left\{ x \in S^{r-1} \mid xt \geq \cos \alpha \right\}$$

with $t \in S^{r-1}$ and $\alpha \geq 0$.

Theorem 6.21 (Quadrature and Covering Radius, Reimer, Yudin). *Let* $r \in \mathbb{N} \setminus \{1\}$, $\mu \in \mathbb{N}$, *and assume* $T = \{t_1, \ldots, t_M\}$ *supports a positive quadrature* $\hat{I}$ *on* $\mathbb{P}_{2\mu}^r(S^{r-1})$. *Then*

$$S^{r-1} \subset \bigcup_{j=1}^{M} C\left(t_j, \chi_{\mu,1}\right)$$

holds, where $\chi_{\mu,1}$ *is the lowest positive zero of* $C_\mu^{\frac{r-2}{2}}(\cos \phi)$. *This is equivalent to* $\rho(T) \leq \chi_{\mu,1}$.

Proof. In view of (4.12), $\eta := \cos \chi_{\mu,1}$ is the greatest zero of $G := G_\mu^r$, i.e., of the reproducing kernel function of $\overset{*}{\mathbb{H}}{}_\mu^r(S^{r-1})$. In particular, $G(\xi)/(\xi - \eta)$ is a polynomial of degree $\mu - 1$. So we get for arbitrary $x \in S^{r-1}$, using orthogonality,

$$\int_{S^{r-1}} G(tx) \frac{G(tx)}{tx - \eta} d\omega(t) = 0.$$

Now assume the first statement to be false. Then an $x_0 \in S^{r-1}$ exists which is contained in none of the caps, i.e., such that $t_j x_0 < \eta$ holds for $j = 1, \ldots, M$. It follows that a neighbourhood $\mathcal{U}_0$ of x_0 in $\mathbb{R}^r$ exists such that

$$t_j x < \eta$$

is valid for $j = 1, \ldots, M$ and all $x \in \mathcal{U}_0 \cap S^{r-1}$. Evaluating the integral with the help of $\hat{I}$, we get for all $x \in \mathcal{U}_0 \cap S^{r-1}$,

$$0 = \sum_{j=1}^{M} A_j \frac{[G(t_j x)]^2}{t_j x - \eta} \leq 0,$$

which implies $G(t_j x) = 0$ for $j = 1, \ldots, M$. For instance, $G(t_1 x_0) = 0$ holds, implying $t_1 \neq x_0$. So there is a

$$u_0 \in S^{r-1} \cap span\{x_0, t_1\}$$

satisfying $u_0 x_0 = 0$ and hence $(t_1 x_0)^2 + (t_1 u_0)^2 = 1$. Now let

$$x(\phi) := x_0 \cdot \cos \phi + u_0 \cdot \sin \phi$$

for $\phi \in \mathbb{R}$. Then there is an $\epsilon > 0$ such that

$$x(\phi) \in \mathcal{U}_0 \cap S^{r-1}$$

holds for $-\epsilon < \phi < \epsilon$, which implies $G(t_1 x(\phi)) = 0$. Therefore,

$$t_1 x(\phi) = (t_1 x_0) \cdot \cos \phi + (t_1 u_0) \cdot \sin \phi$$

is a zero of G, and as a continuous function it must be constant for $-\epsilon < \phi < \epsilon$. This implies $t_1 x_0 = 0 = t_1 u_0$, in contradiction to our result from above. So our assumption was false, and

$$S^{r-1} \subset \bigcup_{j=1}^{M} C\left(t_j, \chi_{\mu,1}\right)$$

is valid. In view of Definition 6.3, this is equivalent to $\rho(T) \leq \chi_{\mu,1}$, and the theorem is proved. $\qquad \square$

Theorem 6.21 can be interpreted in purely geometric terms. Actually, in view of Theorem 6.19 the following corollary holds.

Corollary 6.22 (Number of Small Caps which Cover the Sphere). *For all* $r \in \mathbb{N} \setminus \{1\}$ *there is a number* $c_r > 0$ *with the following property. For all* $\mu \in \mathbb{N}$ *there exists a covering of* S^{r-1} *by* $M(r, \mu)$ *spherical caps of radius* $\frac{\pi}{2\mu}$, *where* $M(r, \mu) \leq c_r \cdot \mu^{r-1}$.

Proof. With the help of Theorem 2.11 we get

$$\lim_{\mu \to \infty} \mu \cdot \chi_{\mu,1} = j_{\alpha,1},$$

where $\alpha = \frac{r-3}{2} \geq -\frac{1}{2}$. It follows that a constant $\gamma_r > 0$ exists such that

$$\mu \cdot \chi_{\mu,1} \leq \gamma_r$$

holds for all $\mu \in \mathbb{N}$. Now we choose $\kappa \in \mathbb{N}$ so large that $\frac{1}{\kappa} \gamma_r \leq \frac{\pi}{2}$ is valid. Then we obtain from the last inequality, by replacing μ by $\kappa\mu$,

$$\chi_{\kappa\mu,1} \leq \frac{\pi}{2\mu}$$

for all $\mu \in \mathbb{N}$. Now let $T = \{t_1, \ldots, t_M\}$ support a product Gauß quadrature on S^{r-1} with $M = M^r_{\kappa\mu}$ nodes. Then Theorem 6.21 yields

$$S^{r-1} \subset \bigcup_{j=1}^{M} C\left(t_j, \chi_{\kappa\mu,1}\right) \subset \bigcup_{j=1}^{M} C\left(t_j, \frac{\pi}{2\mu}\right),$$

where the number of nodes satisfies

$$M^r_{\kappa\mu} \leq 2(\kappa\mu + 1)^{r-1},$$

see Theorem 6.19. Now it is easy to see that a constant $c_r > 0$ exists, such that

$$M^r_{\kappa\mu} \leq c_r \cdot \mu^{r-1}$$

holds for arbitrary $\mu \in \mathbb{N}$. With $M(r, \mu) := M^r_{\kappa\mu}$ the statement of the corollary is true. $\square$

In the following we investigate the distribution of the weights of a positive quadrature, where we consider the nodes to be weighted by the map

$$T \ni t_j \mapsto A_j \in \mathbb{R}_+.$$

Lemma 6.23 (Weight Distribution (I)). *Let $r \in \mathbb{N}\backslash\{1\}$, $\mu \in \mathbb{N}$, $K := \frac{1}{\omega_{r-1}} C^{\frac{r}{2}}_\mu$, and let $\psi_{\mu,1}$ denote the lowest positive zero of $C^{\frac{r}{2}}_\mu (\cos\phi)$. Then the following is valid. If $\hat{I}$ is a positive quadrature on $\mathbb{P}^r_{2\mu}(S^{r-1})$ with nodes $t_1, \ldots, t_M$, and weights $A_1, \ldots, A_M$, then*

$$\sum_{t_j \in C(x,\psi)} A_j \leq \frac{1}{[\tilde{K}(\cos\psi)]^2} \cdot \frac{1}{K(1)} \tag{6.61}$$

holds for $0 < \psi < \psi_{\mu,1}$ and all $x \in S^{r-1}$, where $\tilde{K} := K/K(1) = \tilde{C}^{\frac{r}{2}}_\mu$.

Proof. First recall that a positive quadrature is characterized by positive weights, and that K is the reproducing kernel of $\overset{*}{\mathbb{P}}{}^r_\mu(S^{r-1})$. So we get for all $x \in S^{r-1}$,

$$\sum_{j=1}^{M} A_j K^2(t_j x) = \int_{S^{r-1}} K^2(t_j x) = K(1).$$

By the assumption on ψ, $K(\cos\phi)$ is monotonically decreasing for $0 < \phi \leq \psi$, which yields

$$0 < K^2(\cos\psi) \sum_{t_j \in C(x,\psi)} A_j \leq \sum_{t_j \in C(x,\psi)} A_j K^2(t_j x) \leq K(1),$$

and (6.61) follows immediately. $\square$

Theorem 6.24 (Weight Distribution (II), Reimer). *Let* $r \in \mathbb{N} \setminus \{1\}$. *For* $\mu \in \mathbb{N}$ *let* $\chi_{\mu,1}$ *denote the lowest positive zero of* $C_\mu^{\frac{r-2}{2}}(\cos\phi)$. *Then a constant* $c_r > 0$ *exists such that for all* $\mu \in \mathbb{N} \setminus \{1\}$ *the nodes* t_j *and weights* A_j *of every positive quadrature* $\hat{I}_\mu$ *on* $\mathbb{P}_{2\mu}^r(S^{r-1})$ *satisfy*

$$\sum_{t_j \in C(x,\chi_{\mu,1})} A_j \leq \frac{c_r}{\mu^{r-1}} \tag{6.62}$$

for all $x \in S^{r-1}$.

Proof. By the notation of Section 2.6, $\psi_{\mu,1}^\lambda$ is the lowest positive zero of $C_\mu^\lambda(\cos\phi)$, and $\cos\psi_{\mu,1}^\lambda$ is the greatest zero of C_μ^λ itself. This is a monotonically decreasing function of λ, see Szegö [73], Theorem 6.21.1, (6.21.3). So we get

$$\frac{\pi}{2\mu} = \psi_{\mu,1}^0 \leq \psi_{\mu,1}^{\frac{r-2}{2}} < \psi_{\mu,1}^{\frac{r}{2}},$$

where we used (2.8) in the evaluation of $\psi_{\mu,1}^0$. Now we return to the notation of Lemma 6.23 and of Theorem 6.24, and obtain

$$\frac{\pi}{2\mu} \leq \chi_{\mu,1} < \psi_{\mu,1}. \tag{6.63}$$

In particular we may apply Lemma 6.23 with $\psi := \chi_{\mu,1}$ to get

$$\sum_{t_j \in C(x,\chi_{\mu,1})} A_j \leq R(r,\mu),$$

where

$$R(r,\mu) := \omega_{r-1}\left[\tilde{C}_m^{\frac{r}{2}}\left(\cos\chi_{\mu,1}\right)\right]^{-2}\binom{\mu+r-1}{r-1}^{-1}.$$

Here $K(1)$ has been evaluated by means of (4.6) and of (4.1). Moreover, Theorem 2.11 yields

$$\chi_{\mu,1} \sim \frac{1}{\mu} \cdot j_{\frac{r-3}{2},1} \tag{6.64}$$

for $\mu \to \infty$. Finally we use Theorem 2.8 and Theorem 2.6, and obtain

$$\lim_{\mu \to \infty} \tilde{C}_\mu^{\frac{r}{2}}\left(\cos\chi_{\mu,1}\right) = Z_{\frac{r-1}{2}}\left(j_{\frac{r-3}{2},1}\right) > 0.$$

It follows that

$$\lim_{\mu \to \infty} \mu^{r-1} R(r,\mu) > 0,$$

and since $\mu^{r-1} R(r,\mu)$ is positive for all $\mu \in \mathbb{N}$, a positive constant c_r exists such that

$$\mu^{r-1} \cdot R(r,\mu) \leq c_r$$

holds for all $\mu \in \mathbb{N}$. This finishes the proof. $\qquad\square$

Remark. For spherical designs, Theorem 6.21 was known to Yudin [76]. An inequality like (6.62) occurs as the *regularity condition* in the work of Sloan and Womersley [65]. The first proof that it holds always of itself is given by Reimer [54].

Theorem 6.21 says, in view of (6.64), that a positive quadrature never leaves a larger hole on the sphere free of nodes, while Theorem 6.24 says that there are never too many weighted nodes on it. We make this more configurative by the following corollary.

Corollary 6.25 (Weight Distribution (III)). *Assumptions and c_r as in Theorem 6.24. Then for all $\mu \in \mathbb{N}$ and all $x \in S^{r-1}$, the weights and nodes of all positive quadratures on $\mathbb{P}_{2\mu}^r(S^{r-1})$ satisfy*

$$\sum_{t_j \in C(x, \frac{\pi}{2\mu})} A_j \leq \frac{c_r}{\mu^{r-1}}. \tag{6.65}$$

Proof. Because of (6.63) we get

$$\sum_{t_j \in C(x, \frac{\pi}{2\mu})} A_j \leq \sum_{t_j \in C(x, \chi_{\mu, 1})} A_j,$$

and (6.65) follows from (6.62). □

By a combination of Theorem 6.21 and of Theorem 6.4, we investigate in what follows the sum of the weights which belong to a larger cap. We begin with the following lemma.

Lemma 6.26 (Number of Small Caps Covering a Large Cap). *Let $r \in \mathbb{N} \setminus \{1, 2\}$. Then a constant k_r exists, such that for all $\mu \in \mathbb{N}$ the following holds. Every spherical cap of radius ϕ, $\frac{\pi}{2\mu} < \phi \leq \frac{\pi}{2}$, can be covered by caps of radius $\frac{\pi}{2\mu}$ whose number does not exceed the value $k_r(\mu \sin \phi)^{r-1}$.*

Proof. Let $\mu \in \mathbb{N}$, and assume the cap to be covered has the center $x \in S^{r-1}$, which is, geographically speaking, the 'pole'. The 'equator'

$$S_0^{r-1} := \left\{ u \in S^{r-1} \mid ux = 0 \right\}$$

is a unit sphere S^{r-2}. So we can cover it for every $\kappa \in \{1, \ldots, 2\mu\}$ by $M_\kappa := M\left(r - 1, 3(2\kappa + 1)\right)$ caps of radius $\frac{\pi}{6(2\kappa+1)}$, see Corollary 6.22. Now assume the centers of these caps to be the points

$$u_1^\kappa, \ldots, u_{M_\kappa}^\kappa \in S_0^{r-1}. \tag{6.66}$$

Then the following holds. For fixed $\kappa \in \{1, \ldots, 2\mu\}$ and $u \in S_0^{r-1}$ there is some $k \in \{1, \ldots, M_\kappa\}$ such that

$$\cos \frac{\pi}{6(2\kappa + 1)} \leq uu_k^\kappa \leq 1 \tag{6.67}$$

is valid.

Next we define a partition of the interval $\left[\frac{\pi}{4\mu}, \frac{\pi}{2}\right]$ by introducing the angles

$$\phi_\kappa := \frac{\kappa\pi}{4\mu}, \quad \kappa = 1, \ldots, 2\mu,$$

and the midpoints

$$\bar{\phi}_\kappa := \tfrac{1}{2}\left(\phi_\kappa + \phi_{\kappa+1}\right) = \frac{(2\kappa+1)\pi}{8\mu}, \quad \kappa = 1, \ldots, 2\mu - 1.$$

After that we keep $\kappa \in \{1, \ldots, 2\mu - 1\}$ temporarily fixed. The geographical 'zone'

$$C\left(x, \phi_{\kappa+1}\right) \setminus C\left(x, \phi_\kappa\right) \tag{6.68}$$

consists of all $t \in S^{r-1}$ with a representation

$$t = x \cdot \cos\phi + u \cdot \sin\phi,$$

where $u \in S_0^{r-1}$, $ux = 0$, and

$$\phi_\kappa < \phi \leq \phi_{\kappa+1}.$$

In particular it contains the points

$$t_j^\kappa := x \cdot \cos\bar{\phi}_\kappa + u_j^\kappa \cdot \sin\bar{\phi}_\kappa, \quad j = 1, \ldots, M_\kappa.$$

Now let t be an arbitrary point of the zone (6.68) again. The corresponding u satisfies (6.67) for some $k \in \{1, \ldots, M_\kappa\}$. We want to estimate the trigonometric distance between t and t_k^κ. To this end we introduce the intermediate point

$$\bar{t} := x \cdot \cos\bar{\phi}_\kappa + u \cdot \sin\bar{\phi}_\kappa.$$

In view of the inequality $|\phi - \bar{\phi}_\kappa| \leq \frac{\pi}{8\mu}$ and of (6.67) it satisfies the following inequalities,

$$
\begin{aligned}
|t - \bar{t}|^2 &= 2(1 - t\bar{t}) = 2\left(1 - \cos(\phi - \bar{\phi}_\kappa)\right) = 4\sin^2\frac{\phi-\bar{\phi}_\kappa}{2} \leq 4\sin^2\frac{\pi}{16\mu}, \\
|\bar{t} - t_k^\kappa|^2 &= 2\left(1 - \bar{t}t_k^\kappa\right) = 2\sin^2\bar{\phi}_\kappa \cdot \left(1 - uu_k^\kappa\right) \leq 2\sin^2\bar{\phi}_\kappa \cdot \left(1 - \cos\frac{\pi}{6(2\kappa+1)}\right) \\
&= 4\sin^2\frac{(2\kappa+1)\pi}{8\mu} \cdot \sin^2\frac{\pi}{12(2\kappa+1)} \leq 4\left[\frac{\pi^2}{12} \cdot \frac{2}{\pi}\frac{\pi}{16\mu}\right]^2 \leq 4\sin^2\frac{\pi}{16\mu},
\end{aligned}
$$

and by the help of the triangular inequality we obtain

$$|t - t_k^\kappa| \leq 4\sin\frac{\pi}{16\mu}.$$

This implies

$$
\begin{aligned}
1 - tt_k^\kappa &\leq 4(1 - \cos\tfrac{\pi}{8\mu}) < 4(1 - \cos^2\tfrac{\pi}{8\mu}) \\
&= 2(1 - \cos\tfrac{\pi}{4\mu}) < 2(1 - \cos^2\tfrac{\pi}{4\mu}) \\
&= 1 - \cos\tfrac{\pi}{2\mu},
\end{aligned}
$$

and it follows that $tt_k^\kappa > \cos\frac{\pi}{2\mu}$, and hence $t \in C(t_k^\kappa, \frac{\pi}{2\mu})$.

In other words, the zone (6.68) is covered by the caps $C(t_j^\kappa, \frac{\pi}{2\mu})$, $j = 1, \ldots, M_\kappa$. It follows that

$$C(x, \phi_{\nu+1}) \setminus C(x, \phi_1) = \bigcup_{\kappa=1}^{\nu} \left(C(x, \phi_{\kappa+1}) \setminus C(x, \phi_\kappa) \right)$$

is covered, for $\nu \in \{1, \ldots, 2\mu - 1\}$, by $\sum_{\kappa=1}^{\nu} M_\kappa$ caps of radius $\frac{\pi}{2\mu}$, and since $C(x, \phi_1)$ is contained in $C(x, \frac{\pi}{2\mu})$, we can cover $C(x, \phi_{\nu+1})$ by

$$1 + \sum_{\kappa=1}^{\nu} M_\kappa$$

caps of radius $\frac{\pi}{2\mu}$, in total.

Next we recall the definition of M_κ, and get with the help of Corollary 6.22 the following estimates,

$$\sum_{\kappa=1}^{\nu} M_\kappa \leq 6^{r-2}c_{r-1}\sum_{\kappa=1}^{\nu}(\kappa + \tfrac{1}{2})^{r-2} < 6^{r-2}c_{r-1}\int_1^{\nu+1}\xi^{r-2}d\xi < \frac{6^{r-2}c_{r-1}}{r-1}(\nu+1)^{r-1}.$$

Now it is obvious that a constant $k_r \in \mathbb{N}$ exists such that

$$1 + \sum_{\kappa=1}^{\nu} M_\kappa \leq k_r\left(\frac{\nu}{2}\right)^{r-1}$$

holds for all $\nu \in \{1, \ldots, 2\mu - 1\}$.

We turn to the last step of the proof. For $\mu = 1$, nothing has to be proved. So let $\mu \geq 2$ in what follows, let $\phi \in (\frac{\pi}{2\mu}, \frac{\pi}{2}]$ be given, and define

$$\nu := \lceil \tfrac{4\mu\phi}{\pi} \rceil - 1,$$

which is a number contained in $\{1, \ldots, 2\mu - 1\}$. Because of

$$C(x, \phi) \subset C(x, \phi_{\nu+1}),$$

we can cover $C(x, \phi)$ by at most

$$k_r\left(\frac{\nu}{2}\right)^{r-1} \leq k_r\left(\mu \cdot \frac{2\phi}{\pi}\right)^{r-1} \leq k_r(\mu \sin\phi)^{r-1}$$

caps of radius $\frac{\pi}{2\mu}$, as claimed.   □

Lemma 6.26 enables us to estimate also the weight sum of a positive quadrature in an arbitrary cap, where it is convenient to use the following definition.

Definition 6.6 (Weight Sum Function). *Let $\hat{I}$ be a positive quadrature with nodes t_j and weights A_j. For $x \in S^{r-1}$ the weight sum function $A_x : \mathbb{R} \to \mathbb{R}$ is defined by*

$$A_x(\phi) := \begin{cases} 0 & for \quad \phi < 0, \\ \sum_{t_j \in C(x,\phi)} A_j & for \quad 0 \le \phi \le \pi, \\ A_x(\pi) & for \quad \phi > \pi. \end{cases}$$

The weight sum function A_x is piecewise constant, monotonically nondecreasing, and right-side continuous. In particular, it is integrable in the sense of Riemann–Stieltjes, where we define the inproper Riemann–Stieltjes integral of $f \in C(\mathbb{R})$ on the interval $[\alpha, \beta]$ by

$$\int_\alpha^\beta f(\phi) \, dA(\phi) := \lim_{\epsilon \to 0+} \int_{\alpha-\epsilon}^\beta f(\phi) \, dA(\phi).$$

This definition guarantees that the equation

$$\int_\alpha^\beta f(\phi) \, dA(\phi) = \sum_{\cos \alpha \le xt_j \le \cos \beta} A_j \, f(t_j x)$$

holds for $0 \le \alpha < \beta \le \pi$.

The weight sum function is subject to the following theorem.

Theorem 6.27. *Let $r \in \mathbb{N} \setminus \{1, 2\}$. Then a constant a_r exists such that for all $\mu \in \mathbb{N}$ the following holds. For all positive quadratures on $\mathbb{P}^r_{2\mu}(S^{r-1})$ and all $x \in S^{r-1}$, the weight sum function A_x satisfies*

$$A_x(\phi) \le a_r (\sin \phi)^{r-1}$$

for $\frac{\pi}{2\mu} < \phi \le \frac{\pi}{2}$.

Proof. By Lemma 6.26 we can cover $C(x, \phi)$ with at most $k_r(\mu \sin \phi)^{r-1}$ caps of radius $\frac{\pi}{2\mu}$. The contribution to the weight sum of each of these caps is estimated by Corollary 6.25. Together this yields

$$A_x(\phi) \le k_r(\mu \sin \phi)^{r-1} \cdot c_r \mu^{-r+1} = a_r(\sin \phi)^{r-1},$$

with the constant $a_r := c_r k_r$. $\qquad \square$

Remark. Note that $A_x(\phi) \le \frac{c_r}{\mu^{r-1}}$ holds for $0 \le \phi \le \frac{\pi}{2\mu}$, see Corollary 6.25. Note also that the bound of Theorem 6.27 does not hold for $\phi = 0$, if x is a node.

6.6   Hyperinterpolation on the Sphere

In Section 6.3 we discussed the question whether interpolatory projections Λ_μ : $C(S^{r-1}) \to \mathbb{P}^r_\mu(S^{r-1})$ of all degrees μ exist, whose Lebesgue constants $\|\Lambda\|_\infty$ grow at the order of the minimal projection norms $\|\Pi_\mu\|_\infty$, which is $\mathcal{O}(\mu^{\frac{r-2}{2}})$. We mentioned that for $r \geq 3$ this question is not yet answered. So it is important to know that hyperinterpolation, to be introduced in what follows, is a generalisation of interpolation, which shares with it the advantage of an easy evaluation, but realizes simultaneously the growth order of the minimal projections, though at the price of some additional evaluations. This favourable detour was suggested first by Sloan [63].

The basic idea of hyperinterpolation is to evaluate the orthogonal projection (6.5) by a positive quadrature

$$\hat{I}_\mu = \sum_{j=1}^M A_j \hat{E}_j, \tag{6.69}$$

which is exact on $\mathbb{P}^r_{2\mu}(S^{r-1})$. We do not specify that the node system $T = \{t_1, \ldots, t_M\}$ and the weights $A_1, \ldots, A_M > 0$ depend on μ. With the abbreviation $K = \Gamma^r_\mu$ and the kernel functions $K_j = K(t_j \cdot)$ the resulting linear operator

$$\hat{L}_\mu : C(S^{r-1}) \to \mathbb{P}^r_\mu(S^{r-1})$$

is given by

$$\left(\hat{L}_\mu F\right)(x) := \sum_{j=1}^M A_j F(t_j) K_j(x) \tag{6.70}$$

for $F \in C(S^{r-1})$ and $x \in S^{r-1}$. Note that $\|F\|_\infty \leq 1$ implies

$$\left|\left(\hat{L}_\mu F\right)(x)\right| \leq \sum_{j=1}^M A_j |K(t_j x)|$$

for $x \in S^{r-1}$. Writing $A_j = \sqrt{A_j} \cdot \sqrt{A_j}$ we obtain

$$\left|\left(\hat{L}_\mu F\right)(x)\right|^2 \leq \sum_{j=1}^M A_j \cdot \sum_{j=1}^M A_j K^2(t_j x)$$

$$= \omega_{r-1} \cdot \int_{S^{r-1}} K^2(tx) d\omega(t)$$

$$= \omega_{r-1} \cdot K(1) = N,$$

where we used (4.6) with $N = N^r_\mu = \dim \mathbb{P}^r_\mu(S^{r-1})$. So $\hat{L}_\mu$ is a bounded linear operator with uniform norm

$$\|\hat{L}_\mu\|_\infty \leq \sqrt{N} = \mathcal{O}(\mu^{\frac{r-1}{2}}) \tag{6.71}$$

as $\mu \to \infty$.

Next let $F \in \mathbb{P}_\mu^r(S^{r-1})$. Then the evaluation of the integral (6.5) is exact, and we get $\hat{L}_\mu F = \Pi_\mu F = F$, where we used that Π_μ is a projection. Hence $\hat{L}_\mu$ is also a projection.

Finally assume that $\hat{I}_\mu$ is a Gauß quadrature (though such a quadrature exists, as we know, in particular situations, only). Then, by Theorem 6.16, T is a fundamental system, the weights are all equal to $\frac{1}{K(1)}$, and $K_j = K(1)L_j$ holds. Together this yields

$$\left(\hat{L}_\mu F\right)(x) = \sum_{j=1} F(t_j)L_j(x).$$

In other words, $\hat{L}_\mu = \Lambda_\mu$ is the interpolatory projection belonging to the fundamental system T. This justifies the following definition.

Definition 6.7 (Hyperinterpolation). *A projection $\hat{L}_\mu$, defined in (6.70) by a positive quadrature $\hat{I}_\mu$ on $\mathbb{P}_{2\mu}^r(S^{r-1})$, is called a* hyperinterpolation operator *of degree μ.*

We put emphasis on the fact that hyperinterpolation operators $\hat{L}_\mu$ of all degrees $\mu \in \mathbb{N}_C$ exist. For instance, it is possible to define $\hat{L}_\mu$ by means of a product Gauß quadrature, see Theorem 6.19. The bound occurring in (6.71) holds also for interpolatory projections if they are supported by the nodes of a Gauß quadrature. For, in this case the eigenvalues of the fundamental matrix are all equal, see Theorem 6.16, (iv), and our statement follows from Corollary 5.20. Nevertheless, the problem is just the existence of Gauß nodes, see Theorem 6.18.

It was not difficult to derive the bound of (6.71). But this bound still fails the minimal projection order by $\frac{1}{2}$. To get the optimal order requires a deeper investigation of hyperinterpolation operators.

Theorem 6.28 (Hyperinterpolation Norm, Sloan and Womersley, Reimer). *Let $r \in \mathbb{N} \setminus \{1, 2\}$. Then a positive constant γ_r exists such that for all $\mu \in \mathbb{N}$*

$$\|\hat{L}_\mu\|_\infty \leq \gamma_r \cdot \mu^{\frac{r-2}{2}}$$

holds for all hyperinterpolation operators $\hat{L}_\mu$ of degree μ.

Proof. For $F \in C(S^{r-1})$, $\|F\|_\infty \leq 1$, we get from (6.70)

$$\left|\left(\hat{L}_\mu F\right)(x)\right| \leq \sum_{t_j \in C(x, \frac{\pi}{2})} A_j|K(t_j x)| + \sum_{-t_j \in C(x, \frac{\pi}{2})} A_j|K(t_j x)|. \tag{6.72}$$

The second sum has the form of the first one, except that it belongs to the quadrature where all nodes are replaced by their complement. This is again exact of degree 2μ. So it suffices to estimate the first sum, but this for an arbitrary hyperinterpolation operator.

Let us decompose this sum as follows,

$$\sum_{t_j \in C(x, \frac{\pi}{2})} A_j |K(t_j x)| = \Sigma_0 + \Sigma_1, \tag{6.73}$$

where Σ_0 is defined by

$$\Sigma_0 := \sum_{t_j \in C(x, \frac{\pi}{2\mu})} A_j |K(t_j x)|,$$

while Σ_1 is the corresponding sum which is extended over all $t_j \in C(x, \frac{\pi}{2}) \setminus C(x, \frac{\pi}{2\mu})$. Using (1.23) we can estimate the first sum in the form

$$\Sigma_0 \leq K(1) \sum_{t_j \in C(x, \frac{\pi}{2\mu})} A_j,$$

and in view of (4.6), (4.4), and (6.65), some constant $d_{r,0}$ exists such that

$$\Sigma_0 \leq d_{r,0} \tag{6.74}$$

holds, independently of the value of μ.

With the help of the weight sum function A_x, see Definition 6.6, the second sum can be represented by the Riemann–Stieltjes integral

$$\Sigma_1 = \int_{\frac{\pi}{2\mu}}^{\frac{\pi}{2}} |K(\cos \phi)| \, dA_x(\phi).$$

Now we estimate

$$K = \frac{1}{\omega_{r-1}} \left(C_\mu^{\frac{r}{2}} + C_{\mu-1}^{\frac{r}{2}} \right),$$

see (4.30), by means of Corollary 2.10, and get, with some constant $d_{r,1}$,

$$|K(\cos \phi)| \leq d_{r,1} \cdot \mu^{\frac{r-2}{2}} \cdot (\sin \phi)^{-\frac{r}{2}} \tag{6.75}$$

for $0 < \phi < \pi$. Inserting this above we obtain

$$\Sigma_1 \leq d_{r,1} \cdot \mu^{\frac{r-2}{2}} \int_{\frac{\pi}{2\mu}}^{\frac{\pi}{2}} (\sin \phi)^{-\frac{r}{2}} dA_x(\phi),$$

and integration by parts yields

$$\Sigma_1 \leq d_{r,1} \cdot \mu^{\frac{r-2}{2}} \cdot \left\{ A_x\left(\frac{\pi}{2}\right) + \frac{r}{2} \int_{\frac{\pi}{2\mu}}^{\frac{\pi}{2}} A_x(\phi)(\sin \phi)^{-\frac{r+2}{2}} \cos \phi \, d\phi \right\}.$$

Here we may estimate $A_x(\frac{\pi}{2})$ above by

$$A_x(\pi) = \sum_{j=1}^{M} A_j = \int_{S^{r-1}} d\omega(\phi) = \omega_{r-1},$$

and $A_x(\phi)$ by the bound of Theorem 6.27. Together this yields

$$\Sigma_1 \leq d_{r,1} \cdot \mu^{\frac{r-2}{2}} \cdot \left\{ \omega_{r-1} + \tfrac{r}{2} a_r \int_0^{\frac{\pi}{2}} (\sin\phi)^{\frac{r-4}{2}} \cos\phi \, d\phi \right\},$$

and in view of (6.73) and (6.74) it is now obvious that a constant γ_r exists such that

$$\sum_{t_j \in C(x,\frac{\pi}{2})} A_j |K(t_j x)| \leq \tfrac{1}{2}\gamma_r \, \mu^{\frac{r-2}{2}}.$$

It follows from (6.72) that $|(\hat{L}_\mu F)(x)| \leq \gamma_r \, \mu^{\frac{r-2}{2}}$ holds for all $x \in S^{r-1}$, which finally yields

$$\|\hat{L}_\mu F\|_\infty \leq \gamma_r \, \mu^{\frac{r-2}{2}}$$

for all $F \in C(S^{r-1})$ with $\|F\|_\infty \leq 1$. This finishes the proof. $\qquad\square$

Remark 1. If hyperinterpolation is supported by the nodes of a product Gauß quadrature, then the number of nodes does not exceed the value $2(\mu+1)^{r-1}$, see Theorem 6.19.

Remark 2. In an interpolatory quadrature, the average weight has the value

$$\frac{1}{N} \sum_{j=1}^{N} A_j = \frac{\omega_{r-1}}{N} = \frac{1}{K(1)}.$$

In view of (6.60) it is worthwhile to introduce the *weighted kernel functions* $\tilde{K}_j = K_j/K(1)$ Along the hypercircle $t_j \cos\phi + u_j \sin\phi$, $t_j \perp u_j \in S^{r-1}$, the weighted kernel function $\tilde{K}_j$ takes the form

$$\Gamma_\mu^r(\cos\phi)/\Gamma_\mu^r(1).$$

See Figure 6.1.

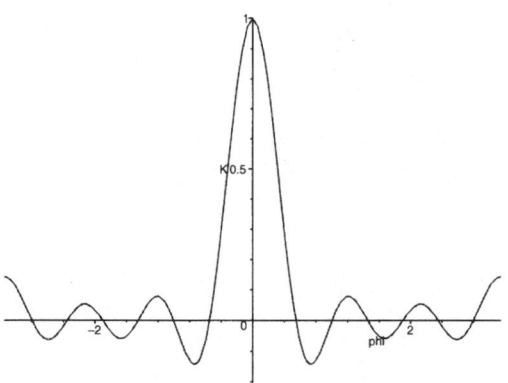

Figure 6.1. Hyperinterpolation: Weighted Kernel Function $\Gamma_6^3(\cos\phi)/\Gamma_6^3(1)$,
in Dependence on the Geodetic Distance ϕ from the Pivot Node.

6.7   Summation of Laplace Series

In the preceding section we considered projections, only. We succeeded in the construction of discrete projections whose Lebesgue constants are growing not faster
than the order of the minimal projection norms. So far they are favourable, though
pointwise convergence in the uniform norm remains excluded by the Theorem of
Banach–Steinhaus. This means that convergence takes place only under additional
assumptions on the approximability, or smoothness, of the function which is to be
approximated. In what follows we depart this setting by making the Laplace series of a spherical function subject to a summation method. The result is linear
operators, but in general no projections, which meet the following three claims,

- they are positive operators,
- their evaluation is easy,
- they converge pointwise in the uniform norm.

The second and third claim need not be explained. The first one is necessary
to make sure that the approximation preserves the order, and eventually even
the shape of the approximated function. Interpolation and hyperinterpolation fail
this aim for $\mu \in \mathbb{N}$, as the Lagrange elements L_j and the kernel functions K_j,
respectively, change the sign.

Positive Polynomial Operators
Let $r \in \mathbb{N} \setminus \{1\}$. For $F \in C(S^{r-1})$ we consider the Laplace series

$$LF = \sum_{\nu=0}^{\infty} \Omega_\nu F,$$

see Section 6.1. It converges to F in the L^2-setting, i.e., in the average. But in $\| \cdot \|_\infty$ the partial sums

$$\Pi_\mu F = \sum_{\nu=0}^{\mu} \Omega_\nu F$$

may happen to diverge, see Theorem 6.6, and it is impossible to get a better result by any other sequence of projections. For this reason, we consider in what follows *mean values* from the orthogonal projections $\Pi_0, \dots, \Pi_\mu$ of the form

$$L_\mu F = \sum_{\nu=0}^{\mu} a_{\mu\nu} \Omega_\nu F \tag{6.76}$$

for $\mu = 0, 1, \dots$. Note that L_μ is formed with the aid of the μ-th row of the subdiagonal infinite real matrix

$$A = \left(a_{\mu\nu} \right)_{\mu,\nu=0,1,\dots}.$$

Obviously, whatever the choice of A may be, by (6.76) a sequence of bounded linear operators

$$L_\mu : C(S^{r-1}) \to \mathbb{P}_\mu^r(S^{r-1}), \quad \mu \in \mathbb{N}_0, \tag{6.77}$$

is defined, and we ask for conditions under which it converges pointwise in the uniform norm.

With the help of (6.4) we get easily the representation

$$(L_\mu F)(x) = \int_{S^{r-1}} F(t) K_\mu(tx) \, d\omega(t), \tag{6.78}$$

where the kernel $K_\mu = K_\mu^A$ is defined by

$$K_\mu := \sum_{\nu=0}^{\mu} a_{\mu\nu} G_\nu \in \mathbb{P}_\mu^1. \tag{6.79}$$

Note that $L_\mu 1 = a_{\mu 0}$ holds. So it is reasonable to assume $a_{\mu 0} = 1$, without restriction of generality if convergence is required. The kernel plays a main role in the following theorem, which meets our claims from above.

Theorem 6.29 (Pointwise Convergence of the L_μ). *Let $r \in \mathbb{N} \setminus \{1\}$, and assume the infinite real matrix $A = (a_{\mu\nu})$ satisfies the following assumptions.*

(i) $a_{\mu\nu} = 0$ *for* $\mu, \nu \in \mathbb{N}_0$, $\nu > \mu$,

(ii) $a_{\mu 0} = 1$ *for* $\mu \in \mathbb{N}_0$, *and* $\lim_{\mu \to \infty} a_{\mu 1} = 1$,

(iii) $K_\mu(\xi) \geq 0$ *for* $-1 \leq \xi \leq 1$, $\mu \in \mathbb{N}_0$.

Then the operators L_μ, $\mu \in \mathbb{N}_0$, defined by (6.76) or (6.78), respectively, are positive and pointwise convergent to the identity in the uniform norm. Moreover,

$$|a_{\mu\nu}| < 1$$

holds for $\nu = 1, \ldots, \mu$, $\mu \in \mathbb{N}$.

Proof. From (iii) it follows that the operators L_μ are positive. Moreover, by Theorem 5.4 the functions 1, x_1, $\ldots$, x_r form a Korovkin set in $C(S^{r-1})$. In view of (4.12) the same holds for the functions G_0, $G_1(ex)$, $e \in S^{r-1}$. But using (6.78) and (6.79) we get

$$\left(L_\mu G_0\right)(x) = \int\limits_{S^{r-1}} G_0(tx)K_\mu(tx)\,d\omega(t) = a_{\mu 0} \cdot G_0(x)$$

for $x \in S^{r-1}$, where we used orthogonality and the reproducing property of G_0 with respect to the constants. Similarly we obtain

$$\left(L_\mu G_1(e\cdot)\right)(x) = \int\limits_{S^{r-1}} G_1(et)K_\mu(tx)\,d\omega(t) = a_{\mu 1} \cdot G_1(ex).$$

Now we use assumption (ii) to get

$$\lim_{\mu\to\infty} L_\mu G_0 = G_0,$$

$$\lim_{\mu\to\infty} L_\mu G_1(e\cdot) = G_1(e\cdot),$$

both in $\|\cdot\|_\infty$. Hence we get uniform convergence to the identity on a Korovkin set, and the remaining follows from Theorem 5.5, the Theorem of Bohman and Korovkin.

Finally, for $\kappa \in \mathbb{N}$ and $-1 \leq \xi \leq 1$, we get from (2.16)

$$1 \pm \tilde{C}_\kappa^{\frac{r-2}{2}}(\xi) \geq 0,$$

where equality is excluded almost everywhere. For fixed $x \in S^{r-1}$ this implies in view of (6.79)

$$0 < \int\limits_{S^{r-1}} K_\mu(tx)\left(1 \pm \tilde{C}_\kappa^{\frac{r-2}{2}}(tx)\right)d\omega(t) = a_{\mu 0} \pm a_{\mu\kappa} = 1 \pm a_{\mu\kappa}$$

for $\kappa = 0, 1, \ldots, \mu$, which finishes the proof. $\square$

Particular Summation Methods

There are various summation methods available to realize the assumptions of Theorem 6.29. For a survey we refer to Beekmann and Zeller, [5]. Here we present two methods of particular interest.

Cesàro Method

To a given real sequence $a_0, a_1, \ldots$ we consider the *formal* power series

$$g(z) = \sum_{\nu=0}^{\infty} a_\nu z^\nu.$$

Formal means that we operate with it algebraically as if it converges in a neighbourhood of $z = 0$. Later the coefficients are to be identified with $G_\nu(\xi)$, for instance.

The partial sums

$$\sigma_\mu^{(0)} := a_0 + \cdots + a_\mu, \tag{6.80}$$

$\mu \in \mathbb{N}_0$, define the new power series

$$g_0(z) = \sum_{\mu=0}^{\infty} \sigma_\mu^{(0)} z^\mu,$$

where

$$(1 - z) g_0(z) = \sigma_0^{(0)} + \sum_{\mu=1}^{\infty} \left\{ \sigma_\mu^{(0)} - \sigma_{\mu-1}^{(0)} \right\} z^\mu = g(z).$$

In other words, the partial sums $\sigma_\mu^{(0)}$ are the coefficients of the power series

$$g_0(z) = \frac{g(z)}{1 - z} = \sum_{\mu=0}^{\infty} \sigma_\mu^{(0)} z^\mu.$$

Likewise we gain the partial sums

$$\sigma_\mu^{(1)} := \sigma_0^{(0)} + \cdots + \sigma_\mu^{(0)}$$

for $\mu \in \mathbb{N}_0$ from the power series

$$g_1(z) = \frac{g(z)}{(1 - z)^2} = \sum_{\mu=0}^{\infty} \sigma_\mu^{(1)} z^\mu,$$

and so on. So we generate the sequences $\{\sigma_\mu^{(k)}\}_{\mu \in \mathbb{N}_0}$, $k = 0, 1, \ldots$, by a repeated summation, where the elements of the k-th sequence are the coefficients of

$$g_k(z) = \frac{g(z)}{(1 - z)^{k+1}} = \sum_{\mu=0}^{\infty} \sigma_\mu^{(k)} z^\mu. \tag{6.81}$$

Now we use (6.81) in order to extend the definition of $\sigma_\mu^{(k)}$ to the case of an arbitrary $k > -1$. Comparing the coefficients of (6.81) and of

$$g_k(z) = \frac{g_0(z)}{(1 - z)^k} = \sum_{\nu=0}^{\infty} \sigma_\nu^{(0)} z^\nu \cdot \sum_{\kappa=0}^{\infty} \frac{(k)_\kappa}{(1)_\kappa} z^\kappa = \sum_{\mu=0}^{\infty} \left(\sum_{\nu=0}^{\mu} \frac{(k)_{\mu-\nu}}{(1)_{\mu-\nu}} \cdot \sigma_\nu^{(0)} \right) z^\mu,$$

we get for arbitrary $k > -1$

$$\sigma_\mu^{(k)} = \sum_{\nu=0}^\mu \frac{(k)_{\mu-\nu}}{(1)_{\mu-\nu}} \cdot \sigma_\nu^{(0)}. \qquad (6.82)$$

Temporarily we put $g(z) := 1$. This yields $\sigma_\nu^{(0)} = 1$ for $\nu \in \mathbb{N}_0$, while (6.81) implies $\sigma_\mu^{(k)} = \frac{(k+1)_\mu}{(1)_\mu}$. Inserting this in (6.82) we get the identity

$$\sum_{\nu=0}^\mu \frac{(k)_{\mu-\nu}}{(1)_{\mu-\nu}} = \frac{(k+1)_\mu}{(1)_\mu},$$

which is of interest itself. It follows that the sum of the weights, which occur in (6.82), is different from unity, except for the case $k = 0$. Therefore we put

$$s_\mu^{(k)} := \frac{(1)_\mu}{(k+1)_\mu} \sigma_\mu^{(k)} \qquad (6.83)$$

in order to get a weighted mean.

Definition 6.8 (Cesàro Means). *For $k > -1$ and $\mu \in \mathbb{N}_0$, $s_\mu^{(k)}$ is called the μ-th Cesàro mean of the partial sums $a_0 + \cdots + a_\nu$ of index k. If*

$$\lim_{\mu \to \infty} s_\mu^{(k)} = s^{(k)}$$

exists, then $s^{(k)}$ is called the (C,k)-limit of the partial sums. It is written in the form

$$s^{(k)} = (C, k) - \sum_{\nu=0}^\infty a_\nu.$$

Note that $s_\mu^{(0)} = \sigma_\mu^{(0)} = a_0 + \cdots + a_\mu$ holds for $\mu \in \mathbb{N}_0$. In the case of a general k we get a similar result by inserting (6.80) in (6.82). It has the form

$$s_\mu^{(k)} = \sum_{\kappa=0}^\mu c_{\mu\kappa}^{(k)} a_\kappa \qquad (6.84)$$

with coefficients $c_{\mu\kappa}^{(k)}$ to be determined in what follows.

For a given $\nu \in \mathbb{N}_0$ we define g by $g(z) := z^\nu$. This implies $a_\kappa = \delta_{\nu\kappa}$ for $\kappa \in \mathbb{N}_0$, and hence

$$c_{\mu\nu}^{(k)} = s_\mu^{(k)} = \frac{(1)_\mu}{(k+1)_\mu} \sigma_\mu^{(k)},$$

where $\sigma_\mu^{(k)}$ is the coefficient occurring with z^μ in the expansion

$$\frac{z^\nu}{(1-z)^{k+1}} = \sum_{\kappa=0}^\infty \frac{(k+1)_\kappa}{(1)_\kappa} z^{\nu+\kappa},$$

see (6.81). This yields $\sigma_\mu^{(k)} = \frac{(k+1)_{\mu-\nu}}{(1)_{\mu-\nu}}$, and hence

$$c_{\mu\nu}^{(k)} = \frac{(\mu - \nu + 1)_\nu}{(\mu - \nu + k + 1)_\nu} \tag{6.85}$$

for $\mu, \nu \in \mathbb{N}_0$, $\nu \leq \mu$. We complete the coefficients by the definition

$$c_{\mu\nu}^{(k)} := 0$$

for $\mu, \nu \in \mathbb{N}_0$, $\nu > \mu$, and define the infinite *Cesàro matrix* of index k by

$$\mathbf{C}^{(k)} := \left(c_{\mu\nu}^{(k)} \right)_{\mu, \nu = 0, 1, \ldots}.$$

Cesàro Kernels and Cesàro Operators

Let $k > -1$. We identify

$$A := \mathbf{C}^{(k)},$$

and define the *Cesàro kernels* $K_\mu = K_\mu^{(r,k)}$ and the *Cesàro operators* $L_\mu = L_\mu^{(r,k)}$ of *index* k by (6.79) and (6.78), respectively, and we ask for conditions to be put on k such that A satisfies the assumptions of Theorem 6.29.

In the present case, the kernels are given by

$$K_\mu^{(r,k)}(\xi) = \sum_{\nu=0}^{\mu} c_{\mu\nu}^{(k)} G_\nu(\xi) \tag{6.86}$$

for $-1 \leq \xi \leq 1$, and $\mu \in \mathbb{N}_0$. We identify a_ν with $G_\nu(\xi)$, and $s_\mu^{(k)}$ with $K_\mu^{(r,k)}(\xi)$, respectively, both for fixed $\xi \in [-1, 1]$. Then (6.86) takes the form (6.84). In particular, we get

$$g_0(z) = \sum_{\mu=0}^{\infty} \sigma_\mu^{(0)} z^\mu = \sum_{\mu=0}^{\infty} \left(\sum_{\nu=0}^{\mu} G_\nu(\xi) \right) z^\mu.$$

The inner sum is well known from (4.28) and (4.30), and we obtain

$$g_0(z) = \frac{1}{\omega_{r-1}} \sum_{\mu=0}^{\infty} \left(C_\mu^{\frac{r}{2}}(\xi) + C_{\mu-1}^{\frac{r}{2}}(\xi) \right) z^\mu = \frac{1}{\omega_{r-1}} \cdot \frac{1+z}{(1 - 2\xi z + z^2)^{\frac{1}{2}}},$$

see (2.3), (2.4), and recall $C_{-1}^{\frac{r}{2}} = 0$. It follows that

$$g_k(z) = \frac{g_0(z)}{(1-z)^k} = \frac{1}{\omega_{r-1}} \cdot \frac{1+z}{(1-z)^k(1 - 2\xi z + z^2)^{\frac{r}{2}}},$$

and we obtain the following generating function for the kernels,

$$\frac{1}{\omega_{r-1}} \cdot \frac{1+z}{(1-z)^k(1 - 2\xi z + z^2)^{\frac{r}{2}}} = \sum_{\mu=0}^{\infty} \frac{(k+1)_\mu}{(1)_\mu} K_\mu^{(r,k)}(\xi) z^\mu. \tag{6.87}$$

The crucial point is now, how can we determine the sign of the coefficients in the power series (6.87). The answer is contained in the following theorem.

Theorem 6.30 (Kogbetliantz). *For* $\lambda \geq 1$, $k \geq 2\lambda - 1$, *the coefficients* $A_\mu^{\lambda,k}(\xi)$, $\mu \in \mathbb{N}_0$, *in the expansion of*

$$f^{\lambda,k}(\xi, z) := \frac{1 + z}{(1 - z)^k (1 - 2\xi z + z^2)^\lambda} = \sum_{\mu=0}^{\infty} A_\mu^{\lambda,k}(\xi)\, z^\mu$$

are nonnegative for $-1 \leq \xi \leq 1$.

Proof. Assume the statement is proved already for $-1 \leq \xi \leq 1$ and $k = \bar{k} := 2\lambda - 1$, and let $k > \bar{k}$. It follows from

$$f^{\lambda,k}(\xi, z) = \frac{1}{(1 - z)^{k - \bar{k}}} \cdot f^{\lambda,\bar{k}}(\xi, z) = \sum_{\mu=0}^{\infty} \frac{(k - \bar{k})_\mu}{(1)_\mu}\, z^\mu \cdot f^{\lambda,\bar{k}}(\xi, z)$$

that $f^{\lambda,k}(\xi, z)$ is the product of two power series with nonnegative coefficients. Therefore its coefficients are also nonnegative.

So it suffices to prove the statement for $k = 2\lambda - 1$, i.e., to prove that the coefficients of

$$f^\lambda(\xi, z) := \frac{1 + z}{(1 - z)^{2\lambda - 1}(1 - 2\xi z + z^2)^\lambda} = \sum_{\mu=0}^{\infty} A_\mu^\lambda(\xi)\, z^\mu$$

are nonnegative for $-1 \leq \xi \leq 1$. For $\lambda = 1$ this takes the form

$$\frac{1 + z}{(1 - z)(1 - 2\xi z + z^2)} = \frac{1}{1 - z} \cdot \sum_{\mu=0}^{\infty} \Big(U_\mu(\xi) + U_{\mu-1}(\xi) \Big)\, z^\mu,$$

see Section 2.1, and we get

$$A_\mu^1(\xi) = \sum_{\nu=0}^{\mu} \Big(U_\nu(\xi) + U_{\nu-1}(\xi) \Big).$$

It follows that

$$A_\mu^1(\cos \phi) = \frac{1}{\sin \phi} \sum_{\nu=0}^{\mu} \Big(\sin(\nu + 1)\phi + \sin \nu\phi \Big) = \left[\frac{\sin(\mu + 1)\frac{\phi}{2}}{\sin \frac{\phi}{2}} \right]^2 \geq 0,$$

see Problem 6.5, and the statement is true. Note that $A_\mu^1(\cos \phi)$ is, apart from a constant factor, the well-known *Fejèr kernel* of degree μ.

Next we assume $\lambda > 1$, and write $f^\lambda(\xi, z)$ in the form

$$f^\lambda(\xi, z) = \frac{1}{(1 - z^2)^{\lambda - 1}} \cdot \left[\frac{1 + z}{(1 - z)(1 - 2\xi z + z^2)} \right]^\lambda.$$

If $\lambda = \nu$ is an integer, then $f^\nu(\xi, z)$ is again a product of power series with nonnegative coefficients, and the statement is true.

So it is left to prove the statement for $\lambda = \nu + \kappa$, where $\nu \in \mathbb{N}$ and $0 < \kappa < 1$. In this case we write

$$f^\lambda(\xi, z) = f^\nu(\xi, z) \cdot h^\kappa(\xi, z),$$

where

$$h^\kappa(\xi, z) = \frac{1}{(1-z)^{2\kappa}} \cdot \frac{1}{(1 - 2\xi z + z^2)^\kappa} = \sum_{\mu=0}^\infty B_\mu^\kappa(\xi) \, z^\mu.$$

By our result from above, the first factor gives rise again to a power series with nonnegative coefficients, and it suffices to prove the same for the second one. So let us investigate $h^\kappa(\xi, z)$, which we write in the form

$$h^\kappa(\xi, z) = \sum_{\nu=0}^\infty C_\nu^\kappa(1) \, z^\nu \cdot \sum_{\nu=0}^\infty C_\nu^\kappa(\xi) \, z^\nu,$$

see (2.3), (2.4). It follows that

$$B_\mu^\kappa(\xi) = \sum_{\nu=0}^\mu C_\nu^\kappa(1) \, C_{\mu-\nu}^\kappa(\xi).$$

In view of $C_0^\kappa = 1$ this yields

$$\tfrac{1}{\kappa} B_\mu^\kappa(\xi) = \tfrac{1}{\kappa}\left[C_\mu^\kappa(1) + C_\mu^\kappa(\xi) \right] + \kappa \sum_{\nu=1}^{\mu-1} \tfrac{1}{\kappa} C_\nu^\kappa(1) \cdot \tfrac{1}{\kappa} C_{\mu-\nu}^\kappa(\xi).$$

Now we use (2.18), and obtain

$$\lim_{k \to 0+} \tfrac{1}{\kappa} B_\mu^\kappa(\xi) = \tfrac{2}{\mu}\left[1 + T_\mu(\xi) \right] > 0$$

for $\mu \in \mathbb{N}$, provided ξ is no zero of the polynomial $1 + T_\mu$.

Now let $\mu \in \mathbb{N}$ be fixed, and let $\xi \in [-1, 1]$ be not the zero of any of the polynomials $1 + T_\nu$, $\nu = 1, \ldots, \mu$. Then a $j \in \mathbb{N}$ exists such that

$$B_\nu^{\kappa/2^j}(\xi) > 0$$

holds for $\nu = 0, 1, \ldots, \mu$. Besides we can use the factorisation

$$h^\kappa(\xi, z) = h^{\frac{\kappa}{2}}(\xi, z) \cdot h^{\frac{\kappa}{2}}(\xi, z)$$

to get

$$B_\mu^\kappa(\xi) = \sum_{\nu=0}^\mu B_{\mu-\nu}^{\frac{\kappa}{2}}(\xi) \cdot B_\nu^{\frac{\kappa}{2}}(\xi).$$

By a repeated application of this formula we find that $B_\mu^\kappa(\xi)$ is a multivariate polynomial with positive coefficients with respect to

$$B_0^{\kappa/2^j}(\xi), \ldots, B_\mu^{\kappa/2^j}(\xi),$$

and we get $B_\mu^\kappa(\xi) > 0$ almost everywhere in $[-1, 1]$. However, $B_\mu^\kappa(\xi)$ is a polynomial. So $B_\mu^\kappa(\xi) \geq 0$ must hold for $-1 \leq \xi \leq 1$. Together with $B_0^\kappa(\xi) = 1$, it follows that $h^\kappa(\xi, z)$ has a power series with nonnegative coefficients for $-1 \leq \xi \leq 1$. This finishes the proof.   □

Remark 1. For the original proof we refer to Kogbetliantz [29], where it is scattered over pp. 159–169. The presented proof follows the lines of [51].

Remark 2. Kogbetliantz proved also that the Lebesgue constants $\|L_\mu^{(r,k)}\|_\infty$ are bounded for $k > \frac{r-2}{2}$, and unbounded for $k \leq \frac{r-2}{2}$.

Now we return to our original problem, and prove the following theorem.

Theorem 6.31 (Positivity and Convergence of Cesàro Operators). *Let* $r \in \mathbb{N} \setminus \{1\}$ *and* $k \geq r - 1$. *Then* $K_\mu^{(r,k)}(\xi) \geq 0$ *holds for* $-1 \leq \xi \leq 1$. *The Cesàro operators* $L_\mu^{(r,k)}$ *are positive and pointwise convergent in the uniform norm to the identity on* $C(S^{r-1})$.

Proof. Let $A = C^{(k)}$. From (6.85) we get

$$a_{\mu 0} = 1 \text{ for } \mu \in \mathbb{N}_0, \quad a_{\mu 1} = 1 - \frac{k}{\mu + k} \text{ for } \mu \in \mathbb{N},$$

such that the assumptions (i) and (ii) of Theorem 6.29 are satisfied. In view of (6.87), (iii) follows from Theorem 6.30 with $\lambda = \frac{r}{2}$. So all assumptions of the Theorem 6.29 are satisfied, and the statement is true.   □

Corollary 6.32 (Operator Norm). *Let* $r \in \mathbb{N} \setminus \{1\}$ *and* $k \geq r-1$. *Then* $\|L_\mu^{(r,k)}\|_\infty = 1$ *holds for* $\mu \in \mathbb{N}_0$.

Proof. In view of (6.78) we get, for $F \in C(S^{r-1})$, $\|F\|_\infty \leq 1$, and $x \in S^{r-1}$,

$$\left| \left(L_\mu^{(r,k)} F\right)(x) \right| \leq \int\limits_{S^{r-1}} K_\mu^{(r,k)}(tx)\, d\omega(t) = L_\mu^{(r,k)} 1 = c_{\mu 0}^{(k)} = 1,$$

see (6.86) and (6.85). This implies $\|L_\mu^{(r,k)}\|_\infty = 1$, as claimed.   □

A Summation Method Based on the Newman–Shapiro Kernels

For $\nu \in \mathbb{N}_0$ let $\eta_{\nu+1} = \cos \chi_{\nu+1}$, $0 < \chi_{\nu+1} < \pi$, be the greatest zero of $G_{\nu+1}$. The *Newman–Shapiro kernels* are defined by

$$K_{2\nu+1}^{(r)}(\xi) := K_{2\nu}^{(r)}(\xi) = g_{\nu+1} \left[\frac{G_{\nu+1}(\xi)}{\xi - \eta_{\nu+1}} \right]^2, \tag{6.88}$$

where the constant is chosen such that

$$\int\limits_{S^{r-1}} K_\mu^{(r)}(tx)\, d\omega(t) = 1 \tag{6.89}$$

holds for $x \in S^{r-1}$ and $\mu \in \mathbb{N}_0$. For arbitrary $\mu \in \mathbb{N}_0$, $K_\mu^{(r)}$ is a univariate polynomial of degree at most μ, which has an expansion

$$K_\mu^{(r)} = \sum_{\nu=0}^{\mu} a_{\mu\nu}^{(r)} G_\nu \tag{6.90}$$

with uniquely determined coefficients $a_{\mu\nu}^{(r)}$ for $\nu = 0, \ldots, \mu$. We extend their definition by putting

$$a_{\mu\nu}^{(r)} := 0 \text{ for } \nu > \mu.$$

So we get a subdiagonal infinite matrix

$$A^{(r)} = \left(a_{\mu\nu}^{(r)} \right)_{\mu,\nu=0,1,\ldots}.$$

It defines a summation method, and by identification $A = A^{(r)}$, a sequence of linear operators $L_\mu = L_\mu^{(r)}$ is defined by (6.78) and (6.79), which we call the *Newman–Shapiro operators*. These operators have been used by Newman and Shapiro [40] in order to prove a *Jackson type inequality* for the sphere, which will occur as a corollary in Section 6.9. However, the operators are also good in our actual setting.

Theorem 6.33 (Newman–Shapiro Operators). *Let $r \in \mathbb{N} \setminus \{1\}$. Then the following holds.*

(i) *The Newman–Shapiro operators $L_\mu^{(r)}$ are positive and in the uniform norm pointwise convergent to the identity on $C(S^{r-1})$.*

(ii) *The elements of the matrix $A^{(r)} = \left(a_{\mu\nu}^{(r)} \right)$ satisfy $a_{\mu0} = 1$, and*

$$0 \leq a_{\mu\nu}^{(r)} < 1$$

for $\nu = 1, \ldots, \mu$, $\mu \in \mathbb{N}$.

(iii) *For $\nu \in \mathbb{N}_0$ the constant in (6.88) is given by*

$$g_{\nu+1}^{-1} = \int_{S^{r-1}} \left[\frac{G_{\nu+1}(tx)}{tx - \eta_{\nu+1}} \right]^2 d\omega(t) = \frac{(2\nu + r)^2}{\omega_{r-1}(\nu+1) \sin^2 \chi_{\nu+1}} \cdot \binom{\nu + r - 2}{r - 2}.$$

Proof. (i) It suffices to consider the case $\mu = 2\nu$. In view of (6.88) and (6.89) the constants $g_{\nu+1}$ are positive, and so the kernels are nonnegative, such that the assumptions (i) and (iii) of Theorem 6.29 are satisfied. In particular the operators $L_\mu^{(r)}$ are positive. Moreover, inserting (6.90) in (6.89) and using $1 \in \overset{*}{\mathbb{H}}_0^r$, we get

$$a_{\mu0}^{(r)} = 1. \tag{6.91}$$

Because of $(\cdot x) \in \overset{*}{\mathbb{H}}_1^r$, we get likewise

$$a_{\mu1}^{(r)} = \int_{S^{r-1}} (tx) K_\mu^{(r)}(tx) \, d\omega(t).$$

Together with (6.89) this yields

$$
\begin{aligned}
a_{\mu 1}^{(r)} - \eta_{\nu+1} &= \int_{S^{r-1}} \left[tx - \eta_{\nu+1} \right] K_{\mu}^{(r)}(tx)\, d\omega(t) \\
&= g_{\nu+1} \int_{S^{r-1}} G_{\nu+1}(tx) \frac{G_{\nu+1}(tx)}{tx - \eta_{\nu+1}}\, d\omega(t) = 0,
\end{aligned}
$$

again by an orthogonality argument. It follows that

$$
a_{\mu 1}^{(r)} = \eta_{\nu+1}. \tag{6.92}
$$

Now recall that $\eta_{\nu+1} = \cos \chi_{\nu+1}$ is the greatest zero of $G_{\nu+1}$, and hence of $C_{\nu+1}^{\frac{r-2}{2}}$. By Theorem 2.11,

$$
\chi_{\nu+1} \sim j_{\frac{r-3}{2},1} \cdot \frac{1}{\nu+1} \tag{6.93}
$$

holds for $\nu \to \infty$. This implies

$$
a_{\mu 1}^{(r)} = 1 - 2 \cdot j_{\frac{r-3}{2},1}^2 \cdot \frac{1}{\mu^2} + o(\mu^{-2}), \tag{6.94}
$$

as $\mu \to \infty$. Together with (6.91) this yields that the assumption (ii) of Theorem 6.29 is also satisfied, and the convergence of the operators follows from this theorem.

(ii) It suffices again to consider the case $\mu = 2\nu$. So let $\lambda := \frac{r-2}{2}$ and $x \in S^{r-1}$. Using (1.27) we get

$$
G_{\nu}(1) = \int_{S^{r-1}} G_{\nu}^2(tx)\, d\omega(t) = \omega_{r-2} \int_{-1}^{1} G_{\nu}^2(\xi)(1-\xi^2)^{\frac{r-3}{2}}\, d\xi = \omega_{r-2}[G_{\nu}, G_{\nu}]_{\lambda},
$$

with the inner product $[\,\cdot\,, \cdot\,]_{\lambda}$ defined in Section 2.3. So the polynomials

$$
P_{\nu} := \sqrt{\frac{\omega_{r-2}}{G_{\nu}(1)}} \cdot G_{\nu}
$$

are orthonormal, and the formula of *Christoffel-Darboux*, see Theorem 2.4, takes the form

$$
\frac{k_{\nu}}{k_{\nu+1}} \cdot \frac{G_{\nu+1}(\xi)G_{\nu}(\eta) - G_{\nu}(\xi)G_{\nu+1}(\eta)}{(\xi - \eta)\sqrt{G_{\nu+1}(1)G_{\nu}(1)}} = \sum_{\kappa=0}^{\nu} \frac{G_{\kappa}(\xi)G_{\kappa}(\eta)}{G_{\kappa}(1)},
$$

where the leading coefficients k_{ν} of P_{ν} are positive. Inserting $\eta := \eta_{\nu+1}$ we get

$$
\frac{G_{\nu+1}(\xi)}{\xi - \eta_{\nu+1}} = \frac{k_{\nu+1}}{k_{\nu}} \cdot \frac{\sqrt{G_{\nu+1}(1)G_{\nu}(1)}}{G_{\nu}(\eta_{\nu+1})} \cdot \sum_{\kappa=0}^{\nu} \frac{G_{\kappa}(\xi)G_{\kappa}(\eta_{\nu+1})}{G_{\kappa}(1)}, \tag{6.95}
$$

where it follows from the interlacing property of the zeros of orthogonal polynomials that

$$G_\kappa(\eta_{\nu+1}) > 0$$

holds for $\kappa = 0, 1, \ldots, \nu$. In other words,

$$\frac{G_{\nu+1}(\xi)}{\xi - \eta_{\nu+1}} = \sum_{\kappa=0}^{\nu} c_{\nu,\kappa} \cdot G_\kappa(\xi)$$

holds with well-known positive coefficients $c_{\nu,\kappa}$, and we get

$$K_\mu^{(r)}(\xi) = g_{\nu+1} \sum_{\iota=0}^{\nu} \sum_{\kappa=0}^{\nu} c_{\nu,\iota} c_{\nu,\kappa} \cdot G_\iota(\xi) G_\kappa(\xi). \tag{6.96}$$

Finally we use the *linearisation formulae of Rogers and Ramanujan*, by which $C_\iota^{\frac{r-2}{2}} \cdot C_\kappa^{\frac{r-2}{2}}$ is a nonnegative linear combination of the polynomials $C_0^{\frac{r-2}{2}}, \ldots,$ $C_{\iota+\kappa}^{\frac{r-2}{2}}$, see Gasper [20]. We bring them to the form

$$
\begin{aligned}
G_\iota G_\kappa \;=\; & \frac{(\iota + \lambda)(\kappa + \lambda)}{\omega_{r-1} \lambda^2} \\
\times\; & \sum_{k=0}^{min\{\iota,\kappa\}} \frac{(\lambda)_k}{(1)_k} \cdot \frac{(\lambda)_{\iota-k}}{(1)_{\iota-k}} \cdot \frac{(\lambda)_{\kappa-k}}{(1)_{\kappa-k}} \cdot \frac{(1)_{\iota+\kappa-2k}}{(2\lambda)_{\iota+\kappa-2k}} \cdot \frac{(2\lambda)_{\iota+\kappa-k}}{(\lambda+1)_{\iota+\kappa-k}} \cdot G_{\iota+\kappa-2k},
\end{aligned}
$$

which is more convenient in our context. Obviously, replacing the products $G_\iota G_\kappa$ in (6.96) with the help of these formulas, it follows, together with $g_{\nu+1} > 0$, that all coefficients in the expansion (6.79) of the kernel $K_\mu = K_\mu^{(r)}$ are nonnegative. The upper bound for the $a_{\mu\nu}^{(r)}$ follows from Theorem 6.29.

(*iii*) By definition,

$$g_{\nu+1}^{-1} = \int_{S^{r-1}} \left[\frac{G_{\nu+1}(tx)}{(tx) - \eta_{\nu+1}} \right]^2 d\omega(t)$$

holds for arbitrary $x \in S^{r-1}$. Using orthogonality and the reproducing property of the G_κ we obtain from (6.95),

$$g_{\nu+1}^{-1} = \left[\frac{k_{\nu+1}}{k_\nu} \cdot \frac{\sqrt{G_{\nu+1}(1) G_\nu(1)}}{G_\nu(\eta_{\nu+1})} \right]^2 \cdot \sum_{\kappa=0}^{\nu} \frac{G_\kappa^2(\eta_{\nu+1})}{G_\kappa(1)}.$$

Using (6.95) once more, but with ξ tending to $\eta_{\nu+1}$, we get

$$g_{\nu+1}^{-1} = \frac{k_{\nu+1}}{k_\nu} \cdot \sqrt{G_{\nu+1}(1) G_\nu(1)} \cdot \frac{G_{\nu+1}'(\eta_{\nu+1})}{G_\nu(\eta_{\nu+1})}.$$

k_ν is the leading coefficient of P_ν, i.e., of

$$P_\nu = \sqrt{\omega_{r-2}G_\nu(1)} \cdot \tilde{C}_\nu^{\frac{r-2}{2}},$$

which we evaluate with the help of (2.13) and of (2.5), if $r \geq 3$, and with the help of (2.8), if $r = 2$. In both cases we obtain for $\nu \in \mathbb{N}$,

$$\frac{k_{\nu+1}}{k_\nu} = \sqrt{\frac{G_{\nu+1}(1)}{G_\nu(1)}} \cdot \frac{2\nu + r - 2}{\nu + r - 2},$$

and hence

$$g_{\nu+1}^{-1} = \frac{2\nu + r - 2}{\nu + r - 2} \cdot G_{\nu+1}(1) \cdot \frac{G'_{\nu+1}(\eta_{\nu+1})}{G_\nu(\eta_{\nu+1})}.$$

For $r \geq 3$ this yields in view of (4.13),

$$g_{\nu+1}^{-1} = \frac{(2\nu + r)^2}{(r-2)\omega_{r-1}} \cdot C_{\nu+1}^{\frac{r-2}{2}}(1) \cdot \frac{(\frac{d}{d\xi}C_{\nu+1}^{\frac{r-2}{2}})(\eta_{\nu+1})}{(\nu+r-2)C_\nu^{\frac{r-2}{2}}(\eta_{\nu+1})}.$$

Finally we use the differential equation

$$(1-\xi^2)\frac{d}{d\xi}C_{\nu+1}^{\frac{r-2}{2}} = -(\nu+1)\xi C_{\nu+1}^{\frac{r-2}{2}} + (\nu+r-2)C_\nu^{\frac{r-2}{2}},$$

which holds for arbitrary r, see, e.g., Problem 2.2. For $\xi := \eta_{\nu+1}$ it provides us with a simple expression for the derivative, and inserting this above we get, together with (2.13),

$$\begin{aligned}
g_{\nu+1}^{-1} &= \frac{(2\nu + r)^2}{(r-2)\omega_{r-1}} \cdot \frac{(r-2)_{\nu+1}}{(1)_{\nu+1}} \cdot \frac{1}{1 - \eta_{\nu+1}^2} \\
&= \frac{(2\nu + r)^2}{\omega_{r-1}\sin^2\chi_{\nu+1}} \cdot \frac{1}{(\nu+1)} \cdot \binom{\nu+r-2}{r-2},
\end{aligned}$$

as claimed. For $r = 2$ we get the same result, again by (4.13), which takes now the form $G_\nu = \frac{\nu}{\omega_{r-1}}C_\nu^0$, see (2.8).

So it is left to consider the case $\nu = 0$, i.e., to evaluate g_1^{-1}. But in this case we get

$$G_1(\xi) = \frac{r}{\omega_{r-1}} \cdot \xi,$$

both for $r = 2$ and for $r \geq 3$, where $\eta_1 = 0$ and $\chi_1 = \frac{\pi}{2}$, and we get

$$g_1^{-1} = \int_{S^{r-1}} \left(\frac{r}{\omega_{r-1}}\right)^2 d\omega(t) = \frac{r^2}{\omega_{r-1}},$$

again as claimed. This finishes the proof.   □

Numerical Evaluation of $A^{(r)}$

If the $A^{(r)}$-summation method is to be applied in practice, all depends on a safe calculation of the submatrix needed.

So it is important that the $a_{\mu\nu}^{(r)}$, which occur as the coefficients in the representation (5.90) of the kernel, can be generated by 'linearisation' of the right side of (6.96) by means of the formulae of Rogers and Ramanujan. This seems to be a rather complicated process. However, in spite of the large number of arithmetic operations, which is required in the evaluation of some of the first submatrices of $A^{(r)}$, this is a numerically very stable method. For, the coefficients occurring in the formulae of Rogers and Ramanujan are large products, whose numerical evaluation has the utmost stability with respect to the relative error. Moreover, the $a_{\mu\nu}^{(r)}$ occur as the sum of many, but positive elements, whose numerical evaluation can be performed again by a very stable algorithm.

A Comparison of the Cesàro and of the Newman–Shapiro Operators

Assume A is one of the Cesàro matrices, or the Newman–Shapiro matrix, while the L_μ are the corresponding operators (6.78). In both cases we get $a_{\mu 0} = 1$ and hence $L_\mu 1 = 1$ for $\mu \in \mathbb{N}_0$. Next let $e \in S^{r-1}$. We define F by $F(x) := ex$ for $x \in S^{r-1}$. Then we get from (6.78) and (6.79),

$$F(x) - \left(L_\mu F\right)(x) = (1 - a_{\mu 1})F(x)$$

for $x \in S^{r-1}$, and hence

$$\|F - L_\mu F\|_\infty = |1 - a_{\mu 1}|.$$

In the Cesàro case $A = C^{(k)}$ we obtain

$$1 - a_{\mu 1} = \frac{k}{\mu + k} \quad \text{for } \mu \in \mathbb{N},$$

see (6.85). In the Newman–Shapiro case $A = A^{(r)}$ we get

$$1 - a_{\mu 1} \sim 2 \cdot j_{\frac{r-3}{2},1}^2 \cdot \frac{1}{\mu^2}$$

for $\mu \to \infty$, see (6.94). So the operators differ in their behaviour essentially, namely by satisfying

$$\|F - L_\mu^{(k)} F\|_\infty \approx \tfrac{1}{\mu}, \quad \text{in the Cesàro case } k \neq 0,$$
$$\|F - L_\mu F\|_\infty \approx \tfrac{1}{\mu^2}, \quad \text{in the Newman–Shapiro case,}$$

both for $\mu \to \infty$. In the Cesàro case $k \neq 0$, and in particular in the case $k \geq r - 1$ where the operators are positive, a simple function such as F cannot be approximated better by $L_\mu^{(k)} F$ than at the order $\frac{1}{\mu}$. Nor does it help to impose further conditions on the function, like smoothness of any order, since F has already all desirable properties. For this reason, the Cesàro methods are said to be $\frac{1}{\mu}$-*saturated*. Likewise the Newman–Shapiro method is $\frac{1}{\mu^2}$-*saturated*.

Remark. By a theorem of Korovkin, it is already in the univariate case impossible, to obtain by positive polynomial operators a better saturation order than $\frac{1}{\mu^2}$, see Lorentz [34], p. 94. We investigate the approximating behaviour of the Newman–Shapiro operators in more detail in Section 6.9.

6.8   Generalized Hyperinterpolation

In this section $A = \left(a_{\mu\nu} \right)_{\mu,\nu=0,1,\ldots}$ is an arbitrary, but fixed matrix which satisfies the assumptions of Theorem 6.29. By (6.78), this means by

$$(L_\mu F)(x) = \int\limits_{S^{r-1}} F(t) K_\mu(tx) \, d\omega(t),$$

it defines a sequence of positive linear operators $L_\mu : C(S^{r-1}) \to \mathbb{P}_\mu^r(S^{r-1})$, $\mu \in \mathbb{N}_0$.

Now let us combine this sequence with a sequence of positive quadratures

$$\hat{I}_\mu = \sum_{j=1}^M A_j \hat{E}_j,$$

$\mu \in \mathbb{N}_0$, where $\hat{I}_\mu$ is exact of degree 2μ, and to be used in the evaluation of the defining integral of L_μ. Then we obtain a sequence of *discrete operators*

$$\hat{L}_\mu : C(S^{r-1}) \to \mathbb{P}_\mu^r(S^{r-1}),$$

which have the form

$$\left(\hat{L}_\mu F\right)(x) = \sum_{j=1}^M A_j F(t_j) K_\mu(t_j x) \tag{6.97}$$

for $\mu \in \mathbb{N}_0$, $F \in C(S^{r-1})$, and $x \in S^{r-1}$. We do not specify that M, the weights A_j, and the nodes $t_j \in S^{r-1}$ depend on μ.

Formally (6.97) agrees with (6.70), i.e., $\hat{L}_\mu$ has the form of a hyperinterpolation operator. In particular, in case of the matrix

$$A = C^{(0)} = \begin{pmatrix} 1 & & & & \\ 1 & 1 & & & \\ 1 & 1 & 1 & & \\ & & & \ddots & \\ 1 & 1 & 1 & & 1 \end{pmatrix},$$

it follows from (6.79) and (4.28) that $K_\mu = \sum_{\nu=0}^{\mu} G_\nu = \Gamma_\mu$ holds, such that the $\hat{L}_\mu$ are, actually, hyperinterpolation operators. This justifies the following definition.

Definition 6.9 (Generalized Hyperinterpolation). *Let* $r \in \mathbb{N} \setminus \{1\}$. *Assume* A *satisfies the assumptions of Theorem 6.29, and* $\{\hat{I}_\mu\}_{\mu \in \mathbb{N}_0}$ *is a sequence of positive quadratures* $\hat{I}_\mu$ *which are exact of degree* 2μ. *Then the operators* $\hat{L}_\mu$ *are called generalized hyperinterpolation operators.*

Theorem 6.34 (Convergence of Generalized Hyperinterpolation). *Every sequence of generalized hyperinterpolation operators on* $C(S^{r-1})$ *converges in the uniform norm to the identity.*

Proof. The proof is necessary since, compared with the L_μ, the $\hat{L}_\mu$ form a quite new operator sequence.

So let $\{\hat{L}_\mu\}_{\mu \in \mathbb{N}_0}$ be a sequence of generalized hyperinterpolation operators. By their definition, they are positive operators. Moreover, for $\mu \in \mathbb{N}$ and $e \in S^{r-1}$ we get, using (6.78) and (6.79),

$$\begin{aligned}
\big(\hat{L}_\mu 1\big)(x) &= \big(L_\mu 1\big)(x) = a_{\mu 0} \cdot 1, \\
\big(\hat{L}_\mu(e\,\cdot)\big)(x) &= \big(L_\mu(e\,\cdot)\big)(x) = a_{\mu 1} \cdot (ex),
\end{aligned}$$

for $x \in S^{r-1}$. In view of the assumption (ii) of Theorem 6.29, this implies

$$\lim_{\mu \to \infty} \hat{L}_\mu F = F$$

in $\| \cdot \|_\infty$ for all $F \in \{1, x_1, \ldots, x_r\}$, which is a Korovkin set in $C(S^{r-1})$, see Theorem 5.4. So Theorem 5.5 yields $\lim_{\mu \to \infty} \hat{L}_\mu F = F$ in $\| \cdot \|_\infty$ for all $F \in C(S^{r-1})$, as claimed. $\qquad\square$

Remark. The evaluation of generalized hyperinterpolation uses the same number of function values as the evaluation of hyperinterpolation or interpolation itself. The unique change is now that the kernel is no longer a pure Jacobi polynomial, though it is still possible to evaluate it by a three-term recurrence relation, as follows.

Evaluation of the Kernels by a Three-Term Recurrence Relation

Using (4.12), we bring the kernel (6.79) of a generalized hyperinterpolation operator to the form

$$K_\mu = \sum_{\nu=0}^{\mu} b_\nu C_\nu^\lambda,$$

where $\lambda := \frac{r-2}{2}$. We do not specify that the coefficients b_ν depend on μ. We are going to use the recurrence relation of the Gegenbauer polynomials, which we write in the form

$$C_\mu^\lambda = \beta_{\mu-1}\, \xi C_{\mu-1}^\lambda + \gamma_{\mu-2} C_{\mu-2}^\lambda,$$

$\mu = 2, 3, \ldots$. Note that the coefficients are well known, see Problem 2.1. We use it in order to replace, step by step, C_μ^λ, $C_{\mu-1}^\lambda$, ... by Gegenbauer polynomials of lower degree. In the first step we get, for instance,

$$K_\mu = (b_\mu \cdot \beta_{\mu-1} \cdot \xi + b_{\mu-1}) C_{\mu-1}^\lambda + (b_\mu \cdot \gamma_{\mu-2} + b_{\mu-2}) C_{\mu-2}^\lambda + \sum_{\nu=0}^{\mu-3} b_\nu C_\nu^\lambda.$$

In the second step, $C_{\mu-1}^\lambda$ is replaced likewise, and so on. This furnishes, finally, the following algorithm.

Algorithm (*Evaluation* $\xi \mapsto K_\mu(\xi)$)

$$s_{\mu+1} := 0, \quad s_\mu := b_\mu,$$

$$s_\nu := b_\nu + \xi \cdot \beta_\nu \cdot s_{\nu+1} + \gamma_\nu \cdot s_{\nu+2}, \quad for\ \nu = \mu - 1, \mu - 2, \ldots, 0,$$

$$K_\mu(\xi) := s_0.$$

Complexity

The algorithm hides the influence of the matrix A. Apparently, in a single evaluation it uses, in the general case, 3μ multiplications and 2μ additions, this means up to at most 2μ multiplications more than the well-known Clenshaw algorithms use. But, most important, the costs are still of order μ.

The evaluation of $\hat{L}_\mu F$ from (6.97) requires M evaluations of the kernel. By means of the fast Fourier transform (FFT), $\mu + 1$ evaluations can be performed simultaneously at $\mathcal{O}(\mu \log \mu)$ arithmetical operations. It follows that the evaluation of (6.97) is possible at $\mathcal{O}\left(M \cdot \log \mu\right) = \mathcal{O}\left(\mu^{r-1} \log \mu\right)$ arithmetical operations, if a product Gauß quadrature is used, see Theorem 6.19.

Stability

In the case $r = 4$, i.e., $\lambda = 1$, where the C_μ^λ are the Chebyshev polynomials U_μ, the evaluation algorithm from above can be identified with the backward directed Clenshaw algorithm of the second kind. This is known to be extremely stable, as its *error norm*, which is a measure for the relative error caused by round-off and by the initial value errors, grows at the order $\mu^2 \log \mu$, only, where μ^2 is a minimal possible order, see Reimer [44]. So there is reason to assume that the error norm grows at a low polynomial order also in the general case.

Final Remarks

Generalized hyperinterpolation is a convergent approximation process by positive operators, which are order preserving, of low complexity, and presumably numerically stable of polynomial order. In the next section we show, that if hyperinterpolation is based on the Newman–Shapiro matrix $A^{(r)}$, convergence takes place, in some sense, even at the best possible order.

6.9 Moduli of Continuity and the Approximation Order

In this section we assume $A = A^{(r)}$, such that the $L_\mu^{(r)}$ and $K_\mu^{(r)}$ are the Newman–Shapiro operators and their kernels, respectively. They are of particular value in what follows.

For $F \in C(S^{r-1})$

$$\lim_{\mu \to \infty} \|F - L_\mu^{(r)} F\|_\infty = 0$$

holds, see Theorem 6.33. We ask how quickly convergence takes place. Because of

$$E_\mu(F) \leq \|F - L_\mu^{(r)} F\|_\infty,$$

every upper bound for the right side provides us also with an estimate of the minimal deviation $E_\mu(F)$ of F in $\mathbb{P}_\mu^r(S^{r-1})$. This is important, in particular, in view of the inequality

$$\|F - \Lambda_\mu F\|_\infty \leq (1 + \|\Lambda_\mu\|_\infty) E_\mu(F),$$

which holds for arbitrary projections Λ_μ onto $\mathbb{P}_\mu^r(S^{r-1})$, see Theorem 5.11.

Moduli of Continuity

Moduli of continuity are used to measure the approximation error. Before we introduce them on $C(S^{r-1})$, we recall their univariate versions for the space $C_{2\pi}$ of 2π-periodic continuous real functions on $\mathbb{R}$.

For $\alpha \geq 0$, $j \in \mathbb{N}$, the j-th forward difference of $f \in C_{2\pi}$ at the point $\xi \in \mathbb{R}$ can be represented in the form

$$\left(\Delta_\alpha^j f\right)(\xi) = \sum_{k=0}^{j} (-1)^{j-k} \binom{j}{k} f(\xi + k\alpha).$$

For $\phi \geq 0$ the *modulus of continuity* of order j is defined by

$$\omega_j(f, \phi) := \max\left\{|(\Delta_\alpha^j f)(\xi)| : \xi \in \mathbb{R},\ |\alpha| \leq \phi\right\}.$$

Note that $\omega_j(f, \phi)$ is a monotonically nondecreasing function of ϕ, which vanishes at $\phi = 0$.

The first two moduli are given also by

$$\omega_1(f; \phi) = \max\left\{|f(\xi) - f(\eta)| : \xi, \eta \in \mathbb{R},\ |\xi - \eta| \leq \phi\right\},$$

$$\omega_2(f; \phi) = \max\left\{|f(\xi - \alpha) - 2f(\xi) + f(\xi + \alpha)| : \xi \in \mathbb{R},\ |\alpha| \leq \phi\right\},$$

respectively, and it is easy to see that

$$\omega_2(f, \phi) \leq 2\omega_1(f, \phi) \tag{6.98}$$

holds, again for $\phi \geq 0$.

Moreover, with the help of a well-known representation formula for differences, we get for $f \in C_{2\pi}^{(j)}$ the inequality

$$\omega_j(f, \phi) \le \phi^j \, \|f^{(j)}\|_\infty, \tag{6.99}$$

where $\| \cdot \|_\infty$ is the maximum norm on $C_{2\pi}$. And finally, the inequalities

$$\omega_j(f, \phi) \le (\mu\phi + 1)^j \, \omega_j(f, \tfrac{1}{\mu}), \tag{6.100}$$

hold for $\phi \ge 0$ and arbitrary $\mu \in \mathbb{N}$. For the proof we refer to Lorentz [34], p. 48. Now let $F \in C(S^{r-1})$, where $r \ge 2$. For arbitrary $u, v \in S^{r-1}$, $u \perp v$, the restriction $F_{u,v}$ of F onto the main circle $S^{r-1} \cap span\{u, v\}$ has the representation

$$F_{u,v}(\phi) = F(u \cos\phi + v \sin\phi), \tag{6.101}$$

$\phi \in \mathbb{R}$, where $F_{u,v}$ is a function belonging to $C_{2\pi}$. This enables us to define the moduli of continuity for $F \in C(S^{r-1})$ by

$$\omega_j(F, \phi) := \max\big\{\omega_j(F_{u,v}, \phi) : u, v \in S^{r-1}, \ u \perp v\big\} \tag{6.102}$$

for $\phi \ge 0$ and $j \in \mathbb{N}$. Note that $\omega_j(F, \phi)$ is monotonically nondecreasing, and vanishing at $\phi = 0$, again. Note also that (6.98) yields

$$\omega_2(F, \phi) \le 2\omega_1(F; \phi). \tag{6.103}$$

Now assume $\phi \ge 0$ and $\mu \in \mathbb{N}$. From (6.100) we get for all $u, v \in S^{r-1}$, satisfying $u \perp v$,

$$\omega_j(F_{u,v}, \phi) \le (\mu\phi + 1)^j \, \omega_j(F_{u,v}, \tfrac{1}{\mu}),$$

which implies immediately

$$\omega_j(F, \phi) \le (\mu\phi + 1)^j \, \omega_j(F, \tfrac{1}{\mu}) \tag{6.104}$$

for $F \in C(S^{r-1})$, as in the univariate case.

Before we extend (6.99) to the multivariate case, we have to replace $C_{2\pi}^{(j)}$ by a proper subspace of $C(S^{r-1})$. To this end we call a function $F \in C(S^{r-1})$ for $j \in \mathbb{N}$ j-times continuously differentiable, if $F_{u,v} \in C_{2\pi}^{(j)}$ holds for arbitrary $u, v \in S^{r-1}$, $u \perp v$. In this case $F_{x,u}^{(j)}(0)$ is the directional derivative of F at the point $x \in S^{r-1}$ in the direction of $u \in S^{r-1}$, $u \perp x$, and $F^{(j)}$ is defined by the set-valued function

$$x \mapsto F^{(j)}(x) := \big\{F_{x,u}^{(j)}(0) : u \in S^{r-1}, \ u \perp x\big\},$$

which is the set of all directional j-th derivatives at the point $x \in S^{r-1}$. Now we may define

$$\|F^{(j)}\|_\infty := \sup\big\{\|F_{u,v}^{(j)}\|_\infty : u, v \in S^{r-1}, \ u \perp v\big\},$$

and may introduce the subspace

$$C^{(j)}(S^{r-1}) := \{F \in C(S^{r-1}) : \|F^{(j)}\|_\infty < \infty\}.$$

Obviously, for $F \in C^{(j)}(S^{r-1})$ and all $u, v \in S^{r-1}$ with $u \perp v$, we get in view of (6.99),

$$\omega_j(F_{u,v}, \phi) \leq \phi^j \|F^{(j)}_{u,v}\|_\infty$$

for $\phi \geq 0$ By the definition of $\omega_j(F, \phi)$ and of $\|F^{(j)}\|_\infty$, this implies

$$\omega_j(F, \phi) \leq \phi^j \|F^{(j)}\|_\infty \tag{6.105}$$

for $F \in C^{(j)}(S^{r-1})$, which is the desired multivariate version of the inequality (6.99).

The Approximation Error

Now we return to our original problem, namely to estimate the error $\|F - L_\mu^{(r)}F\|_\infty$ of the Newman–Shapiro operator $L_\mu^{(r)}$. After that, the error of its discretized version is also investigated. The following lemma is basic in both cases.

Lemma 6.35. *Let* $r \in \mathbb{N} \setminus \{1\}$. *Given a constant* $c_r > 0$, *a constant* d_r *exists such that the following holds. If* $A : [-1, 1] \to \mathbb{R}$ *is an arbitrary monotonically nondecreasing function such that*

$$(i) \quad \int_0^\pi K_{2\nu}^{(r)}(\cos\phi)\, dA(\phi) = \frac{1}{\omega_{r-2}},$$

$$(ii) \quad \int_0^\pi \frac{G_{\nu+1}^2(\cos\phi)}{\cos\phi - \cos\chi_{\nu+1}}\, dA(\phi) = 0,$$

$$(iii) \quad \int_0^{\chi_{\nu+1}} dA(\phi) \leq \frac{c_r}{\nu+1},$$

holds for all $\nu \in \mathbb{N}$, *then*

$$\int_0^\pi K_{2\nu}^{(r)}(\cos\phi)(\mu\phi + 1)^j\, dA(\phi) \leq d_r$$

is valid for $j \in \{1, 2\}$, *and again for arbitrary* $\nu \in \mathbb{N}_0$. *The integrals are defined in the sense of Riemann and Stieltjes.*

Proof. The proof is technical and lengthy, and can be found in the Appendix (C).

The Approximation Error of $\mathbf{L}_\mu^{(r)}$

Theorem 6.36 (Approximation Error of $\mathbf{L}_\mu^{(r)}$). *Let* $r \in \mathbb{N} \setminus \{1\}$. *There is a constant* $k_r > 0$ *such that the following holds. For all* $F \in C(S^{r-1})$ *and* $\mu \in \mathbb{N}$, *the approximation error of the Newman–Shapiro operator* $L_\mu^{(r)}$ *satisfies*

$$\|F - L_\mu^{(r)}F\|_\infty \leq k_r \cdot \omega_2\left(F, \tfrac{1}{\mu}\right).$$

Proof. $L_\mu = L_\mu^{(r)}$ is defined by (6.78), with the kernel (6.88). It suffices to consider the case $\mu = 2\nu$, $\nu \in \mathbb{N}$.

So let $F \in C(S^{r-1})$ and $x \in S^{r-1}$ be given. In view of (6.89) we obtain

$$\left(L_\mu^{(r)} F\right)(x) - F(x) = \int_{S^{r-1}} \left(F(t) - F(x)\right) K_\mu^{(r)}(tx)\, d\omega(t)$$

$$= \int_{-1}^{1} K_\mu^{(r)}(\xi) \int_{S_0^{r-1}} \left(F(\xi x + \sqrt{1-\xi^2}\, u) - F(x)\right)(1-\xi^2)^{\frac{r-3}{2}}\, d\bar\omega(u)d\xi,$$

where $d\bar\omega(u)$ is the surface element of the 'equator'

$$S_0^{r-1} := \{u \in S^{r-1} \mid ux = 0\},$$

which is a unit sphere S^{r-2}. Replacing u by $-u$ we get a similar equation, and taking the average we get

$$\left(L_\mu^{(r)} F\right)(x) - F(x) = \tfrac{1}{2} \int_{-1}^{1} K_\mu^{(r)}(\xi) \int_{S_0^{r-1}}$$
$$\left(F(\xi x + \sqrt{1-\xi^2}\, u) - 2F(x) + F(\xi x - \sqrt{1-\xi^2}\, u)\right)(1-\xi^2)^{\frac{r-3}{2}}\, d\bar\omega(u)d\xi.$$

Substituting $\xi = \cos\phi$ and using (6.101), we bring this to the form

$$\left(L_\mu^{(r)} F - F\right)(x)$$
$$= \tfrac{1}{2} \int_0^\pi K_\mu^{(r)}(\cos\phi) \int_{S_0^{r-1}} \left(F_{x,u}(\phi) - 2F_{x,u}(0) + F_{x,u}(-\phi)\right)(\sin\phi)^{r-2} d\bar\omega(u)\, d\phi,$$

and with the aid of (6.102) and (6.104) we get

$$\left|\left(L_\mu^{(r)} F - F\right)(x)\right| \leq \tfrac{1}{2} \int_0^\pi K_\mu^{(r)}(\cos\phi) \int_{S_0^{r-1}} \omega_2(F,\phi)(\sin\phi)^{r-2} d\bar\omega(u)\, d\phi$$

$$\leq \tfrac{1}{2} \omega_2(F, \tfrac{1}{\mu}) \cdot \int_0^\pi K_\mu^{(r)}(\cos\phi)(\mu\phi + 1)^2(\sin\phi)^{r-2}\, d\phi \cdot \int_{S_0^{r-1}} d\bar\omega(u). \qquad (6.106)$$

It is left to prove that the integrals

$$I_{\nu,2} = \int_0^\pi K_\mu^{(r)}(\cos\phi)(\mu\phi + 1)^2(\sin\phi)^{r-2}\, d\phi$$

are bounded. To this end let A be defined by $A(0) = 0$ and

$$dA(\phi) = (\sin\phi)^{r-2}\, d\phi.$$

We check the assumptions of Lemma 6.35. Apparently, (i) holds in view of (1.27) and of (6.89), and (ii) holds likewise by an orthogonality argument. Moreover,

$$\int_0^{\chi_{\nu+1}} dA(\phi) = \int_0^{\chi_{\nu+1}} (\sin \phi)^{r-2}d\phi \leq \int_0^{\chi_{\nu+1}} \phi^{r-2}d\phi = \frac{1}{r-1} \cdot (\chi_{\nu+1})^{r-1}$$

is valid, and in view of (6.93) a constant c_r exists such that (iii) holds.

Now that the assumptions of Lemma 6.35 are satisfied, a constant d_r exists such that $I_{\nu,\cdot} \leq d_r$ is valid for $\nu \in \mathbb{N}_0$. Finally we define the constant k_r by

$$k_r := \frac{1}{2} \cdot d_r \cdot \int_{S_0^{r-1}} d\bar{\omega}(u),$$

and the statement of the theorem follows immediately from (6.106).   □

Corollary 6.37 (Jackson's Inequality on S^{r-1}, Newman–Shapiro). *Assumptions and $k_r > 0$ as in Theorem 6.36. Then*

$$\|F - L_\mu^{(r)}F\|_\infty \leq 2\,k_r\,\omega_1(F, \tfrac{1}{\mu})$$

holds for all $F \in C(S^{r-1})$ and arbitrary $\mu \in \mathbb{N}$.

Proof. In view of (6.103), the statement follows from Theorem 6.36.   □

Corollary 6.38. *Assumptions and $k_r > 0$ as in Theorem 6.36. For $j \in \{1,2\}$ and $F \in C^{(j)}(S^{r-1})$,*

$$\|F - L_\mu^{(r)}F\|_\infty \leq j k_r \cdot \|F^{(j)}\|_\infty \cdot \frac{1}{\mu^j}$$

holds for $\mu \in \mathbb{N}$.

Proof. In view of (6.105), the statement follows from Theorem 6.36 and Corollary 6.37, respectively.   □

Remark. We remarked at the end of Section 6.7 that the approximation by positive polynomial operators cannot be better than of order $\mathcal{O}(\frac{1}{\mu^2})$. In this sense the $L_\mu^{(r)}$ are approximating smooth functions at the best possible order.

The Approximation Error of $\hat{L}_\mu^{(r)}$

In the preceding, the approximation error of the Newman–Shapiro operators $L_\mu^{(r)}$ is estimated by the moduli of continuity of the first and of the second order, respectively. If we evaluate $L_\mu^{(r)}$ by a positive quadrature which is exact of degree 2μ to get the generalized hyperinterpolation operator $\hat{L}_\mu^{(r)}$,

$$\left(\hat{L}_\mu^{(r)}F\right)(x) = \sum_{j=1}^M A_j F(t_j) K_\mu^{(r)}(t_j x), \tag{6.107}$$

$F \in C(S^{r-1})$, $x \in S^{r-1}$, then the estimation is disturbed, and it is quite uncertain whether similar bounds hold for $\hat{L}_\mu^{(r)}$ instead of $L_\mu^{(r)}$. Fortunately we can give a positive answer for the modulus of continuity of the first order.

Theorem 6.39. *Approximation Error of $\hat{L}_\mu^{(r)}$, Reimer] Let $r \in \mathbb{N} \setminus \{1\}$. There is a constant $\hat{k}_r$ such that the following holds. For all $F \in C(S^{r-1})$ and $\mu \in \mathbb{N}$,*

$$\|F - \hat{L}_\mu^{(r)} F\|_\infty \leq \hat{k}_r \omega_1(F, \tfrac{1}{\mu})$$

is valid. The constant does not depend on the choice of the quadratures used in the definition of $\hat{L}_\mu^{(r)}$.

Proof. It suffices to consider the case $\mu = 2\nu$, $\nu \in \mathbb{N}$. By the assumption on the quadratures we get

$$\sum_{j=0}^{M} A_j K_\mu^{(r)}(t_j x) = \int_{S^{r-1}} K_\mu^{(r)}(tx) d\omega(t) = 1$$

for $x \in S^{r-1}$, see (6.89), and for arbitrary $F \in C(S^{r-1})$ we obtain

$$\left(\hat{L}_\mu^{(r)} F\right)(x) - F(x) = \sum_{j=0}^{M} A_j \left(F(t_j) - F(x)\right) K_\mu^{(r)}(t_j x).$$

Every $t \in S^{r-1}$ can be written in the form $t = (tx)x + \sqrt{1 - (tx)^2}\, u$, where $u \in S^{r-1}$, $u \perp x$. Hence we get, with $tx = \cos\phi$,

$$|F(t) - F(x)| = |F_{xu}(\phi) - F_{xu}(0)| \leq \omega_1(F, \phi).$$

Using the weight sum function from Definition 6.6, we get

$$\left|\left(\hat{L}_\mu^{(r)} F - F\right)(x)\right| \leq \sum_{j=0}^{M} A_j \omega_1\left(F, \arccos(t_j x)\right) K_\mu^{(r)}(t_j x)$$

$$= \int_0^\pi \omega_1(F, \phi) K_\mu^{(r)}(\cos\phi)\, dA_x(\phi)$$

$$\leq \omega_1(F, \tfrac{1}{\mu}) \cdot \int_0^\pi K_\mu^{(r)}(\cos\phi)(\mu\phi + 1)\, dA_x(\phi), \quad (6.108)$$

where we used (6.104). We want to apply Lemma 6.35 with $A := A_x$, so we check its assumptions.

(i) Using (1.27) and (6.89), again, we obtain

$$\omega_{r-2} \int_0^\pi K_\mu^{(r)}(\cos\phi)\, dA_x(\phi) = \sum_{j=0}^{M} A_j K_\mu^{(r)}(t_j x) = \int_{S^{r-1}} K_\mu^{(r)}(tx)\, d\omega(t) = 1,$$

such that (i) is satisfied.

(ii) Likewise we obtain, again using orthogonality,

$$\omega_{r-2}\int_0^{\vec{\pi}}\frac{G_{\nu+1}^2(\cos\phi)}{\cos\phi-\cos\chi_{\nu+1}}\,dA_x(\phi) \;=\; \sum_{j=0}^M A_j\,\frac{G_{\nu+1}^2(t_jx)}{t_jx-\eta_{\nu+1}}$$

$$=\;\int_{S^{r-1}}G_{\nu+1}(tx)\cdot\frac{G_{\nu+1}(tx)}{tx-\eta_{\nu+1}}\,d\omega(t)\;=\;0,$$

and (ii) is also satisfied.

(iii) Finally we get from Theorem 6.24, with $\chi_{\nu+1}=\chi_{\nu+1,1}$,

$$\int_0^{\chi_{\nu+1}}dA_x(\phi)\;=\;\sum_{t_jx\ge\cos\chi_{\nu+1}}A_j\;\le\;\frac{c_r}{(\nu+1)^{r-1}},$$

with some constant c_r, and (iii) holds at this choice of c_r.

So, all assumptions of Lemma 6.35 are satisfied, and a constant $\hat{k}_r$ exists such that the factor occurring with $\omega_1(F,\frac{1}{\mu})$ in (6.108) is bounded by $\hat{k}_r$. As the constant does not depend on x, this finishes the proof. □

Remark 1. Corollary 6.37 yields $\lim_{\mu\to\infty}\|F-L_\mu^{(r)}F\|_\infty=0$. In view of $L_\mu F\in\mathbb{P}_\mu^r(S^{r-1}$, this is a direct proof of the Theorem of Weierstrass for the sphere $D=S^{r-1}$.

Remark 2. An estimate of the error with the help of $\omega_2(F,\frac{1}{\mu})$ would require that if

$$t_j\;=\;(t_jx)x+\sqrt{1-(t_jx)^2}\,u_j,$$

$u_j\in S^{r-1}$, $u_j\perp x$, is a quadrature node, then its counterpart

$$t_j\;=\;(t_jx)x-\sqrt{1-(t_jx)^2}\,u_j$$

is also, and this for arbitrary $x\in S^{r-1}$. In general this condition cannot be satisfied.

Remark 3. Apart from the particular shape of the kernel, (6.107) has exactly the form of (6.70). So it is worthwhile to compare the weighted kernels. The weighted kernel $\Gamma_\mu^r/\Gamma_\mu^r(1)$ has been discussed in Remark 2 at the end of Section 6,6. We refer in particular to Figure 6.1. Now let

$$\hat{K}_\mu^{(r)}(\xi)\;:=\;K_\mu^{(r)}(\xi)/\Gamma_\mu^r(1)$$

be the corresponding weighted Newman–Shapiro kernel, which is, of course, nonnegative for $-1\le\xi\le1$. It follows from (6.88), (4.13), Theorem 6.33, and Theorem 2.11, that

$$J\;:=\;\lim_{\mu\to\infty}\hat{K}_\mu^{(3)}(1)\;=\;\frac{1}{j_{0,1}^2}\;\approx\;0.1724$$

holds. The behaviour of these kernels is demonstrated by Figure 6.2. Actually, they are rapidly decreasing to zero left from unity, thus making visible a strong localisation property of the discretized Newman–Shapiro operators.

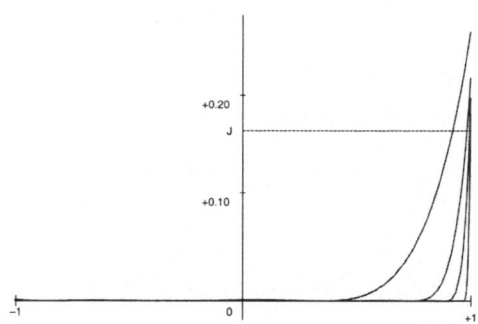

Figure 6.2. Weighted Newman–Shapiro Kernels
$(r = 3,\ \mu = 6,\ 12,\ 20,\ \text{and}\ 40)$.

Remark 4. $L_\mu^{(r)}F$ takes advantage of the *local* behaviour of F, and so does $\hat{L}_\mu^{(r)}F$. So it is reasonable to truncate the sum (6.97) by omitting all contributions of t_j-s which do not belong to a given neighbourhood of x (*truncated generalized hyperinterpolation*). This method reduces the arithmetical cost essentially.

6.10   Truncated Generalized Hyperinterpolation

Let $r \in \mathbb{N} \setminus \{1, 2\}$ be fixed. In what follows, we explain the concept of   truncated generalized hyperinterpolation in case of the discretized Newman–Shapiro operators.

For $\nu \in \mathbb{N}$ let $\mu := 2\nu$, and note that $\eta_{\nu+1} = \cos\chi_{\nu+1} > 0$ holds. We introduce *truncation radii* $\beta_{\nu+1}$, which are assumed to satisfy

$$\chi_{\nu+1} < \beta_{\nu+1} \leq \tfrac{\pi}{2}, \tag{6.109}$$

and put $\xi_{\nu+1} := \cos\beta_{\nu+1}$, such that

$$0 \leq \xi_{\nu+1} < \eta_{n+1} < 1$$

is valid. On the right side of (6.107) we omit all the terms which belong to a node

$$t_j \notin C(x, \beta_{\nu+1}),$$

and obtain the *truncated* discretized Newman–Shapiro operator $\hat{M}_\mu^{(r)}$ defined by

$$\left(\hat{M}_\mu^{(r)} F\right)(x) := \sum_{t_j \in C(x,\beta_{\nu+1})} A_j F(t_j) K_\mu^{(r)}(t_j x)$$

for $F \in C(S^{r-1})$ and $x \in S^{r-1}$, whose evaluation is much cheaper than the evaluation of $\hat{L}_\mu^{(r)}$ itself. In what follows we investigate, first, the error caused by truncation. After that we compare the result with the cost reduction.

To begin with, let us write (6.107) in the form

$$\left(\hat{L}_\mu^{(r)} F\right)(x) = \left(\hat{M}_\mu^{(r)} F\right)(x) + R + S, \tag{6.110}$$

where

$$R = \sum_{0 \le t_j x < \xi_{\nu+1}} A_j F(t_j) K_\mu^{(r)}(t_j x),$$

$$S = \sum_{-1 \le t_j x < 0} A_j F(t_j) K_\mu^{(r)}(t_j x).$$

In view of the kernel definition (6.88) we obtain

$$|S| \le \|F\|_\infty \cdot g_{\nu+1} \cdot \sum_{-1 \le t_j x < 0} A_j \left[\frac{G_{\nu+1}(t_j x)}{t_j x - \eta_{\nu+1}} \right]^2$$

$$\le \|F\|_\infty \cdot g_{\nu+1} \cdot \frac{1}{\eta_{\nu+1}^2} \cdot \sum_{j=1}^{M} A_j \left[G_{\nu+1}(t_j x) \right]^2$$

$$= \|F\|_\infty \cdot g_{\nu+1} \cdot \frac{1}{\eta_{\nu+1}^2} \cdot \int_{S^{r-1}} \left[G_{\nu+1}(t_j x) \right]^2 d\omega(t)$$

$$= \|F\|_\infty \cdot g_{\nu+1} \cdot \frac{1}{\eta_{\nu+1}^2} \cdot G_{\nu+1}(1),$$

where we used the reproducing property of the kernel in the last step. By the well-known interlacing property of the zeros of orthogonal polynomials we obtain

$$0 < \eta_2 < \eta_3 < \cdots < 1,$$

and hence

$$|S| \le \|F\|_\infty \cdot g_{\nu+1} \cdot \frac{1}{\eta_2^2} \cdot G_{\nu+1}(1),$$

where $g_{\nu+1} \approx (\nu+1)^{-(r+1)}$ follows from Theorem 6.33, and $G_{\nu+1}(1) \approx (\nu+1)^{r-2}$ from (4.12) and (4.2). Together this yields that a constant $\gamma_{r,1} > 0$ exists such that

$$|S| \le \gamma_{r,1} \cdot (\nu+1)^{-3} \cdot \|F\|_\infty \tag{6.111}$$

holds for all $\nu \in \mathbb{N}$, independently of the choice of x.

Likewise we get for $|R|$ the not so optimistic estimate

$$|R| \leq \|F\|_\infty \cdot g_{\nu+1} \cdot \frac{1}{[\xi_{\nu+1} - \eta_{\nu+1}]^2} \cdot \sum_{0 \leq t_j x < \xi_{\nu+1}} A_j \left[G_{\nu+1}(t_j x)\right]^2$$

$$= \|F\|_\infty \cdot g_{\nu+1} \cdot \frac{1}{[\xi_{\nu+1} - \eta_{\nu+1}]^2} \cdot \int_{\beta_{\nu+1}}^{\frac{\pi}{2}} \left[G_{\nu+1}(\cos\phi)\right]^2 dA_x(\phi),$$

with the weight sum function A_x of Definition 6.6.

Now it follows from (4.12), (1.23) and Theorem 2.9 that a constant $\gamma_{r,2} > 0$ exists such that

$$|R| \leq \|F\|_\infty \cdot g_{\nu+1} \cdot \frac{\gamma_{r,2}}{[\xi_{\nu+1} - \eta_{\nu+1}]^2} \cdot \left[G_{\nu+1}(1)\right]^2 \int_{\beta_{\nu+1}}^{\frac{\pi}{2}} \left[\frac{1}{(\nu+2)\sin\phi}\right]^{r-2} dA_x(\phi)$$

holds for all $\nu \in \mathbb{N}$ and $x \in S^{r-1}$, and integration by parts yields

$$|R| \leq \|F\|_\infty \cdot g_{\nu+1} \cdot \frac{\gamma_{r,2}}{[\xi_{\nu+1} - \eta_{\nu+1}]^2} \cdot \frac{[G_{\nu+1}(1)]^2}{(\nu+2)^{r-2}}$$

$$\cdot \left\{A_x\left(\frac{\pi}{2}\right) + (r-2) \int_{\beta_{\nu+1}}^{\frac{\pi}{2}} A_x(\phi) \cdot \frac{\cos\phi \, d\phi}{(\sin\phi)^{r-1}}\right\}.$$

Now we use Theorem 6.27, and obtain

$$|R| \leq \|F\|_\infty \cdot g_{\nu+1} \cdot \frac{\gamma_{r,2}}{[\xi_{\nu+1} - \eta_{\nu+1}]^2} \cdot \frac{[G_{\nu+1}(1)]^2}{(\nu+2)^{r-2}} \cdot \left\{\omega_{r-1} + (r-2)\, a_r \int_{\beta_{\nu+1}}^{\frac{\pi}{2}} \cos\phi \cdot d\phi\right\}.$$

Moreover, using the growth orders of $g_{\nu+1}$ and $G_\nu(1)$ obtained already above we get

$$g_{\nu+1} \cdot \frac{[G_{\nu+1}(1)]^2}{(\nu+2)^{r-2}} = \mathcal{O}\left(\frac{1}{(\nu+1)^3}\right),$$

and it is now obvious that a constant $\gamma_{r,3} > 0$ exists such that

$$|R| \leq \frac{4 \cdot \gamma_{r,3}}{[\xi_{\nu+1} - \eta_{\nu+1}]^2} \cdot \frac{1}{(\nu+1)^3} \cdot \|F\|_\infty$$

$$= \frac{\gamma_{r,3}}{\left[\sin\frac{\beta_{\nu+1}+\chi_{\nu+1}}{2} \cdot \sin\frac{\beta_{\nu+1}-\chi_{\nu+1}}{2}\right]^2} \cdot \frac{1}{(\nu+1)^3} \cdot \|F\|_\infty$$

is valid for all $\nu \in \mathbb{N}$ and $x \in S^{r-1}$.

Now all depends on the choice of the truncation radii $\beta_{\nu+1}$, where we assume that they are chosen such that $\chi_{\nu+1} = o(\beta_{\nu+1})$ holds for $\nu \to \infty$. Then a constant $\gamma_{r,4} > 0$ exists such that

$$|R| \leq \frac{\gamma_{r,4}}{(\nu+1)^3} \cdot \frac{1}{\beta_{\nu+1}^4} \cdot \|F\|_\infty \qquad (6.112)$$

holds for $\nu \in \mathbb{N}$ and $x \in S^{r-1}$, and summarizing (6.110)–(6.112) we see that a constant $\gamma_r > 0$ exists such that

$$\|F - \hat{M}_\mu^{(r)} F\|_\infty \leq \|F - \hat{L}_\mu^{(r)} F\|_\infty + \frac{\gamma_r}{(\nu+1)^3}\left(1 + \frac{1}{\beta_{\nu+1}^4}\right) \cdot \|F\|_\infty \qquad (6.113)$$

is valid.

Next let us choose a *truncation order* α, satisfying $0 \leq \alpha < 1$, and let us define the truncation radii by

$$\beta_{\nu+1} := \frac{1}{(2\nu+1)^\alpha}.$$

In view of (6.93), the assumption $\chi_{\nu+1} = o(\beta_{\nu+1})$ is satisfied, and the assumption (6.109) is valid for all sufficiently large $\nu \in \mathbb{N}$, say for $\nu \geq \nu(\alpha)$. Moreover, a constant $\gamma_r(\alpha) > 0$ exists such that

$$\|F - \hat{M}_\mu^{(r)} F\|_\infty \leq \|F - \hat{L}_\mu^{(r)} F\|_\infty + \gamma_r(\alpha) \cdot (\mu+1)^{4\alpha-3} \cdot \|F\|_\infty \qquad (6.114)$$

is valid for $\mu = 2\nu \geq 2\nu(\alpha)$. Note that the constant $\gamma_r(\alpha)$ depends neither on F, nor on the choice of the positive quadrature used in the definition of $\hat{L}_\mu^{(r)}$.

Before we discuss this result in detail, let us consider the expected value $E(\alpha, \mu)$ of nodes ξ_j which are contained in the cap $C(x, (\mu+1)^{-\alpha})$. If $M = M_\mu$ is the total number of nodes in the quadrature, then

$$\frac{E(\alpha, \mu)}{M_\mu} \sim \frac{|C(x, (\mu+1)^{-\alpha})|}{\omega_{r-1}} \sim \frac{1}{r-1} \cdot \frac{\omega_{r-2}}{\omega_{r-1}} \cdot \left(\frac{1}{\mu+1}\right)^{\alpha(r-1)}$$

holds for $\mu \to \infty$, see Problem 6.4.

Now assume that product Gauß quadratures are used, such that

$$M_\mu \lesssim (r-1)! \, N_\mu \sim 2(\mu+1)^{r-1}$$

holds with $N_\mu = \dim \mathbb{P}_\mu^r(S^{r-1})$, see (6.59) and (4.4). Inserting this above we obtain

$$E(\alpha, \mu) \lesssim \frac{2^\alpha}{r-1} \cdot \frac{\omega_{r-2}}{\omega_{r-1}} \cdot \left[(r-1)!\right]^{1-\alpha} \cdot N_\mu^{1-\alpha} \qquad (6.115)$$

for $\mu = 2\nu \to \infty$. In the important case $r = 3$ this takes the form

$$E(\alpha, \mu) \stackrel{<}{\sim} \frac{1}{2} \cdot N_\mu^{1-\alpha}. \tag{6.116}$$

Now we turn to the estimate (6.114). In view of Corollary 6.38 and of the Remark following, the best possible approximation order is $\mathcal{O}(\mu^{-2})$. It occurs for $F \in C^{(2)}(S^{r-1})$ at exact integration. On the other hand, for $\alpha := \frac{1}{4}$ we get

$$\|F - \hat{M}_\mu^{(r)} F\|_\infty \leq \|F - \hat{L}_\mu^{(r)} F\|_\infty + \mathcal{O}(\mu^{-2}) \cdot \|F\|_\infty,$$

which is saying that the approximation order is not destroyed by truncation, whereas the cost reduces to an expected value of

$$E(\tfrac{1}{4}, \mu) = \mathcal{O}(N_\mu^{\frac{3}{4}})$$

function evaluations.

For $\alpha = \frac{1}{2}$ the effect is more dramatic, since we get from (6.114), in view of Theorem 6.39,

$$\|F - \hat{M}_\mu^{(r)} F\|_\infty \leq \hat{k}_r \omega_1(F, \tfrac{1}{\mu}) + \mathcal{O}(\mu^{-1}) \cdot \|F\|_\infty,$$

at the expected value of

$$E(\tfrac{1}{2}, \mu) = \mathcal{O}(N_\mu^{\frac{1}{2}})$$

function evaluations. Again, the order of the estimate is not destroyed. In particular, for $F \in C^{(1)}(S^{r-1})$ we get

$$\|F - \hat{M}_\mu^{(r)} F\|_\infty = \mathcal{O}(\mu^{-1}) \cdot \|F\|_\infty,$$

see (6.105).

Finally, for arbitrary $\epsilon \in (0, \frac{3}{4})$ and $\alpha := \frac{3}{4} - \epsilon$ we still get pointwise convergence, i.e., (6.114) implies

$$\lim_{\mu \to \infty} \hat{M}_\mu^{(r)} F = F$$

for all $F \in C(S^{r-1})$ in $\| \cdot \|_\infty$, at the expected value of

$$E(\alpha, \mu) = \mathcal{O}(N_\mu^{\frac{1}{4} + \epsilon})$$

function evaluations. These results are summarized in Table 6.1.

α	function class	approximation order	order of $E(\alpha,\mu)$	upper bound for $E(\alpha,\mu)$, $r=3$
$0+$	$C^{(2)}(S^{r-1})$	μ^{-2} *)	N_μ^{1-}	$(\mu+1)^2+1$
$\frac{1}{4}$	$C^{(2)}(S^{r-1})$	μ^{-2} *)	$N_\mu^{\frac{3}{4}}$	$\frac{1}{2}(\mu+1)^{\frac{3}{2}}$
$\frac{1}{2}$	$C^{(1)}(S^{r-1})$	μ^{-1}	$N_\mu^{\frac{1}{2}}$	$\frac{1}{2}(\mu+1)$
$\frac{3}{4}-$	$C(S^{r-1})$	$o(1)$	$N_\mu^{\frac{1}{4}+}$	$\frac{1}{2}(\mu+1)^{\frac{1}{2}+}$

Table 6.1. Truncated Discretized Newman–Shapiro Operators,
Approximation Order and Cost (Product Gauß Quadrature).
*) As yet proved for the undiscretized operators, only.

Remarks. If the truncation order α is chosen suitably, then the approximation order is not destroyed by truncation, but the evaluation cost reduces significantly.

To give a numerical example. Let $r = 3$, and assume product Gauß quadratures are applied. For $\mu = 1024$ the evaluation of $\left(\hat{L}_\mu^{(3)}F\right)(x)$ uses 1050626 function values, while $\left(\hat{M}_\mu^{(3)}F\right)(x)$ requires, on the average, at most 16408 evaluations for $\alpha = \frac{1}{4}$, and at most 513 evaluations for $\alpha = \frac{1}{2}$, where we recall that the guaranteed approximation order is not destroyed in either case. Moreover, if the approximated function is just continuous, the parameter α should be chosen close to the value $\frac{3}{4}$, where only about 16 evaluations are needed in the average.

If a function behaves differently in different regions of S^{r-1}, an adaptive method can be used which changes the value of α in dependence on the local order of smoothness.

Reduction of the Truncation Cost

It is very expensive, in general, to determine the quadrature nodes which are located in a given spherical cap with center $x \in S^{r-1}$. Therefore it is important to know that the problem can be inverted, in some sense, by solving it just once, for instance at the pole e_3, and applying locally a quadrature which is gained from the original one by a rotation, which maps e_3 to the point x of interest. Actually, we did not yet decide which quadrature should be used, and all quadratures which do not differ apart from a rotation, are equal in our sense. In particular they agree in their degree of exactness.

Now let $t_1, \ldots, t_M$ be the nodes of a (high precision) quadrature, which we assume to be ordered in advance by their distance from e_r. Say

$$-1 \leq t_{M,r} \leq \ldots \leq t_{2,r} \leq t_{1,r} \leq 1$$

holds, where $t_{j,r} = t_j e_r$. Moreover, let us assume that a truncated approximant to

$$\left(L_\mu^{(r)} F\right)(x), \quad x \in S^{r-1},$$

is wanted, which uses m nodes of a quadrature, nearest to x. Of course, depending on x, a rotation $B = B_x$ exists such that

$$B e_r = x$$

is valid. And since it does not matter which quadrature is applied, we decide to use the quadrature whose nodes and weights are given by

$$\bar{t}_j := B t_j, \quad \bar{A}_j := A_j, \quad j = 1, \dots, M,$$

respectively. In view of

$$\bar{t}_j x = t_j e_r = t_{j,r},$$

it is obvious that, among the new nodes,

$$\bar{t}_1, \dots, \bar{t}_m$$

are located nearest to x, such that truncation means that (6.107) has to be replaced by the sum

$$\left(\hat{M}_{\mu,m}^{(r)} F\right)(x) = \sum_{j=1}^m A_j F(\bar{t}_j) K_\mu^{(r)}(\bar{t}_j x).$$

In other words, using the fixed numbers

$$c_j := A_j \cdot K_\mu^{(r)}(t_{j,r}), \quad j = 1, \dots, m, \tag{6.117}$$

which can be calculated also in advance, we obtain the approximation

$$\left(\hat{M}_{\mu,m}^{(r)} F\right)(x) = \sum_{j=1}^m c_j \cdot F(B_x t_j), \tag{6.118}$$

to $(\hat{L}_\mu^{(r)} F)(x)$, whose evaluation is quite easy. Note that the operator $\hat{M}_\mu^{(r)}$ is positive again. The choice of m should conform with Table 6.1. For instance, in the case $r = 3$, a good choice would be $m = \mu + 1$, if a $C^{(1)}$–function is to be approximated.

Numerical Example

Figure 6.3 shows the restriction to S_+^2 of a truncated approximant $\hat{M}_{\mu,m}^{(3)} F$ of degree $\mu = 160$, which uses only $m = 41$ quadrature nodes. Figure 6.4 presents the corresponding approximation error. The approximation is based on a product Gauß quadrature of degree of exactness 161, whose poles are located on the equator. This implies that the node density is minimal at the north pole e_3, such that the method does not use unnecessarily many points. The serious question, how a spherical function should be represented figuratively, is solved by means of an area preserving parametrisation of the sphere, given by the map $x : 2B^2 \to S^2$,

$$x_1 = \sqrt{1 - \frac{u_1^2 + u_2^2}{4}} \cdot u_1, \quad x_2 = \sqrt{1 - \frac{u_1^2 + u_2^2}{4}} \cdot u_2, \quad x_3 = 1 - \frac{u_1^2 + u_2^2}{2},$$

for $u_1^2 + u_2^2 \leq 4$. Note that the distortion is modest in the northern hemisphere S_+^2, which corresponds to the parameter domain $u_1^2 + u_2^2 \leq 2$. The approximated function is given by

$$F(x(u)) = \tfrac{1}{2}\big(1 + \sin[5 \cdot (u_1 - u_2) \cdot (u_2 - \tfrac{3}{5})]\big).$$

The (relative) error has an absolute value up to about $5 \cdot 10^{-3}$, but is far less in areas with a minor curvature — in correspondence to our theory. Note that, in comparison to the number $m = 41$, the complete product Gauß quadrature uses 13122 nodes, half of them being located on a hemisphere. This makes the efficiency of our cost reduction method visible.

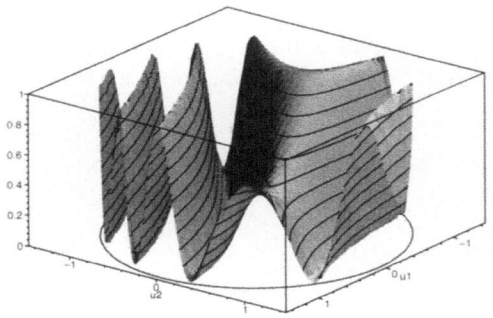

Figure 6.3. Truncated Approximant $\hat{M}_{\mu,m}^{(3)}F$, $\mu = 160$, $m = 41$. Restriction to S_+^2, under an Area Preserving Parametrisation.

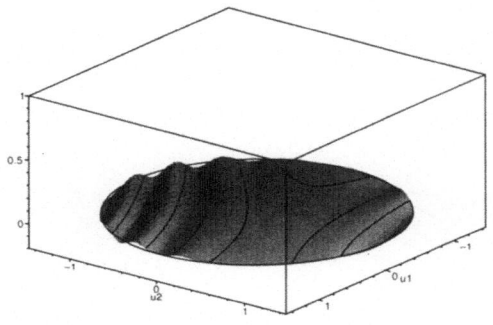

Figure 6.4. Error Function belonging to Figure 6.3.

6.11   Problems

Problem 6.1. Prove:

$$\int_{S^1} F(x)d\omega(x) = \frac{2\pi}{2\mu + 1} \sum_{j=0}^{2\mu} F(t_j)$$

holds for $F \in \mathbb{P}^2_{2\mu}(S^1)$, if the nodes $t_0, t_1, \ldots, t_{2\mu}$ are equidistantly distributed on S^1.

Problem 6.2. Prove:

$$\int_{-1}^{1} f(\xi)\frac{d\xi}{\sqrt{1 - \xi^2}} = \frac{\pi}{\mu + 1} \sum_{\nu=0}^{\mu} f\left(\cos\frac{2\nu + 1}{2\mu + 2}\,\pi\right)$$

holds for all $f \in \mathbb{P}^1_{2\mu+1}$.

Problem 6.3. Let $r \in \mathbb{N} \setminus \{1\}$ be arbitrary, and assume the vertices of a regular simplex are located on S^{r-1}. Then they support a Gauß quadrature, which is exact of degree 2.

Problem 6.4. Calculate the content of a spherical cap $C(x, \phi)$ asymptotically for $\phi \to 0+$.

Problem 6.5. Prove the identity

$$\frac{1}{\sin\phi} \sum_{\nu=0}^{\mu} \left(\sin(\nu + 1)\phi + \sin\nu\phi\right) = \left[\frac{\sin(\mu + 1)\frac{\phi}{2}}{\sin\frac{\phi}{2}}\right]^2.$$

Chapter 7

Approximation on the Ball

In this chapter the domain is the unit ball B^r, $r \geq 2$. Of course, all that has been said in Section 5 with respect to an arbitrary compact domain is valid also for B^r. But there are particulars now which require a separate consideration.

7.1 Orthogonal Projections and Appell Series

Let $r \in \mathbb{N} \setminus \{1\}$, $s \in \mathbb{N}_0$. By $L_s^2(B^r)$ we denote the space of all measurable functions $B^r \to \mathbb{R}$ for which

$$\int\limits_{B^r} F^2(x)(1 - |x|^2)^{\frac{s-1}{2}} dx$$

exists, with the usual identification of equivalent functions. $L_s^2(B^r)$ is provided with the inner product

$$[E, F]_{r,s} := \int\limits_{B^r} E(x)F(x)(1 - |x|^2)^{\frac{s-1}{2}} dx, \tag{7.1}$$

$E, F \in L_s^2(B^r)$, which extends the definition (4.89) for $E, F \in C(B^r)$. In both spaces $\| \cdot \|_{r,s} := \sqrt{[\cdot, \cdot]_{r,s}^2}$ is the induced norm. Note that $F \in C(B^r)$ implies

$$\|F\|_{r,s} \leq \|F\|_\infty \cdot \|1\|_{r,s}. \tag{7.2}$$

In view of Theorem 5.8, the Theorem of Weierstrass, this implies that $\mathbb{P}^r(B^r)$ is dense in $C(B^r)$, both in $\| \cdot \|_\infty$ and in $\| \cdot \|_{r,s}$.

B^r contains interior points, so we are allowed to identify polynomials and their restrictions onto B^r. In particular we get by identification

$$\mathbb{P}^r = \mathbb{P}^r(B^r), \quad \mathbb{P}_\mu^r = \mathbb{P}_\mu^r(B^r), \quad \mathbf{V}_\mu^{r,s} = \mathbf{V}_\mu^{r,s}(B^r).$$

The Appell spaces $\mathbf{V}_\mu^{r,s}$ were introduced in Definition 4.2.

In what follows, X is either the space $C(B^r)$ or one of the spaces $L_s^2(B^r)$, respectively. We are interested in the orthogonal projections

$$\Omega_\nu^{r,s} : X \to \mathbf{V}_\nu^{r,s}$$

and

$$\Pi_\mu^{r,s} : X \to \mathbb{P}_\mu^r,$$

with respect to the inner product $[\,\cdot\,,\,\cdot\,]_{r,s}$ for $\nu, \mu \in \mathbb{N}_0$. Note that (4.99) yields

$$\Pi_\mu^{r,s} = \sum_{\nu=0}^{\mu} \Omega_\nu^{r,s}. \tag{7.3}$$

The $\Pi_\mu^{r,s}$ are called *Appell projections*. The following theorem corresponds to Theorem 6.1.

Theorem 7.1 (Convergence of the Appell Projections). *In $\|\cdot\|_{r,s}$, the orthogonal projections $\Pi_\mu^{r,s}$, $\mu = 0, 1, \ldots$, are pointwise convergent to the identity on X.*

Proof. The proof follows exactly the lines of the proof of Theorem 6.1. $\square$

The result of Theorem 7.1 can also be written in the form

$$\sum_{\nu=0}^{\infty} \Omega_\nu^{r,s} F = F, \tag{7.4}$$

which holds in $\|\cdot\|_{r,s}$ for all $F \in X$.

Definition 7.1 (Appell Series). *The* Appell series operator $\Pi^{r,s} : X \to X$ *is defined by*

$$\Pi^{r,s} := \sum_{\nu=0}^{\infty} \Omega_\nu^{r,s},$$

in the sense of pointwise convergence in $\|\cdot\|_{r,s}$. For $F \in X$ the series (7.4) is called the Appell series of F.

Note that convergence of an Appell series means that the partial sums converge to F *in the quadratic mean*, which is the same as

$$\lim_{\mu \to \infty} \int_{B^r} \left[F(x) - \sum_{\nu=0}^{\mu} \left(\Omega_\nu^{r,s} F \right)(x) \right]^2 (1 - |x|^2)^{\frac{s-1}{2}} \, dx = 0.$$

Moreover, using Theorem 4.27, we confirm that

$$\left[F - \sum_{|n|=\nu} [\hat{U}_n^{r,s}, F]_{r,s} \cdot V_n^{r,s} , \hat{U}_m^{r,s} \right]_{r,s} = 0$$

holds for $|m| = \nu \in \mathbb{N}_0$. In view of (4.93) this implies

$$\left[F - \sum_{|n|=\nu} [\hat{U}_n^{r,s}, F]_{r,s} \cdot V_n^{r,s} , V \right]_{r,s} = 0$$

for all $V \cong \mathbf{V}_\nu^{r,s}$, and hence

$$\Omega_\nu^{r,s} F = \sum_{|n|=\nu} [\hat{U}_n^{r,s}, F]_{r,s} \cdot V_n^{r,s},$$

see Definition 5.6 and Theorem 5.12. Similarly we get

$$\Omega_\nu^{r,s} F = \sum_{|n|=\nu} [F, V_n^{r,s}]_{r,s} \cdot \hat{U}_n^{r,s}.$$

We formulate these results in the symmetric form

$$\sum_{|n|=\nu} [\hat{U}_n^{r,s}, F]_{r,s} \cdot V_n^{r,s} = \Omega_\nu^{r,s} F = \sum_{|n|=\nu} [F, V_n^{r,s}]_{r,s} \cdot \hat{U}_n^{r,s}. \tag{7.5}$$

Because of (7.3) this yields

$$\sum_{|m|\leq\mu} [\hat{U}_m^{r,s}, F]_{r,s} \cdot V_m^{r,s} = \Pi_\mu^{r,s} F = \sum_{|m|\leq\mu} [F, V_m^{r,s}]_{r,s} \cdot \hat{U}_m^{r,s}, \tag{7.6}$$

and, correspondingly, the Appell series (7.4) takes the two different forms

$$\sum_{\nu=0}^\infty \sum_{|n|=\nu} [\hat{U}_n^{r,s}, F]_{r,s} \cdot V_n^{r,s} = F = \sum_{\nu=0}^\infty \sum_{|n|=\mu} [F, V_n^{r,s}]_{r,s} \cdot \hat{U}_n^{r,s}, \tag{7.7}$$

where convergence holds in both cases in $\| \cdot \|_{r,s}$ for $F \in X$.

7.2 Summation of Appell Series

We are interested in summation methods which make the Appell series of all continuous functions convergent to F in the uniform norm. Our knowledge of the summability of Laplace series will help us solve this problem.

Let $r \in \mathbb{N} \setminus \{1\}$, $s \in \mathbb{N}_0$. We define the linear map $\Phi : C(B^r) \to C(S^{r+s})$ by

$$\left(\Phi F \right)(x_1, \ldots, x_r, x_{r+1}, \ldots, x_{r+s+1}) := F(x_1, \ldots, x_r) \tag{7.8}$$

for $F \in C(B^r)$ and $x_1^2 + \cdots + x_r^2 + x_{r+1}^2 + \cdots + x_{r+s+1}^2 = 1$, which implies, obviously, $x_1^2 + \cdots + x_r^2 \leq 1$. Φ is injective, and maps $C(B^r)$ bijectively onto the subspace $\overset{\circ}{C}(S^{r+s})$ of $C(S^{r+s})$ which consists of the functions that do not depend on $x_{r+1}, \ldots, x_{r+s+1}$. These functions are invariant under all rotations of $\mathbb{R}^{r+s+1}$ which keep $e_1, \ldots, e_r$ fixed.

In what follows we consider Φ as a map

$$\Phi \ : \ C(B^r) \to \overset{\circ}{C}(S^{r+s}),$$

which is an isometry. For $(m, 0) \in \mathbb{N}_0^r \times \mathbb{N}_0^{s+1}$ we get, in particular,

$$\Phi V_m^{r,s} \ = \ V_{m,0}^{r+s+1,-1},$$

see (3.58). These polynomials are linearly independent. But in view of (4.86) and of Theorem 4.24, the subfamily with $|m| = \mu$ is even a basis for the $(e_1, \ldots, e_r)$-kernel of the space $\overset{*}{\mathbb{H}}_{\mu}^{r+s+1}(S^{r+s})$, which we denote by $\overset{\circ}{\mathbb{H}}_{\mu}^{r+s+1}$. It follows that Φ maps $\mathbf{V}_{\mu}^{r,s}$ bijectively onto $\overset{\circ}{\mathbb{H}}_{\mu}^{r+s+1}$, and we get

$$\mathbf{V}_{\mu}^{r,s} \ \cong \ \overset{\circ}{\mathbb{H}}_{\mu}^{r+s+1}$$

for $\mu \in \mathbb{N}_0$. Moreover, using (1.25) we obtain

$$
\begin{aligned}
\langle \Phi F_1, \Phi F_2 \rangle_{S^{r+s}} \ &= \ \int_{S^{r+s}} \left(\Phi F_1 \cdot \Phi F_2 \right)(x_1, \ldots, x_{r+s+1}) \, d\tilde{\omega}(x_1, \ldots, x_{r+s+1}) \\
&= \ \int_{S^{r+s}} \left(F_1 \cdot F_2 \right)(x_1, \ldots, x_r) \, d\tilde{\omega}(x_1, \ldots, x_{r+s+1}) \\
&= \ \omega_s \cdot \int_{B^r} \left(F_1 \cdot F_2 \right)(x) \, (1 - |x|^2)^{\frac{s-1}{2}} \, dx
\end{aligned}
$$

for $F_1, F_2 \in C(B^r)$, where $d\tilde{\omega}(\,\cdot\,)$ is the surface element of S^{r+s+1}. Because of (7.1) we may write this result in the form

$$\langle \Phi F_1, \Phi F_2 \rangle_{S^{r+s}} \ = \ \omega_s \cdot [F_1, F_2]_{r,s}. \tag{7.9}$$

In particular, Φ preserves orthogonality relations. For instance it maps orthogonal spaces onto orthogonal spaces.

For $\nu, \mu \in \mathbb{N}_0$ we consider now the orthogonal projections

$$\Omega_\nu \ : \ C(S^{r+s}) \to \overset{*}{\mathbb{H}}_{\nu}^{r+s+1}(S^{r+s})$$

and

$$\Pi_\mu \ : \ C(S^{r+s}) \to \mathbb{P}_{\mu}^{r+s+1}(S^{r+s}),$$

which belong to the inner product $\langle \,\cdot\, , \,\cdot\, \rangle_{S^{r+s}}$, and note that

$$\Pi_\mu \ = \ \sum_{\nu=0}^{\mu} \Omega_\nu \tag{7.10}$$

holds, see (6.1) with r replaced by $r + s + 1$.

We investigate the action of Ω_ν on ΦF for $F \in C(B^r)$. By definition, ΦF does not depend on $x_{r+1}, \ldots, x_{r+s+1}$. So it is natural to ask whether $\Omega_\nu \Phi F$ does. We decide this question as follows.

Let $\nu \in \mathbb{N}_0$. For $\tilde{x} = (x_1, \ldots, x_r, x_{r+1}, \ldots, x_{r+s+1})' \in S^{r+s}$ we put $x := (x_1, \ldots, x_r)' \in B^r$, and so on. Using (6.4), with r again replaced by $r + s + 1$, we get, returning to full notation,

$$\left(\Omega_\nu \Phi F\right)(\tilde{x}) = \int\limits_{S^{r+s}} G_\nu^{r+s+1}(\tilde{t}\tilde{x}) \cdot \left(\Phi F\right)(\tilde{t}\tilde{x}) \, d\tilde{\omega}(\tilde{t})$$

$$= \int\limits_{S^{r+s}} G_\nu^{r+s+1}(\tilde{t}\tilde{x}) \cdot F(t) \, d\tilde{\omega}(\tilde{t}).$$

Now assume $\tilde{A}$ is an arbitrary rotation in $\mathbb{R}^{r+s+1}$ with fixed points $e_1, \ldots, e_r$. Using the substitution $\tilde{t} = \tilde{A}\tilde{u}$ we obtain

$$\left(\Omega_\nu \Phi F\right)(\tilde{A}\tilde{x}) = \int\limits_{S^{r+s}} G_\nu^{r+s+1}(\tilde{t}'\tilde{A}\tilde{x}) F(t) \, d\tilde{\omega}(\tilde{t})$$

$$= \int\limits_{S^{r+s}} G_\nu^{r+s+1}(\tilde{u}'\tilde{x}) F(u) \, d\tilde{\omega}(\tilde{u}) = \left(\Omega_\nu \Phi F\right)(\tilde{x}).$$

Therefore $\Omega_\nu \Phi F$ belongs to the $(e_1, \ldots, e_r)$-kernel of $\overset{*}{\mathbb{H}}{}_\mu^{r+s+1}(S^{r+s})$, which is, in view of Corollary 1.5, the same as

$$\Omega_\nu \Phi F \in \overset{\circ}{\mathbb{H}}{}_\nu^{r+s+1}.$$

It follows that $\Phi^{-1} \Omega_\nu \Phi$ is a map $C(B^r) \to \mathbf{V}_\nu^{r,s}$ with the property that

$$\left[F - \Phi^{-1}\Omega_\nu \Phi F, V\right]_{r,s} = \frac{1}{\omega_s}\left\langle \Phi F - \Omega_\nu \Phi F, \Phi V\right\rangle_{S^{r+s}} = 0$$

holds for all $F \in C(B^r)$ and $V \in \mathbf{V}_\nu^{r,s}$. Here we used that Ω_ν is a projection, together with $\Phi V \in \overset{\circ}{\mathbb{H}}{}_\nu^{r+s+1} \subset \overset{*}{\mathbb{H}}{}_\nu^{r+s+1}(S^{r+s})$. Therefore, $\Phi^{-1}\Omega_\nu \Phi$ has the defining properties of the orthogonal projection $\Omega_\nu^{r,s}$, and we get

$$\Phi^{-1}\Omega_\nu \Phi = \Omega_\nu^{r,s}. \tag{7.11}$$

Moreover, by summation we obtain

$$\Phi^{-1}\Pi_\mu \Phi = \Pi_\mu^{r,s} \tag{7.12}$$

for $\mu \in \mathbb{N}_0$.

Theorem 7.2 (Summation of an Appell Series). *Let $r \in \mathbb{N} \setminus \{1\}$, $s \in \mathbb{N}_0$. Assume* $A = \left(a_{\mu,\nu} \right)_{\mu,\nu=0,1,\ldots}$ *is a real subdiagonal matrix such that the operators*

$$L_\mu := \sum_{\nu=0}^{\mu} a_{\mu,\nu} \Omega_\nu,$$

$\mu \in \mathbb{N}_0$, converge for $\mu \to \infty$ in $\| \cdot \|_\infty$ pointwise to the identity on $C(S^{r+s})$. Then the operators

$$L_\mu^{r,s} := \sum_{\nu=0}^{\mu} a_{\mu,\nu} \Omega_\nu^{r,s},$$

$\mu \in \mathbb{N}_0$, converge in $\| \cdot \|_\infty$ to the identity on $C(B^r)$.

Proof. Let $F \in C(B^r)$, which implies $\Phi F \in C(S^{r+s})$ and hence, by assumption,

$$\lim_{\mu \to \infty} \| \Phi F - L_\mu \Phi F \|_\infty = 0.$$

Since Φ is isometric, we obtain, with the help of (7.11),

$$
\begin{aligned}
\| F - L_\mu^{r,s} F \|_\infty &= \Big\| F - \sum_{\nu=0}^{\mu} a_{\mu,\nu} \Omega_\nu^{r,s} F \Big\|_\infty \\
&= \Big\| \Phi F - \sum_{\nu=0}^{\mu} a_{\mu,\nu} \Phi \Omega_\nu^{r,s} F \Big\|_\infty \\
&= \Big\| \Phi F - \sum_{\nu=0}^{\mu} a_{\mu,\nu} \Omega_\nu \Phi F \Big\|_\infty = \| \Phi F - L_\mu \Phi F \|_\infty,
\end{aligned}
$$

where the right side tends to zero for $\mu \to \infty$. $\square$

Provided the assumptions are satisfied, the result of Theorem 7.2 can be written in the form

$$\sum_{\nu=0}^{\infty} \sum_{|n|=\nu} a_{\mu,\nu} \left[F, V_n^{r,s} \right]_{r,s} \cdot \hat{U}_n^{r,s} = F = \sum_{\nu=0}^{\infty} \sum_{|n|=\nu} a_{\mu,\nu} \left[\hat{U}_n^{r,s}, F \right]_{r,s} \cdot V_n^{r,s}, \quad (7.13)$$

which holds now for all $F \in C(B^r)$ in $\| \cdot \|_\infty$. Here we used (7.5).

Corollary 7.3 (Positive Operators). *For $r \in \mathbb{N} \setminus \{1\}$, $s \in \mathbb{N}_0$, the matrices $C^{(k)}$, $k \geq r+s$, and $A^{(r+s+1)}$ satisfy the assumptions on A of Theorem 7.2. So the operators $L_\mu^{r,s}$ are positive, in their case.*

Proof. Using (7.11) we get

$$L_\mu^{r,s} = \Phi^{-1} L_\mu \Phi, \quad (7.14)$$

where, in the present case, all three factors are positive operators, the operator L_μ by the particular assumptions on A, see Theorem 6.31 and Theorem 6.33, respectively. $\square$

Corollary 7.4 (Order of Convergence). *Assume* $A = A^{(r+s+1)}$ *is the Newman–Shapiro matrix in* $\mathbb{R}^{r+s+1}$, $r \in \mathbb{N} \setminus \{1\}$, $s \in \mathbb{N}_0$, *and assume* $F \in C(B^r)$ *is* j-times continuously differentiable in a neighbourhood of B^r, where $j \in \{1, 2\}$. *Then*

$$\|F - L_\mu^{r,s} F\|_\infty = \mathcal{O}(\mu^{-j})$$

holds for $\mu \to \infty$.

Proof. Let $j \in \{1, 2\}$, and let $F \in C(B^r)$ satisfy the assumptions. Then we get

$$\Phi F \in C^{(j)}(S^{r+s}),$$

and by Corollary 6.38 we obtain

$$\|\Phi F - L_\mu \Phi F\|_\infty = \mathcal{O}(\mu^{-j})$$

for $\mu \to \infty$. Now we use (7.14), and get

$$
\begin{aligned}
\|F - L_\mu^{r,s} F\|_\infty &= \|\Phi F - \Phi L_\mu^{r,s} F\|_\infty \\
&= \|\Phi F - L_\mu \Phi F\|_\infty = \mathcal{O}(\mu^{-j}),
\end{aligned}
$$

as claimed. $\qquad\square$

7.3 Interpolation on the Ball

In Section 5.4 we discussed interpolatory projections onto an arbitrary finite-dimensional subspace of $C(D)$, where D is compact. It was favourable to base the theory on reproducing kernels, whatever the inner product has been.

In the case $D = S^{r-1}$, the inner product has been induced by the invariant measure in a quite natural way, but now that we consider the case $D = B^r$, we are more or less free in its choice. Nor is it indispensable to use reproducing kernels, provided there are plain functions with comparable properties available. It is quite impossible to discuss all thinkable settings, so we consider in what follows four examples, only.

(A) Fundamental Systems Distributed on Certain Hyperplanes

Temporarily we forget that we are considering the ball. Actually, every fundamental system for $\mathbb{P}_\mu^r$ gives rise to various fundamental systems for $\mathbb{P}_\mu^r(B^r)$, just by mapping it to B^r by means of an affine-linear transform.

In what follows we construct a fundamental system for $\mathbb{P}_\mu^r$ from known fundamental systems for the spaces $\mathbb{P}_0^{r-1}, \ldots, \mathbb{P}_\mu^{r-1}$. The construction follows the lines of the construction of Theorem 6.9, where we begin by recalling the well-known fact, that every $\mu + 1$ pairwise different points in $\mathbb{R}^1$ form a fundamental system for $\mathbb{P}_\mu^1$.

So let $r \in \mathbb{N} \setminus \{1\}$ and $\mu \in \mathbb{N}_0$, in what follows, and assume that for every $\nu \in \{0, \ldots, \mu\}$, a fundamental system

$$T^{r-1,\nu} = \left\{ \bar{t}_j^{r-1,\nu} \mid j = 1, \ldots, M_\nu^{r-1} \right\}$$

for $\mathbb{P}_\nu^{r-1}$ is known. Then, in particular,

$$\bar{t}_j^{r-1,\nu} \in \mathbb{R}^{r-1}$$

is valid for $j = 1, \ldots, M_\nu^{r-1}$, where

$$M_\nu^{r-1} = \dim \mathbb{P}_\nu^{r-1} = \binom{\nu + r - 1}{r - 1},$$

see (3.8). Now assume that the abscissae

$$\xi_0, \xi_1, \ldots, \xi_\mu \in \mathbb{R} \tag{7.15}$$

are pairwise different. With their help we define the nodes

$$t_j^{r,\nu} := \begin{pmatrix} \bar{t}_j^{r-1,\nu} \\ \xi_\nu \end{pmatrix} \in \mathbb{R}^r \tag{7.16}$$

for $j = 1, \ldots, M_\nu^{r-1}$, $\nu = 0, \ldots, \mu$. Their number is

$$\sum_{\nu=0}^{\mu} M_\nu^{r-1} = M_\mu^r = \dim \mathbb{P}_\mu^r,$$

such that there is hope that they form a fundamental system for $\mathbb{P}_\mu^r$. Actually, the following holds.

Theorem 7.5 (Recursive Construction of Fundamental Systems). *Let* $r \in \mathbb{N} \setminus \{1\}$, $\mu \in \mathbb{N}_0$. *Assume that for every* $\nu \in \{0, \ldots, \mu\}$,

$$T^{r-1,\nu} = \left\{ \bar{t}_j^{r-1,\nu} \,\middle|\, j = 1, \ldots, M_\nu^{r-1} \right\}$$

is a fundamental system for $\mathbb{P}_\nu^{r-1}$. *Then the system*

$$T^{r,\mu} := \left\{ t_j^{r,\nu} \,\middle|\, j = 1, \ldots, M_\nu^{r-1}, \ \nu = 0, \ldots, \mu \right\},$$

whose members are defined by (7.16), is a fundamental system for $\mathbb{P}_\mu^r$.

Proof. The proof is similar to the proof of Theorem 6.9. We have to show that the interpolation problem

$$F(t_j^{r,\nu}) = y_j^\nu,$$

$j = 1, \ldots, M_\nu^{r-1}$, $\nu = 0, \ldots, \mu$, has a solution $F \in \mathbb{P}_\mu^r$ for arbitrary data $y_j^\nu \in \mathbb{R}$ on the right side. So, assume these data are given.

In the first step, we solve the system

$$\bar{Q}_\mu(\bar{t}_j^{r-1,\mu}) = y_j^\mu, \quad j = 1, \ldots, M_\mu^{r-1},$$

by a polynomial $\bar{Q}_\mu \in \mathbb{P}_\mu^{r-1}$. This is possible by the assumption on $T^{r-1,\mu}$. After we define $Q_\mu \in \mathbb{P}_\mu^r$ by $Q_\mu(x) := \bar{Q}_\mu(\bar{x})$, where $x = (\bar{x}, x_r)' \in \mathbb{R}^{r-1} \times \mathbb{R}^1$. Obviously, in view of (7.16), this is a solution of the system

$$Q_\mu(t_j^{r,\mu}) = y_j^\mu, \quad j = 1, \ldots, M_\mu^{r-1},$$

since Q_μ does not depend on x_r. It follows that $P_\mu := Q_\mu$ is an element of $\mathbb{P}_\mu^r$ which solves the equations

$$P_\mu(t_j^{r,\mu}) = y_j^\mu, \quad j = 1, \ldots, M_\mu^{r-1}.$$

In the second step, let $\bar{Q}_{\mu-1} \in \mathbb{P}_{\mu-1}^{r-1}$ solve the equations

$$\bar{Q}_{\mu-1}(\bar{t}_j^{r-1,\mu-1}) = y_j^{\mu-1} - P_\mu(t_j^{r,\mu-1}), \quad j = 1, \ldots, M_{\mu-1}^{r-1}.$$

This is possible by the assumption on $T^{r-1,\mu-1}$. After that we define $Q_{\mu-1} \in \mathbb{P}_{\mu-1}^r$ by $Q_{\mu-1}(x) := \bar{Q}_{\mu-1}(\bar{x})$, it solves the system

$$Q_{\mu-1}(t_j^{r,\mu-1}) = y_j^{\mu-1} - P_\mu(t_j^{r,\mu-1}), \quad j = 1, \ldots, M_{\mu-1}^{r-1}.$$

Now we define

$$P_{\mu-1}(x) := P_\mu(x) + \frac{x_r - \xi_\mu}{\xi_{\mu-1} - \xi_r} \cdot Q_{\mu-1}(x).$$

Then $P_{\mu-1}$ is an element of $\mathbb{P}_\mu^r$ which solves the equations

$$P_{\mu-1}(t_j^{r,\nu}) = y_j^\nu, \quad j = 1, \ldots, M_\nu^{r-1},$$

for $\nu \in \{\mu, \mu - 1\}$.

We continue this construction by defining $\bar{Q}_\kappa$ for $\kappa \in \{\mu - 2, \ldots, 0\}$ such that $Q_\kappa \in \mathbb{P}_\kappa^r$ is a solution of the system

$$Q_\kappa(t_j^{r,\kappa}) = y_j^\kappa - P_{\kappa+1}(t_j^{r,\kappa}), \quad j = 1, \ldots, M_\kappa^{r-1},$$

and by putting

$$P_\kappa := P_{\kappa+1} + \prod_{\lambda=\kappa+1}^{\mu} \frac{x_r - \xi_\lambda}{\xi_\kappa - \xi_\lambda} \cdot Q_\kappa.$$

Then P_κ is a polynomial of $\mathbb{P}_\mu^r$ which solves the equations

$$P_\kappa(t_j^{r,\nu}) = y_j^\nu, \quad j = 1, \ldots, M_\nu^{r-1},$$

for $\nu \in \{\mu, \mu - 1, \ldots, \kappa\}$.

Obviously, $F := P_0 \in \mathbb{P}_\mu^r$ solves the problem posed in the beginning, and since there exists a solution $F \in \mathbb{P}_\mu^r$ for arbitrary data on the right side, $T^{r,\mu}$ is a fundamental system for $\mathbb{P}_\mu^r$. $\qquad\square$

Remarks. As indicated, we may assume that the nodes of $T^{r,\mu}$ are all located in B^r. Moreover, because of $M_0^{r-1} = M_0^r = 1$, we may assume $T^{r-1,0} = \{0\}$ and $T^{r,0} = \{e_r\}$, which implies $\xi_0 = 1$. The remaining ξ_ν may be assumed to be in the alternating order defined by

$$-1 < \xi_1 < \xi_3 < \cdots < \xi_2 < \xi_0 = 1.$$

The advantage is now that the nodes can be put, in the increasing numbers $M^{r-1,\nu}$, at intersections of B^r and hyperplanes $x_r = \xi_\nu$, which have increasing content. It is even possible to get a more or less regular node distribution by a proper choice of the abscissae.

A distinguished choice for $T^{1,\mu}$ is the zeros of the polynomial $(1 - \xi^2)P_\mu'(\xi)$, where P_μ is the Legendre polynomial of degree μ. Actually, by a result of Fejér,[18], they form an extremal fundamental system for $\mathbb{P}_\mu^1(B^1)$.

For a radial distribution of the nodes see Sündermann, [72].

(B) The Reproducing Kernel of $\mathbb{P}_\mu^r(B^r)$ Belonging to $[\,\cdot\,,\,\cdot\,]_{r,s}$

Let $r \in \mathbb{N} \setminus \{1\}$, again. We use the notation of Section 7.2, and are interested in the kernel function $\Gamma_\mu^{r,s}$ of $\mathbb{P}_\mu^r(B^r)$ with respect to the inner product $[\,\cdot\,,\,\cdot\,]_{r,s}$.

In view of (4.99), the kernel is given by

$$\Gamma_\mu^{r,s}(x,y) = \sum_{\nu=0}^{\mu} G_\nu^{r,s}(x,y),$$

and (4.100) yields

$$\sum_{|n|\leq\mu} \hat{U}_n^{r,s}(x)V_n^{r,s}(y) = \Gamma_\mu^{r,s}(x,y) = \sum_{|n|\leq\mu} \hat{U}_n^{r,s}(y)V_n^{r,s}(x) \qquad (7.17)$$

for $x, y \in B^r$. Of course, this representation of the kernel by means of the bi-orthonormal Appell systems is of great theoretical interest. However, the space $\mathbb{P}_\mu^r(B^r)$ is rotation-invariant, therefore Theorem 1.2 says that it must be possible to represent the kernel also in the form

$$\Gamma_\mu^{r,s}(x,y) = h_\mu^{r,s}(|x|, |y|, xy), \qquad (7.18)$$

where $h_\mu^{r,s}(\xi, \eta, \zeta)$ is a continuous function of three real variables. Actually, in the applications, where complexity plays a dominant role, the representation (7.18) seems to be more useful than (7.17) is. For this reason, we derive the explicit form of (7.18), in what follows, but only in the most important case $s = 0$.

So let $s := 0$. In particular, we define $x \in B^r$ for $\tilde{x} = (x_1, \ldots, x_r, x_{r+1})' \in S^r$ by $x := (x_1, \ldots, x_r)'$, and for $F \in C(B^r)$ and $(x, x_{r+1})' \in S^r$ we put

$$\tilde{F}(x, x_{r+1}) := (\Phi F)(x, x_{r+1}) = F(x).$$

Now let $\mu \in \mathbb{N}_0$ and $F \in \mathbb{P}^r_\mu(B^r)$, such that $\tilde{F} \in \mathbb{P}^{r+1}_\mu(S^r)$ holds. Using our knowledge about the space $\mathbb{P}^{r+1}_\mu(S^r)$, we represent $\tilde{F}(x, x_{r+1})$ in the form

$$\tilde{F}(x, x_{r+1}) = \int_{S^r} \tilde{F}(t, t_{r+1}) \Gamma^{r+1}_\mu(tx + t_{r+1}x_{r+1}) \, d\tilde{\omega}(t, t_{r+1}),$$

which yields

$$F(x) = \int_{S^r} F(t) \Gamma^{r+1}_\mu(tx + t_{r+1}x_{r+1}) \, d\tilde{\omega}(t, t_{r+1}).$$

This is no longer a reproducing kernel representation. However, by replacing x_{r+1} by $-x_{r+1}$, we get an additional equation for $F(x)$, and taking the average we obtain

$$F(x) = \frac{1}{2} \int_{S^r} F(t) \left\{ \Gamma^{r+1}_\mu(tx + t_{r+1}x_{r+1}) + \Gamma^{r+1}_\mu(tx - t_{r+1}x_{r+1}) \right\} d\tilde{\omega}(t, t_{r+1}).$$

But for $(t, t_{r+1})' \in S^r$, either

$$t_{r+1} = +\sqrt{1 - t^2} \quad or \quad t_{r+1} = -\sqrt{1 - t^2}$$

holds, and we get in both cases, together with $|x_r| = \sqrt{1 - \bar{x}^2}$,

$$F(x) = \frac{1}{2} \int_{S^r} F(t) \left\{ \Gamma^{r+1}_\mu\left(tx + \sqrt{1 - t^2}\sqrt{1 - x^2}\right) \right.$$
$$\left. + \Gamma^{r+1}_\mu\left(tx - \sqrt{1 - t^2}\sqrt{1 - x^2}\right) \right\} d\tilde{\omega}(t, t_{r+1}).$$

In view of (1.24) this implies

$$F(x) = \int_{B^r} F(t) \Gamma^{r,0}_\mu(t, x) \frac{dt}{\sqrt{1 - t^2}} \tag{7.19}$$

with

$$\Gamma^{r,0}_\mu(x, y) = \frac{1}{2} \left\{ \Gamma^{r+1}_\mu\left(xy + \sqrt{1 - x^2}\sqrt{1 - y^2}\right) + \Gamma^{r+1}_\mu\left(xy - \sqrt{1 - x^2}\sqrt{1 - y^2}\right) \right\} \tag{7.20}$$

for $x, y \in B^r$. Actually, it is easy to see that the right side of (7.20) has all defining properties of the reproducing kernel of $\mathbb{P}^r_\mu(B^r)$ with respect to the inner product

$[\,\cdot\,,\,\cdot\,]_{r,0}$, see Definition 1.1, such that the identification (7.20) is correct. Moreover, (7.18) holds with $h_\mu^{r,0}$ defined by

$$h_\mu^{r,0}(\xi,\eta,\zeta) := \frac{1}{2}\Big\{\Gamma_\mu^{r+1}\Big(\zeta + \sqrt{1-\xi^2}\sqrt{1-\eta^2}\Big) + \Gamma_\mu^{r+1}\Big(\zeta - \sqrt{1-\xi^2}\sqrt{1-\eta^2}\Big)\Big\}.$$
(7.21)

The case $s \geq 1$ would require a symmetrisation with respect to further arguments.

Remarks

a) The evaluation of $\Gamma_\mu^{r,0}(x,y)$ is still possible at reasonable cost.

b) Since we know the kernel of $\mathbb{P}_\mu^r(B^r)$ with respect to the inner product $[\,\cdot\,,\,\cdot\,]_{r,0}$ explicitly, the general results of Section 5.4 and of Section 5.5 on interpolation are applicable in practice at a reasonable order of complexity.

c) If $t_1,\ldots,t_N \in B^r$ form a fundamental system for $\mathbb{P}_\mu^r(B^r)$, then the diagonal of the fundamental matrix is occupied by the positive numbers $\frac{1}{2}\{\Gamma_\mu^{r+1}(1) + \Gamma_\mu^{r+1}(2t_j^2 - 1)\}$, $j = 1,\ldots,N$. It follows that the weights of a Gauß quadrature, if it exists, are no longer equal, except all of the nodes are located on a unique sphere, see Corollary 5.34.

(C) Interpolation in the Spaces $\mathbf{V}_\nu^{r,s}$

As in the preceding example, we can write the kernel in the less complex form

$$G_\nu^{r,s}(x,y) = g_\nu^{r,s}(|x|,|y|,xy),$$

instead of (4.100). See Theorem 1.2 again. Another way to reduce the complexity is indicated by the following theorem.

Theorem 7.6 (Basis in $\mathbf{V}_\nu^{r,s}$). *Let $r \in \mathbb{N}\setminus\{1,2\}$, $s \in \mathbb{N}_0$, and $\nu \in \mathbb{N}_0$. And assume that the points $t_1,\ldots,t_{N_\nu} \in S^{r-1}$ form a fundamental system for $\overset{*}{\mathbb{P}}_\nu^r(S^{r-1})$. Then*

$$\mathbf{V}_\nu^{r,s} = span\Big\{G_\nu^{r+s+1}(t_j\,\cdot\,)|\,j = 1,\ldots,N_\nu\Big\}$$

holds, where $N_\nu = \dim \overset{*}{\mathbb{P}}_\nu^r = \dim \mathbf{V}_\nu^{r,s}$.

Proof. Of course, $N_\nu = \dim \overset{*}{\mathbb{P}}_\nu^r$ holds by the definition of the fundamental system.

Now recall that G_ν^{r+s+1} is the reproducing kernel of $\overset{*}{\mathbb{H}}_\nu^{r+s+1}(S^{r+s})$. It follows from (4.13) and (3.51) that

$$G_\nu^{r+s+1}(tx) = \frac{2\nu + r + s - 1}{(r+s-1)\,\omega_{r+s}} \cdot \sum_{|n|=\nu} V_n^{r,s}(x)\,t^n$$

holds for $(t_1,\ldots,t_r,0,\ldots,0)' \in S^{r+s}$ and $(x_1,\ldots,x_r,*,\ldots,*)' \in B^{r+s+1}$, i.e., for $x \in B^r$ and $t \in S^{r-1}$, respectively. So we obtain

$$\sum_{|n|=\nu} V_n^{r,s}(x)\,t_j^n = \frac{(r+s-1)\,\omega_{r+s}}{2\nu + r + s - 1} \cdot G_\nu^{r+s+1}(t_j x)$$

for $j = 1, \ldots, N_\nu$. The matrix $\left(t_j^n \right)$ is regular, see Theorem 5.14. So there are coefficients $c_{n,j}^\nu$, such that

$$V_n^{r,s} = \sum_{j=1}^{N_\nu} c_{n,j}^\nu \cdot G_\nu^{r+s+1}(t_j \cdot)$$

holds for $|n| = \nu$, and the first statement is true because of (4.93).

It follows that $\dim \mathbf{V}_n^{r,s} \leq N_\nu$. But in view of (3.54), the $V_n^{r,s}$, $|n| = \nu$, are linearly independent, while their number is N_ν, which implies

$$N_\nu \leq \dim \mathbf{V}_\nu^{r,s}.$$

Together this yields $\dim \mathbf{V}_\nu^{r,s} = N_\nu$, as claimed. □

Remark. Theorem 7.6 closes the gap which occurred in the beginning of Section 4.4. In particular, in view of this theorem, Corollary 4.30 now gives notice how the average operator $\Pi_T^{r,s+1}$, $T = \{t_1, \ldots, t_k\}$ orthonormal, $1 \leq k \leq r - 1$, acts on the space $\mathbf{V}_\mu^{r,s}$, where the result is formulated by means of the reproducing kernel $G_\mu^{k,s}$ of $\mathbf{V}_\mu^{k,s}$.

(D) Interpolation in the Vertices of a Regular Simplex

In what follows, the spaces $\mathbb{P}_1^r(B^r)$ are treated for $r \in \mathbb{N}$ in formula, as one of the few examples where this is possible. Actually,

$$\dim \mathbb{P}_1^r(B^r) = r + 1$$

holds, see (3.8), and this is exactly the number of vertices of a simplex in $\mathbb{R}^r$.

So, assume $t_1, \ldots, t_{r+1} \in S^{r-1}$ are the vertices of a *regular simplex*. Then

$$t_j t_k = \begin{cases} 1 & , \text{ if } k = j, \\ -\frac{1}{r} & , \text{ if } k \neq j, \end{cases} \tag{7.22}$$

holds for $j, k = 1, \ldots, r + 1$, see Solution of Problem 6.3. For example, the *regular tetrahedron* can be moved into a position where

$$t_1 = \begin{bmatrix} 1 \\ 0 \\ 0 \end{bmatrix}, \quad t_2 = \begin{bmatrix} -\frac{1}{3} \\ \frac{\sqrt{8}}{3} \\ 0 \end{bmatrix}, \quad t_3 = \begin{bmatrix} -\frac{1}{3} \\ -\frac{\sqrt{2}}{3} \\ \sqrt{\frac{2}{3}} \end{bmatrix}, \quad t_4 = \begin{bmatrix} -\frac{1}{3} \\ -\frac{\sqrt{2}}{3} \\ -\sqrt{\frac{2}{3}} \end{bmatrix}.$$

The reproducing kernel with respect to the inner product $[\cdot, \cdot]_{r,0}$ is given by (7.20), and using (4.30) together with (2.5), we obtain

$$\Gamma_1^{r+1}(\xi) = \frac{1}{\omega_r}\left(C_1^{\frac{r+1}{2}}(\xi) + C_0^{\frac{r+1}{2}}(\xi) \right) = \frac{1}{\omega_r}\left((r+1)\xi + 1 \right).$$

It follows that the elements of the fundamental matrix are given by

$$\Gamma_1^{r,0}(t_j, t_k) = \frac{1}{r\omega_r} \cdot \left\{ \begin{array}{ll} r(r+2) & , \ if \ k = j, \\ -1 & , \ if \ k \neq j, \end{array} \right.$$

for $j, k = 1, \ldots, r+1$. For instance, in case of the regular tetrahedron, the fundamental matrix takes the form

$$\left(\Gamma_1^{3,0}(t_j t_k) \right)_{j,k=1,\ldots,4} = \frac{1}{3\omega_3} \cdot \left(\begin{array}{cccc} 15 & -1 & -1 & -1 \\ -1 & 15 & -1 & -1 \\ -1 & -1 & 15 & -1 \\ -1 & -1 & -1 & 15 \end{array} \right).$$

The inverse is given by

$$\left(\left[L_j, L_k \right]_{3,0} \right)_{j,k=1,\ldots,4} = \frac{\omega_3}{64} \cdot \left(\begin{array}{cccc} 13 & 1 & 1 & 1 \\ 1 & 13 & 1 & 1 \\ 1 & 1 & 13 & 1 \\ 1 & 1 & 1 & 13 \end{array} \right),$$

where $L_1, \ldots, L_4$ are the Lagrange elements belonging to $t_1, \ldots, t_4$. See (5.34) and (5.30). The fundamental matrix is not diagonal, so the vertices of a regular simplex *do not support* a Gauß quadrature for $\mathbb{P}_1^r(B^r)$, see Theorem 6.16, though they support a Gauß quadrature for $\mathbb{P}_1^r(S^{r-1})$, see Problem 6.3, again.

It follows from (7.22) that the Lagrangians are given by

$$L_j(x) = \frac{1}{r+1} \left(r(t_j x) + 1 \right)$$

for $j = 1, \ldots, r+1$. Moreover,

$$\sum_{j=1}^{r+1} L_j^2(x) \leq 1 \tag{7.23}$$

holds for all $x \in B^3$, which implies

$$\|L_1\|_\infty = \cdots = \|L_{r+1}\|_\infty = 1.$$

The proof is left to the reader, see Problem 7.3.

7.4 Quadrature on Sphere and Ball are Related Topics

In this section we restrict ourselves to the discussion of a principle, by which results on quadratures obtained for the sphere can be used to derive quadratures for the ball.

Let $r \in \mathbb{N}$ and $s \in \mathbb{N}_0$, and assume a quadrature on S^{r+s} is known, such that

$$\int_{S^{r+s}} \tilde{F}(\tilde{t}) \, d\tilde{\omega}(\tilde{t}) = \sum_{j=1}^{M} \tilde{A}_j \tilde{F}(\tilde{t}_j) \tag{7.24}$$

holds for all $\tilde{F} \in \mathbb{P}_\mu^r(S^{r+s})$, where $\mu \in \mathbb{N}_0$, and where $\tilde{t}_j \in S^{r+s}$ is valid for $j = 1, \ldots, M$.

Every polynomial $F \in \mathbb{P}_\mu^r(B^r)$ can be understood to be a spherical polynomial $\tilde{F} \in \mathbb{P}_\mu^{r+s+1}(S^{r+s})$ by the definition

$$\tilde{F}(x_1, \ldots, x_{r+s+1}) := F(x_1, \ldots, x_r)$$

for $(x_1, \ldots, x_{r+s+1})' \in S^{r+s}$. With the help of (1.25) and of (7.24) it follows that

$$\int_{\overline{E}^r} F(x)(1 - |x|^2)^{\frac{s-1}{2}} \, dx = \frac{1}{\omega_s} \int_{S^{r+s}} \tilde{F}(\tilde{t}) \, d\tilde{\omega}(\tilde{t}) = \frac{1}{\omega_s} \sum_{j=1}^{M} \tilde{A}_j \tilde{F}(\tilde{t}_j).$$

Now we assume the nodes $\tilde{t}_j$ to have the representation

$$\tilde{t}_j = (t_{j,1}, \ldots, t_{j,r+s+1})' \in S^{r+s},$$

and define the points

$$t_j := (t_{j,1}, \ldots, t_{j,r})' \in B^r$$

for $j = 1 \ldots, M$. Then we obtain, finally,

$$\int_{B^r} F(x)(1 - |x|^2)^{\frac{s-1}{2}} \, dx = \sum_{j=1}^{M} A_j F(t_j) \tag{7.25}$$

with weights defined by

$$A_j := \frac{1}{\omega_s} \cdot \tilde{A}_j$$

for $j = 1 \ldots, M$. The quadrature (7.25) inherits the degree of exactness from the quadrature (7.24). For instance, if it is gained from a product Gauß quadrature, then it is exact of degree $2\mu + 1$ and provided with positive weights. The only disadvantage of this method is that the constructed quadrature might use more nodes than necessary, though their number is restricted, in the present case, by

$$M \leq 2(\mu + 1)^{r+s},$$

see Theorem 6.19. For $s = 0$ this is the order of $\dim \mathbb{P}_\mu^r$.

To get rid of the factor 2, we make a new and more general approach to the quadrature problem on the ball in what follows. The method is similar to the method used in case of the sphere.

Product Gauß Quadratures

Let $r \in \mathbb{N}$, $\lambda > -\frac{1}{2}$. For $F \in C(B^r)$ we define the integral

$$I_\lambda^r F := \int_{B^r} F(x)(1 - |x|^2)^{\lambda - \frac{1}{2}} dx. \tag{7.26}$$

In particular, we get

$$I_\lambda^1 F = \int_{-1}^{1} F(\xi)(1 - \xi^2)^{\lambda - \frac{1}{2}} d\xi = \int_0^\pi F(\cos\phi)(\sin\phi)^{2\lambda} d\phi.$$

A quadrature with $\mu+1$ nodes and positive weights, which is exact on $\mathbb{P}_{2\mu+1}^1$, is already known, for instance the Gauß–Gegenbauer quadrature (6.49). It corresponds to the cosine Gauß quadrature (6.53).

So we may assume, in what follows, that a *product Gauß quadrature* on $\mathbb{P}_{2\mu+1}^{r-1}$, $r \geq 2$, is known, and we construct a similar quadrature on $\mathbb{P}_{2\mu+1}^r$ with its help.

Theorem 7.7 (Product Gauß Quadrature on B^r). *For* $r \geq 1$, $\lambda > -\frac{1}{2}$, *and arbitrary* $\mu \in \mathbb{N}_0$, *there exists a product Gauß quadrature satisfying*

$$I_\lambda^r F = \sum_{j=1}^{M} A_j^{r,\lambda} F(x_j^{r,\lambda})$$

for $F \in \mathbb{P}_{2\mu+1}^r$ *with* $M = (\mu + 1)^r$ *nodes* $x_j^{r,\lambda} \in B^r$ *and positive weights* $A_j^{r,\lambda}$.

Proof. We use mathematical induction. Of course, for $r = 1$ the statement is true, see formula (6.49), again.

Next let $r \in \mathbb{N} \setminus \{1\}$, and assume

$$I_\lambda^{r-1,\lambda} \bar{F} = \sum_{j=1}^{\bar{M}} A_j^{r-1,\lambda} F(\bar{x}_j^{r-1,\lambda}) \tag{7.27}$$

holds for all $\bar{F} \in \mathbb{P}_{2\mu+1}^{r-1}$, where $\bar{M} = (\mu + 1)^{r-1}$, $\bar{x}_j^{r-1,\lambda} \in B^{r-1}$, and where the weights $A_j^{r-1,\lambda}$ are positive.

After that let $F \in \mathbb{P}_{2\mu+1}^r$. We write the integral (7.26) in the form

$$I_\lambda^r F = \int_{-1}^{1} \int_{B^{r-1}} F(\sqrt{1 - x_r^2}\, \bar{u}, x_r)(1 - |\bar{u}|^2)^{\lambda - \frac{1}{2}}(1 - x_r^2)^{\lambda + \frac{r-1}{2}} d\bar{u} dx_r$$

$$= \int_0^\pi \left(\int_{B^{r-1}} F(\sin\phi \cdot \bar{u}, \cos\phi)(1 - |\bar{u}|^2)^{\lambda - \frac{1}{2}} d\bar{u} \right) (\sin\phi)^{2\lambda + r} d\phi.$$

If we replace in the inner integral $\bar{u}$ by $-\bar{u}$, then we see that this integral is an even trigonometric polynomial of degree $2\mu + 1$ with respect to ϕ, such that we may evaluate the outer integral with the help of the cosine Gauß quadrature (6.53), with $\kappa := \lambda + \frac{r}{2}$ instead of λ. It follows that

$$I_\lambda^r F = \sum_{\nu=1}^{\mu+1} c_\nu^\kappa \int_{B^{r-1}} F(\sin \psi_\nu^\kappa \cdot \bar{u}, \cos \psi_\nu^\kappa)(1 - |\bar{u}|^2)^{\lambda - \frac{1}{2}} d\bar{u}.$$

Now we apply (7.27) to the integrals occurring on the right side, and get

$$I_\lambda^r F = \sum_{\nu=1}^{\mu+1} \sum_{j=1}^{\bar{M}} c_\nu^\kappa A_j^{r-1,\lambda} F(\sin \psi_\nu^\kappa \cdot \bar{x}_j^{r-1,\lambda}, \cos \psi_\nu^\kappa),$$

which is a positive quadrature with $M = (\mu + 1)\bar{M} = (\mu + 1)^r$ nodes in B^r, and which is exact on $\mathbb{P}_{2\mu+1}^r$. We write it in the form

$$I_\lambda^r F = \sum_{j=1}^{M} A_j^{r,\lambda} F(x_j^{r,\lambda}),$$

with nodes $x_j^{r,\lambda} \in B^r$ and weights $A_j^{r,\lambda} > 0$ for $j = 1, \ldots, M$, and mathematical induction finishes the proof. □

Remark 1. For $\lambda := \frac{1}{2}$ it follows from Theorem 7.7, in particular, that for every $\mu \in \mathbb{N}_0$ there is a positive quadrature which is exact on $\mathbb{P}_{2\mu+1}^r(B^r)$ with respect to the standard integral

$$IF := \int_{B^r} F(x)dx,$$

which is supported by at most $(\mu+1)^r$ nodes. In other words, positive quadratures of all degrees exist.

Remark 2. In view of Remark 1 it follows now from Theorem 5.31 that for all $\mu \in \mathbb{N}_0$ a positive quadrature exists, which is exact of degree μ with respect to I, and which is supported by at most $N_\mu^r = \dim \mathbb{P}_\mu^r = \binom{\mu+r}{r}$ nodes.

We do not investigate quadratures on B^r in more detail. Our main purpose has been to show that positive quadratures of all degrees exist, and in particular that the degree $2\mu + 1$ is attainable with $(\mu + 1)^r$ nodes. For more sophisticated quadratures we refer to the literature.

7.5 Hyperinterpolation and Generalized Hyperinterpolation

Evaluating the orthogonal projection

$$\left(\Pi_\mu^{r,s} F\right)(x) = \int_{B^r} F(t)\Gamma_\mu^{r,s}(t,x)dt$$

by means of a positive quadrature of degree 2μ, we obtain a linear operator

$$\hat{L}_\mu : C(B^r) \to \mathbb{P}_\mu^r(B^r)$$

of the form

$$\left(\hat{L}_\mu F\right)(x) = \sum_{j=1}^M A_j F(t_j)\Gamma_\mu^{r,s}(t_j,x) \tag{7.28}$$

for $F \in C(B^r)$, $x \in B^r$, which corresponds to (6.70). We call $\hat{L}_\mu$, which is again a projection onto $\mathbb{P}_\mu^r(B^r)$, a *hyperinterpolation* operator on B^r.

Similarly we may treat the operators $L_\mu^{r,s}$ of Theorem 7.2 to get *generalized hyperinterpolation* operators of the form

$$\left(\hat{L}_\mu^{r,s} F\right)(x) = \sum_{j=1}^M A_j F(t_j)K_\mu^{r,s}(t_j,x), \tag{7.29}$$

$F \in C(B^r)$, $x \in B^r$, where the kernel is defined by means of a proper subdiagonal matrix A in the form

$$K_\mu^{r,s} := \sum_{\nu=0}^\mu a_{\mu,\nu} G_\nu^{r,s}. \tag{7.30}$$

In general, $\hat{L}_\mu^{r,s}$ is no projection, but it is a positive operator, if the kernel (7.30) is nonnegative. This is valid, for instance, in the situations of Corollary 7.3.

We do not go into further details, since it is more convenient to understand generalized hyperinterpolation on the ball B^r from its originating from the sphere S^{r+s}, where the whole theory of Section 6 is ready for application.

7.6 Evaluation of Multivariate Orthogonal Expansions

If we replace $G_\nu^{r,s}$ in (7.30) by means of (4.100), then (7.29) takes the form

$$\sum_{|n|\le\mu} a_n(F)\hat{U}_n(x) = \left(L_\mu^{r,s} F\right)(x) = \sum_{|n|\le\mu} b_n(F)V_n^{r,s}(x),$$

and the question arises, how can such expansions be evaluated. We give a quite general answer, which uses the following definition.

Definition 7.2 (Multivariate Orthogonal Polynomials). *Let $\mathbb{P}^r$, $r \in \mathbb{N}$ be furnished with the inner product $[\cdot, \cdot]$, which has the property that $[FG, H] = [F, GH]$ holds for arbitrary $F, G, H \in \mathbb{P}^r$. Then a polynomial $P \in \mathbb{P}^r_\mu \setminus \mathbb{P}^r_{\mu-1}$, $\mu \in \mathbb{N}_0$, is called an orthogonal polynomial of degree μ, if $[P, Q] = 0$ holds for all $Q \in \mathbb{P}_{\mu-1}$, with $P^r_{-1} = [0]$.*

Remark. For $r \geq 2$ and $\mu > 0$, orthogonal polynomials are not uniquely determined, apart from a constant factor, by their degree. For example, let $[\cdot, \cdot] := [\cdot, \cdot]_{r,s}$, where $s \in \mathbb{N}_0$. Then the polynomials $V^{r,s}_m$ and $U^{r,s}_m$ are orthogonal, see (4.99), (4.93), and (4.88). But by no means are they uniquely determined apart from a constant factor, except for $\mu = 0$.

It is of great importance that multivariate orthogonal polynomials satisfy a three-term recurrence relation — as in the univariate case.

Theorem 7.8 (Three-Term Recurrence Relation). *Let $r \in \mathbb{N}$, and let $\{W_m\}_{m \in \mathbb{N}^r_0}$ be a family of linearly independent orthogonal polynomials with the property that*

$$\mathbb{P}^r_\mu = span\{W_m \mid |m| \leq \mu\}$$

holds for all $\mu \in \mathbb{N}_0$. Then a recurrence relation

$$W_m = \sum_{|n|=|m|-1} \left(a_{m,n} x + b_{m,n} \right) W_n + \sum_{|n|=|m|-2} c_{m,n} W_n \qquad (7.31)$$

holds for $n \in \mathbb{N}^r_0$, $|m| \geq 2$, with coefficients $a_{m,n} \in \mathbb{R}^r$ and $b_{m,n}, c_{m,n} \in \mathbb{R}$, where W_n is defined by $W_n := 0$ for $n \notin \mathbb{N}^r_0$.

Proof. Let $\mu \in \mathbb{N} \setminus \{1\}$ and $|m| = \mu$. By assumption,

$$W_m = \sum_{|l|=\mu} w_{m,l}\, x^l + TLD$$

holds with real coefficients $w_{m,l}$, where TLD is some term of lower degree. For every $l \in \mathbb{N}^r_0$ with $|l| = \mu$, a component $l_\nu \geq 1$, $\nu \in \{1, \ldots, r\}$, exists, such that

$$x^l = x_\nu \cdot x^{l-e_\nu}$$

is valid with x^{l-e_ν} contained in $\mathbb{P}^r_{\mu-1}$. This monomial is a linear combination of the W_n, $|n| \leq \mu - 1$, by assumption. Using this we obtain for W_m a representation of the form

$$W_m = \sum_{|n|=\mu-1} (a_{m,n} x) W_n + TLD,$$

with coefficients $a_{m,n} \in \mathbb{R}^r$, and TLD denoting some polynomial of degree $\mu - 1$, which is a linear combination of the W_n, $|n| \leq \mu - 1$, again. Together this yields that W_m has a representation

$$W_m = \sum_{|n|=\mu-1} (a_{m,n} x + b_{m,n}) W_n + \sum_{|n|=\mu-2} c_{m,n} W_n + Y_m$$

with additional coefficients $b_{m,n}, c_{m,n} \in \mathbb{R}$, and with a remainder $Y_m \in \mathbb{P}^r_{\mu-3}$. In the case $\mu = 2$ we get $Y_m = 0$, and (7.31) is true. In the case $\mu \geq 3$ we obtain, using orthogonality,

$$
\begin{aligned}
[Y_m, Y_m] &= - \sum_{|n|=\mu-1} [(a_{m,n}x)W_n, Y_m] \\
&= - \sum_{|n|=\mu-1} [W_n, (a_{m,n}x)Y_m] = 0,
\end{aligned}
$$

since $(a_{m,n}x)Y_m$ is a polynomial of degree at most $\mu - 2$. It follows that $Y = 0$, and (7.31) is true, again. This finishes the proof. $\square$

Since the Appell polynomials are orthogonal polynomials, their three-term recurrence relation can be used in a low cost evaluation of $V_m^{r,s}$ or $\hat{U}_m^{r,s}$ expansions, respectively.

7.7 Problems

Problem 7.1. Let $t_1, \ldots, t_{r+1} \in S^{r-1}$, $r \in \mathbb{N} \setminus \{1\}$, be the vertices of a regular simplex. Determine the eigenvalues of the fundamental matrix $\left(\Gamma_1^{r,0}(t_j t_k)\right)_{j,k=1,\ldots,r+1}$.

Problem 7.2. Use the result of Problem 7.1 to get an upper bound for

$$
\sum_{j=1}^{r+1} L_j^2(x), \quad x \in B^r,
$$

where the L_j are the Lagrange elements in $\mathbb{P}_1^r(B^r)$ with respect to $t_1, \ldots, t_{r+1}$.

Problem 7.3. Let $L_1, \ldots, L_{r+1} \in \mathbb{P}_1^r(B^r)$ be the Lagrange elements belonging to the vertices $t_1, \ldots, t_{r+1} \in S^{r-1}$ of a regular simplex. Give a direct proof for

$$
\max \left\{ \sum_{j=1}^{r+1} L_j^2(x) \,\bigg|\, x \in B^r \right\} = 1.
$$

Problem 7.4. The nodes of Problem 7.3 support a Gauß quadrature.

Part IV

Applications

Chapter 8

Tomography

In this section we consider a recovery problem for real functions F, which are hidden in a given function space X and which are to be reconstructed from the values λF, where λ varies in a family Λ of linear functionals on X. In practice, F will be some density function, while the values λF are accessible to measurement.

Of great importance is the case where Λ consists of integrals which are extended over a certain affine hyperplane of dimension k, $k \in \{0, \ldots, r-1\}$. The case $k = 0$ corresponds to point evaluation, and hence with traditional interpolation. For $k = 1$ the integrals are stretched over a line and can be measured by X-ray, for instance (computer tomography, CT). For $k = r - 1$ we get the important case where the integrals are extended over hyperplanes of co-dimension 1. In the case $r = 3$ they can be made subject to nuclear-spin magnetic-resonance tomography (MRT).

Not even in interpolation is it self-evident that reconstruction is possible at all, and it was necessary to prove Theorem 5.4 on the existence of a fundamental system. This is the framework within we have to judge the work of *J. K. A. Radon*, [42], who showed already in 1917 that tomography is possible in rather general geometric situations.

Integration of a function F over a hyperplane of dimension $k \geq 1$ requires that $|F|$ decreases rapidly enough at infinity. This holds in so-called *Schwartz spaces*. A stronger, but nevertheless realistic assumption is that F has compact support. In this case we may assume without restriction of generality that the support is contained in the unit ball B^r. More serious is that we assume F to be continuous, where we are aware, however, that $C(B^r)$ stands even for more general spaces, as $L^2(B^r)$, for instance.

8.1   Radon Transform

In this section we are concerned with the case $r \in \mathbb{N} \setminus \{1\}$, $k = r - 1$, which is ruled by the *Radon transform*.

Let Z^r denote the cylinder $Z^r := [-1, 1] \times S^{r-1}$. We consider the traditional Radon transform $\mathcal{R}_0$ as a map

$$\mathcal{R}_0 \ : \ C(B^r) \to C(Z^r),$$

defined by

$$(\mathcal{R}_0 F)(\sigma, t) := \int\limits_{\substack{v \perp t \\ v^2 \leq 1 - \sigma^2}} F(\sigma t + v)\, dv \tag{8.1}$$

for $F \in C(B^r)$ and $(\sigma, t) \in Z^r$. The integral is extended over the intersection of B^r and the affine hyperplane $tx = \sigma$.

Substituting $v = \sqrt{1 - \sigma^2}\, u$, we obtain

$$(\mathcal{R}_0 F)(\sigma, t) = (1 - \sigma^2)^{\frac{r-1}{2}} \int\limits_{\substack{u \perp t \\ |u| \leq 1}} F(\sigma t + \sqrt{1 - \sigma^2}\, u)\, du, \tag{8.2}$$

where, in particular,

$$(\mathcal{R}_0 1)(\sigma, t) = \Omega_{r-1}(1 - \sigma^2)^{\frac{r-1}{2}} \tag{8.3}$$

holds for $(\sigma, t) \in Z^r$. Obviously, $\mathcal{R}_0$ is a positive linear operator with uniform norm

$$\|\mathcal{R}_0\|_\infty = \Omega_{r-1}. \tag{8.4}$$

In practice, F will occur as a density function, which is provided with a certain physical dimension. This dimension changes by integration over the $(r-1)$-dimensional space area. We restore it by considering the ratio $\mathcal{R}F := \mathcal{R}_0 F / \mathcal{R}_0 1$. From (8.2) and (8.3) it follows that

$$(\mathcal{R}F)(\sigma, t) = \frac{1}{\Omega_{r-1}} \int\limits_{\substack{u \perp t \\ |u| \leq 1}} F(\sigma t + \sqrt{1 - \sigma^2}\, u)\, du \tag{8.5}$$

holds for $F \in C(B^r)$ and $(\sigma, t) \in Z^r$, thus defining a map

$$\mathcal{R} \ : \ C(B^r) \to C(Z^r)$$

which we call the *normalized Radon transform*. $\mathcal{R}$ is again a positive linear operator, its uniform norm is given by

$$\|\mathcal{R}\|_\infty = 1. \tag{8.6}$$

The normalized Radon transform is provided with the pleasant property of mapping polynomials onto polynomials. This is a consequence of the following theorem, which is basic for the whole theory.

Theorem 8.1 (Davison and Grünbaum). *Let* $r \in \mathbb{N} \setminus \{1\}$ *and* $\kappa \in \mathbb{N}$. *For* $\mu \in \mathbb{N}_0$
and $a \in S^{r-1}$

$$\left[\mathcal{R}G_\mu^{r+\kappa}(a \cdot)\right](\sigma, t) = \tilde{G}_\mu^{r+\kappa}(\sigma) G_\mu^{r+\kappa}(at)$$

holds for $(\sigma, t) \in Z^r$, *where* $\tilde{G}_\mu^{r+\kappa} = G_\mu^{r+\kappa}/G_\mu^{r+\kappa}(1) = \tilde{C}_\mu^{\frac{r+\kappa-2}{2}}$.

Proof. In view of (8.5) we obtain

$$\left[\mathcal{R}G_\mu^{r+\kappa}(a \cdot)\right](\sigma, t) = \frac{1}{\Omega_{r-1}} \int\limits_{\substack{u \perp t \\ |u| \leq 1}} G_\mu^{r+\kappa}\left(a'[\sigma t + \sqrt{1 - \sigma^2}\, u]\right) du.$$

Replacing the integrator u by Au, where A is a rotation in $\mathbb{R}^r$ with fixed point t, we obtain

$$\left[\mathcal{R}G_\mu^{r+\kappa}(a \cdot)\right](\sigma, t) = \frac{1}{\Omega_{r-1}} \int\limits_{\substack{u \perp t \\ |u| \leq 1}} G_\mu^{r+\kappa}\left(a'A[\sigma t + \sqrt{1 - \sigma^2}\, u]\right) du.$$

For fixed σ, t and u, the integrand is a continuous function of A. So let us apply the average operator $\Pi_T^{r,\kappa}$, $T := (t)$, from Section 4.4 to both sides. Because of (4.108), the left side remains unchanged. On the right side we may commute the average operator and the integral by a continuity argument, and using Corollary 4.31 we obtain, with $x = [\sigma t + \sqrt{1 - \sigma^2}\, u]$,

$$\left[\mathcal{R}G_\mu^{r+\kappa}(a \cdot)\right](\sigma, t) = \frac{1}{\Omega_{r-1}} \int\limits_{\substack{u \perp t \\ |u| \leq 1}} \tilde{G}_\mu^{r+\kappa}(at) G_\mu^{r+\kappa}(\sigma) du = \tilde{G}_\mu^{r+\kappa}(\sigma) G_\mu^{r+\kappa}(at),$$

as claimed. $\qquad\qquad\qquad\qquad\qquad\qquad\qquad\qquad\qquad\qquad\qquad\qquad \Box$

For the original work see Davison and Grünbaum [14].

Corollary 8.2 (Action of $\mathcal{R}$ on the spaces $\mathbf{V}_\mu^{r,s}$). *Let* $r \in \mathbb{N} \setminus \{1\}$, $s \in \mathbb{N}_0$. *Then*

$$(\mathcal{R}F_\mu)(\sigma, t) = \tilde{C}_\mu^{\frac{r+s-1}{2}}(\sigma) F_\mu(t)$$

holds for $F_\mu \in \mathbf{V}_\mu^{r,s}$, $(\sigma, t) \in Z^r$, *and* $\mu \in \mathbb{N}_0$.

Proof. Put $\kappa := s + 1$, such that

$$\tilde{G}_\mu^{r+\kappa} = \tilde{C}_\mu^{\frac{r+s-1}{2}}.$$

Theorem 7.6 says that every $F_\mu \in \mathbf{V}_\mu^{r,s}$ is a linear combination of functions

$$G_\mu^{r+s+1}(t_j \cdot), \quad t_j \in S^{r-1},$$

whose images under $\mathcal{R}$ are known from Theorem 8.1. Together this yields the statement. $\qquad\qquad\qquad\qquad\qquad\qquad\qquad\qquad\qquad\qquad\qquad\qquad\qquad \Box$

8.2   Adjoint Operator and the Inverse

Action on the Spaces $\mathbf{V}_\mu^{r,1}$, and the Adjoint Operator

We consider the particular case $s := 1$, i.e., we assume $F_\mu \in \mathbf{V}_\mu^{r,1}$, where $\mu \in \mathbb{N}_0$. Then we have the marvellous situation where the factor $\tilde{C}_\mu^{\frac{r+s-1}{2}} = \tilde{C}_\mu^{\frac{r}{2}}$, which occurs in Corollary 8.2, is apart from a constant factor the reproducing kernel of $\overset{*}{\mathbb{P}}{}_\mu^r(S^{r-1})$, and moreover the following lemma is valid.

Lemma 8.3. *Let $r \in \mathbb{N} \setminus \{1\}$, $\mu \in \mathbb{N}_0$. Then*

$$F_\mu(x) = \tfrac{1}{\omega_{r-1}} \int\limits_{S^{r-1}} F_\mu(t) C_\mu^{\frac{r}{2}}(tx) d\omega(t)$$

holds for all $F_\mu \in \mathbf{V}_\mu^{r,1}$ and $x \in B^r$.

Proof. We begin the proof by considering the statement for an arbitrary $s \in \mathbb{N}_0$ instead of $s = 1$. Because of $G_\mu^{r+s+1} = const \cdot C_\mu^{\frac{r+s-1}{2}}$, and of Theorem 7.6, $F_\mu \in \mathbf{V}_\mu^{r,s}$ has the representation

$$F_\mu(x) = \sum_{j=1}^N \gamma_j C_\mu^{\frac{r+s-1}{2}}(t_j x)$$

for $x \in B^r$ with some points $t_j \in S^{r-1}$ and real coefficients γ_j. For fixed x we get

$$C_\mu^{\frac{r+s-1}{2}}(-tx) = (-1)^\mu C_\mu^{\frac{r+s-1}{2}}(tx),$$

which says that $C_\mu^{\frac{r+s-1}{2}}(tx)$ is an element of $\overset{*}{\mathbb{P}}{}_\mu^r(S^{r-1})$ with respect to the variable $t \in S^{r-1}$. It follows that

$$\frac{1}{\omega_{r-1}} \int\limits_{S^{r-1}} F_\mu(t) C_\mu^{\frac{r+s-1}{2}}(tx) d\omega(t) = \frac{1}{\omega_{r-1}} \sum_{j=1}^N \gamma_j \int\limits_{S^{r-1}} C_\mu^{\frac{r+s-1}{2}}(t_j t) C_\mu^{\frac{r+s-1}{2}}(tx) d\omega(t)$$

$$= \sum_{j=1}^N \gamma_j \int\limits_{S^{r-1}} \overset{*}{\Gamma}{}_\mu^{r+s-1}(t_j t) C_\mu^{\frac{r+s-1}{2}}(tx) d\omega(t).$$

We arrived at a point where we have to assume $s = 1$. At this choice, only, the first factor of the integrand is the reproducing kernel of $\overset{*}{\mathbb{P}}{}_\mu^r(S^{r-1})$, and it follows that

$$\frac{1}{\omega_{r-1}} \int\limits_{S^{r-1}} F_\mu(t) C_\mu^{\frac{r}{2}}(tx) d\omega(t) = \sum_{j=1}^N \gamma_j C_\mu^{\frac{r}{2}}(t_j x) = F_\mu(x),$$

as claimed.   □

It is worthwhile mentioning that $\overset{*}{\Gamma}{}^{r+2}_{\mu}(xy)$ *is not* the reproducing kernel of $\mathbf{V}^{r,1}_{\mu}$.

Next we restore in Corollary 8.2 the original meaning of $\sigma = tx$, this means we consider the equation

$$(\mathcal{R}F_{\mu})(tx,t) = \tilde{C}^{\frac{r}{2}}_{\mu}(tx)F_{\mu}(t),$$

which holds for $F_{\mu} \in \mathbf{V}^{r,1}_{\mu}$, $x \in B^r$ and $t \in S^{r-1}$. In view of Lemma 8.3 we get

$$\int\limits_{S^{r-1}} (\mathcal{R}F_{\mu})(tx,t)d\omega(t) = \frac{\omega_{r-1}}{C^{\frac{r}{2}}_{\mu}(1)} \cdot F_{\mu}(x), \tag{8.7}$$

such that $F_{\mu}(x)$ is reconstructed from $\mathcal{R}F_{\mu}$, apart from a constant factor, which depends on μ, unfortunately. For else we would know the inverse of $\mathcal{R}$. Nevertheless, we are inspired to investigate the operator

$$\mathcal{R}^* : C(Z^r) \to C(B^r)$$

defined by

$$(\mathcal{R}^*G)(x) := \int\limits_{S^{r-1}} G(tx,t)d\omega(t) \tag{8.8}$$

for $G \in C(Z^r)$ and $x \in B^r$. Actually, with its help, (8.7) takes the form

$$\mathcal{R}^*\mathcal{R}\,F_{\mu} = \frac{\omega_{r-1}}{C^{\frac{r}{2}}_{\mu}(1)} \cdot F_{\mu}. \tag{8.9}$$

To be more precise, we have defined already the inner product

$$[F_1, F_2]_{r,1} = \int\limits_{B^r} F_1(x)F_2(x)dx$$

for $F_1, F_2 \in C(B^r)$, see (4.89). Now let us provide the image space, and more generally $C(Z^r)$ with the inner product defined by

$$(G_1, G_2)_r := \int\limits_{Z^r} G_1(\sigma,t)G_2(\sigma,t)d(\sigma,t)$$

for $G_1, G_2 \in C(Z^r)$, with $d(\sigma,t) := d\sigma d\omega(t)$. Then we get for arbitrary $F \in C(B^r)$

and $G \in C(Z^r)$,

$$
\begin{aligned}
\left[F, \mathcal{R}^*G\right]_{r,1} &= \int_{B^r} F(x)\left(\int_{S^{r-1}} G(tx,t)\, d\omega(t)\right) dx \\
&= \int_{S^{r-1}} \left(\int_{B^r} F(x)\, G(tx,t)\, dx\right) d\omega(t) \\
&= \int_{S^{r-1}} \left(\int_{-1}^{1} \int_{\substack{u \perp t \\ u^2 \leq 1-\sigma^2}} F(\sigma t + u)\, G(\sigma,t)\, du\, d\sigma\right) d\omega(t) \\
&= \int_{S^{r-1}} \int_{-1}^{1} (\mathcal{R}_0 F)(\sigma,t)\, G(\sigma,t)\, d\sigma\, d\omega(t) \\
&= \left(\mathcal{R}_0 F, G\right)_r.
\end{aligned}
$$

In other words, $\mathcal{R}^*$ is the adjoint operator of $\mathcal{R}_0$. Besides, replacing $\mathcal{R}_0 F$ by $\mathcal{R}F \cdot \mathcal{R}_0 1$, we find that $\mathcal{R}^*$ is also the adjoint of $\mathcal{R}$, but with respect to a slightly modified inner product in $C(Z^r)$. So we get a better understanding of formula (8.9), which says that the spaces $\mathbf{V}_\mu^{r,1}$ are *eigenspaces* of the operator $\mathcal{R}^*\mathcal{R}$ with respect to the eigenvalue

$$
\sigma_\mu = \left(\overset{*}{\varGamma}^r_\mu(1)\right)^{-1}.
$$

In this setting the σ_μ are called the *singular values* of $\mathcal{R}$. Naturally, the spaces must be orthogonal with respect to $[\cdot, \cdot]_{r,1}$, — which has been anticipated by Theorem 4.28. For a general s we refer to Rosier [60].

The Inverse

Equation (8.9) shows that $\mathcal{R}^*$ is not far off the inverse, except that the singular values are not all equal. So we change the operator a bit, and consider, instead of (8.7), the following integral, with $D_\sigma := \frac{\partial}{\partial \sigma}$,

$$
\int_{S^{r-1}} \left(D_\sigma^{r-1} \mathcal{R}_0 F_\mu\right)(tx,t)\, d\omega(t) = \Omega_{r-1} \int_{S^{r-1}} \left(D_\sigma^{r-1}[(1-\sigma^2)^{\frac{r-1}{2}} \mathcal{R} F_\mu]\right)(tx,t)\, d\omega(t)
$$

$$
= \Omega_{r-1} \int_{S^{r-1}} \left(D_\sigma^{r-1}[(1-\sigma^2)^{\frac{r-1}{2}} \tilde{C}_\mu^{\frac{r}{2}}(\sigma)]\right)_{\sigma=tx} F_\mu(t)\, d\omega(t).
$$

Here we used the definition of $\mathcal{R}$ together with (8.3) and Corollary 8.2.
Now we assume r to be odd, such that the identity

$$
\Omega_{r-1} D_\sigma^{r-1}\left[(1-\sigma^2)^{\frac{r-1}{2}} \tilde{C}_\mu^{\frac{r}{2}}(\sigma)\right] = 2(2\pi i)^{r-1} \frac{1}{\omega_{r-1}} C_\mu^{\frac{r}{2}}(\sigma)
$$

holds, which is proved in the Appendix (D). Inserting this above we obtain

$$\int\limits_{S^{r-1}} \left(D_\sigma^{r-1}\mathcal{R}_0 F_\mu\right)(tx,t)\, d\omega(t) = 2(2\pi i)^{r-1}\int\limits_{S^{r-1}} \frac{1}{\omega_{r-1}} C_\mu^{\frac{r}{2}}(tx) F_\mu(t)\, d\omega(t)$$

$$= 2(2\pi i)^{r-1}\cdot F_\mu(x)$$

for arbitrary $F_\mu \in \mathbf{V}_\mu^{r,1}$, $\mu \in \mathbb{N}_0$, and $x \in B^r$, where we used Lemma 8.3 in the last step. But in view of (4.99), every polynomial is a linear combination of F_μ-s, and we get

$$\frac{1}{2(2\pi i)^{r-1}}\int\limits_{S^{r-1}} \left(D_\sigma^{r-1}\mathcal{R}_0 F\right)(tx,t)\, d\omega(t) = F(x)$$

for arbitrary $F \in \mathbb{P}^r$ and $x \in B^r$. It follows that

$$\frac{1}{2(2\pi i)^{r-1}}\mathcal{R}^* D_\sigma^{r-1}\mathcal{R}_0 = id\,(\mathbb{P}^r), \tag{8.10}$$

and we obtain

$$\mathcal{R}_0^{-1} = \frac{1}{2(2\pi i)^{r-1}}\mathcal{R}^* D_\sigma^{r-1}, \tag{8.11}$$

but for $r \in \{3,5,\dots\}$, only.

For $r \in \{2,4,\dots\}$ an identity similar to (8.10) exists. It has the form

$$\frac{i}{2(2\pi i)^{r-1}}\mathcal{R}^*\mathcal{H}D_\sigma^{r-1}\mathcal{R}_0 = id(\mathbb{P}^r), \tag{8.12}$$

where $\mathcal{H}$ denotes the *Hilbert transform*, which we need not explain here. Of course, in this case we get

$$\mathcal{R}_0^{-1} = \frac{i}{2(2\pi i)^{r-1}}\mathcal{R}^*\mathcal{H}D_\sigma^{r-1}. \tag{8.13}$$

The equations (8.11) and (8.13) are well known as the *Lorentz–Radon inversion formulae*.

We do not go into further details, except for the following remarks. If

$$F(x) = f(tx),$$

$f \in C[-1,1]$, $t \in S^{r-1}$, $x \in B^r$, is a zonal function, then we obtain from (8.2)

$$(\mathcal{R}_0 F)(\sigma,t) = (1-\sigma^2)^{\frac{r-1}{2}}\int\limits_{\substack{u\perp t \\ |u|\le 1}} f(\sigma)\, du$$

$$= \Omega_{r-1}(1-\sigma^2)^{\frac{r-1}{2}} f(\sigma).$$

This formula says, that if f is *not differentiable* at the point $\sigma \in (-1,1)$, then $\mathcal{R}_0 F$ is *not partially differentiable* with respect to the first argument at (σ,t). So the Lorentz–Radon inversion formulae, which hold on the image of $\mathbb{P}^r$, are *not even applicable* on the full image space of $C(B^r)$. But even if $D_\sigma^{r-1}\mathcal{R}_0 F$ exists, by suitable smoothness assumptions or in a generalized sense, a finite approximation to this partial derivative is required, for instance by a divided difference. The price for safe convergence is even smoothness assumptions of an order higher than $r-1$, which are unacceptable in higher dimensional spaces. For this reason we look for an alternative reconstruction method.

8.3   Reconstruction by Approximation

In what follows, $s \in \mathbb{N}_0$ is arbitrary, again. Theorem 4.28 says that

$$\mathbb{P}^r = \bigoplus_{\mu=0}^{\infty} \mathbf{V}_\mu^{r,s}$$

holds, and since $\mathbb{P}^r$ is dense on $C(B^r)$, Corollary 8.2 describes the action of the bounded linear operator $\mathcal{R}$ onto $C(B^r)$ completely by its action onto the polynomials.

To be more concrete, let us assume that the summation matrix

$$A = \left(a_{\mu,\nu} \right)_{\mu,\nu=0,1,\ldots}$$

satisfies the assumptions of Theorem 7.2. Realisations of this assumption are described by Corollary 7.3.

Now let $F \in C(B^r)$ be the function to be reconstructed from its image $\mathcal{R}F$, which is assumed to be known. We introduce the Appell projections

$$F_\nu := \Omega_\nu^{r,s} F \in \mathbf{V}_\nu^{r,s}$$

for $\nu \in \mathbb{N}_0$, such that Theorem 7.2 yields

$$F = \lim_{\mu \to \infty} \sum_{\nu=0}^{\mu} a_{\mu,\nu} F_\nu,$$

where convergence takes place in the uniform norm on B^r. Since $\mathcal{R}$ is bounded, this implies

$$(\mathcal{R}F)(\sigma,t) = \lim_{\mu \to \infty} \sum_{\nu=0}^{\mu} a_{\mu,\nu} (\mathcal{R}F_\nu)(\sigma,t),$$

and using Corollary 8.2 we obtain

$$(\mathcal{R}F)(\sigma,t) = \lim_{\mu \to \infty} \sum_{\nu=0}^{\mu} a_{\mu,\nu} \tilde{C}_\nu^{\frac{r+s-1}{2}}(\sigma) F_\nu(t), \tag{8.14}$$

uniformly for $(\sigma, t) \in Z^r$. Actually, this formula describes the action of $\mathcal{R}$ on $C(B^r)$, as indicated.

In the following we try to regain F from $\mathcal{R}F$, at least approximately, by constructing the approximants

$$L_\mu^{r,s} F = \sum_{\nu=0}^{\mu} a_{\mu,\nu} F_\nu \qquad (8.15)$$

to F for $\mu \in \mathbb{N}_0$.

First let $r \in \mathbb{N}_0$, and choose an arbitrary nonzero element $F_\nu \in \mathbf{V}_\nu^{r,s}$. By Theorem 7.2 we get

$$F_\nu = \lim_{\mu \to \infty} L_\mu^{r,s} F_\nu = \lim_{\mu \to \infty} \sum_{\kappa=0}^{\mu} a_{\mu,\kappa} \Omega_\kappa^{r,s} F_\nu = \lim_{\mu \to \infty} a_{\mu,\nu} \cdot F_\nu,$$

and it follows that

$$\lim_{\mu \to \infty} a_{\mu,\nu} = 1 \qquad (8.16)$$

for arbitrary $\nu \in \mathbb{N}_0$.

Now it is easy to regain $F_\nu(t)$ from (8.14). Actually, by the orthogonality of the Gegenbauer polynomials, see Theorem 2.3, we obtain

$$\int_{-1}^{1} (\mathcal{R}F)(\sigma, t) C_\nu^{\frac{r+s-1}{2}}(\sigma)(1 - \sigma^2)^{\frac{r+s-2}{2}} d\sigma = \lim_{\mu \to \infty} a_{\mu,\nu} c_\nu^{r,s} F_\nu(t) = c_\nu^{r,s} \cdot F_\nu(t)$$

$$(8.17)$$

for $\nu \in \mathbb{N}_0$ and $t \in S^{r-1}$, where the constants are defined by

$$c_\nu^{r,s} := \int_{-1}^{1} \tilde{C}_\nu^{\frac{r+s-1}{2}}(\sigma) C_\nu^{\frac{r+s-1}{2}}(\sigma)(1 - \sigma^2)^{\frac{r+s-2}{2}} d\sigma. \qquad (8.18)$$

They are positive. But note that by formula (8.17) the restriction of F_ν onto S^{r-1} is reconstructed from $\mathcal{R}F$, not F_ν itself. Here we have arrived at a point where we have to distinguish the case $s = 1$ and the general case.

The Case $s = 1$

Let $s = 1$. Then Lemma 8.3 provides us with just the information needed for the full reconstruction of F_ν. Actually, using (8.17) and Lemma 8.3 we obtain, now for $x \in B^r$,

$$F_\nu(x) = \frac{1}{\omega_{r-1} c_\nu^{r,1}} \int_{S^{r-1}} \int_{-1}^{1} (\mathcal{R}F)(\sigma, t) C_\nu^{\frac{r}{2}}(tx) C_\nu^{\frac{r}{2}}(\sigma)(1 - \sigma^2)^{\frac{r-1}{2}} d\sigma d\omega(t).$$

Replacing $\mathcal{R}$ with $\mathcal{R}_0/\mathcal{R}_0\,1$ we get, using (8.3) again,

$$F_\nu(x) = \lambda_\nu \int\limits_{Z^r} (\mathcal{R}_0 F)(\sigma, t)\, C_\nu^{\frac{r}{2}}(tx)\, C_\nu^{\frac{r}{2}}(\sigma)\, d(\sigma, t) \tag{8.19}$$

with

$$\lambda_\nu := \left(\omega_{r-1} \cdot \Omega_{r-1} \cdot c_\nu^{r,1}\right)^{-1}.$$

We evaluate the parameter as follows. From (8.18) we get in view of (4.13)

$$c_\nu^{r,1} = \frac{r\,\omega_{r+1}}{2\nu + r} \int\limits_{-1}^{1} \tilde{C}_\nu^{\frac{r}{2}}(\sigma)\, G_\nu^{r+2}(\sigma)\, (1 - \sigma^2)^{\frac{r-1}{2}}\, d\sigma.$$

With an arbitrary $\tilde{x} \in S^{r+1}$ we can write this equation in the form

$$c_\nu^{r,1} = \frac{r}{2\nu + r} \cdot \frac{\omega_{r+1}}{\omega_r} \int\limits_{S^{r+1}} \tilde{C}_\nu^{\frac{r}{2}}(\tilde{t}\tilde{x})\, G_\nu^{r+2}(\tilde{t}\tilde{x})\, d\tilde{\omega}(\tilde{t}),$$

where $d\tilde{\omega}(\tilde{t})$ is the surface element of S^{r+1}, see (1.27) with $r + 2$ instead of r. Here $\tilde{C}_\nu^{\frac{r}{2}}(\,\cdot\,\tilde{x})$ is an element of $\overset{*}{\mathbb{H}}{}_\nu^{r+2}(S^{r+1})$, where G_ν^{r+2} is the reproducing kernel function. Together this yields

$$c_\nu^{r,1} = \frac{r}{2\nu + r} \cdot \frac{\omega_{r+1}}{\omega_r}. \tag{8.20}$$

Inserting this above we obtain, together with (1.8),

$$\lambda_\nu = \frac{2\nu + r}{\omega_{r-1}^2}. \tag{8.21}$$

We summarize our results in the following theorem.

Theorem 8.4 (Reconstruction of Approximants, s=1). *Let $r \in \mathbb{N} \setminus \{1\}$, and assume the summation matrix* A *satisfies the assumptions of Theorem 7.2. Then*

$$\left(L_\mu^{r,1} F\right)(x) = \int\limits_{Z^r} (\mathcal{R}_0 F)(\sigma, t)\, K_\mu^A(\sigma, tx)\, d(\sigma, t) \tag{8.22}$$

is valid for $F \in C(B^r)$, $x \in B^r$, and $\mu \in \mathbb{N}_0$, where the kernel is defined by

$$K_\mu^A(\sigma, \tau) := \sum_{\nu=0}^{\mu} \lambda_\nu\, a_{\mu,\nu}\, C_\nu^{\frac{r}{2}}(\sigma)\, C_\nu^{\frac{r}{2}}(\tau) \tag{8.23}$$

for $(\sigma, \tau) \in [-1, 1]^2$. In particular,

$$\lim_{\mu \to \infty} L_\mu^{r,1} F = F$$

holds in $\|\cdot\|_\infty$ for arbitrary $F \in C(B^r)$.

Proof. We obtain the first statement by inserting (8.19) in (8.15). Convergence follows from Theorem 7.2. $\square$

Remark 1. The approximants of Theorem 8.4 are of particular value if A is the Newman–Shapiro matrix $A^{(r+2)}$, see Corollary 7.3 and 7.4. In this case the operators $L_F^{r,1}$ are positive, and convergence takes place up to the best possible order $\mathcal{O}(\mu^{-2})$, which depends on the smoothness of F.

Remark 2. The operators (8.22) approximate the inverse of the positive operator $\mathcal{R}_0$. So not all of the kernels can be nonnegative. See Figure 8.2, for example.

The Case s arbitrary

Let $s \in \mathbb{N}_0$ be arbitrary, again. In this case, Lemma 8.3 is not available, such that we need a different method for the reconstruction of F_ν from its restriction onto S^{r-1}. We indicate in advance that the result will be rather unsatisfactory because of its complexity, but for completeness, and to be able to judge the extra role of the parameter $s = 1$ more precisely, we do not omit it.

We begin by recalling (4.93), which says, for instance, that $F_\nu \in \mathbf{V}_\nu^{r,s}$, $\nu \in \mathbb{N}_0$, has a uniquely determined expansion

$$F_\nu(x) = \sum_{|n|=\nu} \gamma_n \hat{U}_n^{r,s}(x), \tag{8.24}$$

$x \in B^r$, with real coefficients γ_n. Moreover, from (4.91) we obtain

$$\hat{U}_n^{r,s}(t) = d_\nu^{r,s} \cdot t^n \tag{8.25}$$

for $t \in S^{r-1}$ and $|n| = \nu$, with the positive constants

$$d_\nu^{r,s} := \left(\nu + \tfrac{r+s-1}{2}\right) \cdot \pi^{-\frac{r}{2}} \cdot \frac{\Gamma(\frac{r+s-1}{2})}{\Gamma(\frac{s+1}{2})}. \tag{8.26}$$

It follows that the restriction of F_ν onto S^{r-1} is given by

$$F_\nu(t) = d_\nu^{r,s} \sum_{|n|=\nu} \gamma_n t^n \tag{8.27}$$

for $t \in S^{r-1}$, and our aim is to reconstruct (8.24) from (8.27).

In (8.27), $F_\nu(t)$ occurs as the restriction of a homogeneous polynomial, whose coefficients are determined by their representer P_n, see (4.44). By a comparison of the expansions (3.51) and (4.36) we get

$$P_n = \tfrac{1}{\omega_{r-1}} \cdot V_n^{r,1},$$

and (8.27) yields

$$\gamma_n = \frac{1}{\omega_{r-1} d_\nu^{r,s}} \cdot \int_{S^{r-1}} F_\nu(t) V_n^{r,1}(t) \, d\omega(t),$$

again for $|n| = \nu$. It follows that (8.24) takes the form

$$F_\nu(x) = \frac{1}{\omega_{r-1} \, d_\nu^{r,s}} \cdot \int\limits_{S^{r-1}} F_\nu(t) \Big(\sum_{|n|=\nu} \hat{U}_n^{r,s}(x) V_n^{r,1}(t) \Big) \, d\omega(t),$$

now for $x \in B^r$. This finishes the reconstruction of F_ν from its restriction onto S^{r-1}, which is given by (8.17). Inserting this we obtain finally, using (8.3) again,

$$F_\nu(x) = \int\limits_{Z^r} (\mathcal{R}_0 F)(\sigma, t) \, K_\nu^{r,s}(\sigma, t, x) \, d(\sigma, t), \tag{8.28}$$

with the kernel defined by

$$K_\nu^{r,s}(\sigma, \tau, x) := \frac{(1-\sigma^2)^{\frac{s-1}{2}}}{\omega_{r-1} \, \Omega_{r-1} \, c_\nu^{r,s} \, d_\nu^{r,s}} \cdot C_\nu^{\frac{r+s-1}{2}}(\sigma) \cdot \sum_{|n|=\nu} V_n^{r,1}(t) \, \hat{U}_n^{r,s}(x) \tag{8.29}$$

for $(\sigma, t, x) \in Z^r \times B^r$.

Theorem 8.5 (Reconstruction of Approximants, s Arbitrary). *Let $r \in \mathbb{N} \setminus \{1\}$, $s \in \mathbb{N}_0$, and assume the summation matrix A satisfies the assumption of Theorem 7.2. Then*

$$(L_\mu^{r,s} F)(x) = \int\limits_{Z^r} (\mathcal{R}_0 F)(\sigma, t) \, K_\mu^{A,r,s}(\sigma, t, x) \, d(\sigma, t) \tag{8.30}$$

holds for $F \in C(B^r)$, $x \in B^r$, and $\mu \in \mathbb{N}_0$, where the kernel is defined for $(\sigma, t, x) \in Z^r \times B^r$ by

$$K_\mu^{A,r,s}(\sigma, t, x) := \sum_{\nu=0}^{\mu} a_{\mu,\nu} \, K_\nu^{r,s}(\sigma, t, x). \tag{8.31}$$

Proof. The statement follows by inserting (8.28) in (8.15).   □

Remark. In comparison with (8.23), the kernel (8.31) is rather complex. But just for $s = 1$ we get, in view of (4.100) and of (8.25), and for $t \in S^{r-1}$ and $x \in B^r$, the following reductions,

$$\sum_{|n|=\nu} V_n^{r,1}(t) \, \hat{U}_n^{r,s}(x) = \sum_{|n|=\nu} V_n^{r,1}(x) \, \hat{U}_n^{r,1}(t)$$

$$= d_\nu^{r,s} \sum_{|n|=\nu} V_n^{r,1}(x) \, t^n$$

$$= d_\nu^{r,1} \cdot C_\nu^{\frac{r}{2}}(tx),$$

where we used (3.51) to get the last equality. If this is inserted in (8.29), the kernel (8.31) breaks down and takes the form

$$K_\mu^{A,r,1}(\sigma, t, x) = K_\mu^{A}(\sigma, tx),$$

exactly, see (8.23) and (8.21).

8.4 Complexity and Stability

By the choice of the Appell index $s \in \mathbb{N}_0$ we decide on an embedding of the given $\mathbb{R}^r$-problem in the larger space $\mathbb{R}^{r+s+1}$, this means on the treatment of $C(B^r)$-functions as $C(S^{r+s})$-functions, with some advantages. It seems to be advisable to keep s as small as possible, so the first guess would be $s = 0$. But in this case we would be forced to use the reconstruction formulae of Theorem 8.5, which are rather complex in comparison with the formulae of Theorem 8.4, which hold however only for $s = 1$. So it is quite natural to use only the parameter value $s = 1$ in practice.

Complexity

So let $s := 1$, again. The evaluation of $\left(L_\mu^{r,1} F\right)(x)$ from (8.22) at a single point $x \in B^r$ requires the evaluation of an integral over $Z^r = [-1,1] \times S^{r-1}$, for instance by the product of a Gauß quadrature on $[-1,1]$ and a product Gauß quadrature on S^{r-1}, with nodes (σ_j, t_k). In this case the number of $\mathcal{R}F$-evaluations is bounded by $2(\mu + 1)^r = \mathcal{O}(\mu^r)$, see Theorem 6.19. In comparison, the evaluation of the *Lorentz–Radon formulae* requires the evaluation of an integral over S^{r-1} by at most $2(\mu + 1)^{r-1} = \mathcal{O}(\mu^{r-1})$ evaluations of $\mathcal{R}F$, only, however at the price of an additional numerical differentiation, — if this makes sense at all.

We must not forget the necessary kernel evaluations at the points $(\sigma_j, t_k) \in Z^r$. They require evaluation of the $C_\nu^{\frac{r}{2}}(\sigma_j)$ and of the $C_\nu^{\frac{r}{2}}(t_k x)$ for $\nu = 0, \ldots, \mu$ together at $(\mu + 1) + 2(\mu + 1)^{r-1}$ points (at most), with a need for $\mathcal{O}(\mu^r)$ arithmetical operations in total, if the recurrence relation of the Gegenbauer polynomials is applied, see Problem 2.1. But apart from their number, the kernel evaluations may be much cheaper than the evaluations of $\mathcal{R}F$, and could also be organized in tabular form. Moreover, for every fixed $j \in \{0, \ldots, \mu\}$, $K_\mu^A(\sigma_j, \cdot)$ is a polynomial of degree μ, which can be evaluated simultaneously at the points $t_k x$ by means of the fast Fourier transform (FFT) at $\mathcal{O}(\mu^{r-1} \log \mu)$ arithmetical operations, such that the total arithmetical costs are of the order $\mathcal{O}(\mu^{r-1} \log \mu)$, only, instead of $\mathcal{O}(\mu^r)$. Here we neglected the necessary organisation work.

In practice we may even argue that all quadrature points (σ_j, t_k) may be omitted, which define an affine hyperplane $\{x \in \mathbb{R}^r \mid x t_k = \sigma_j\}$, which is not intersecting a given neighbourhood of the evaluation point $x_0 \in B^r$. For instance, if this neighbourhood is the open ball $\{x \in \mathbb{R}^r : |x - x_0| < d\}$, then all quadrature points could be omitted which satisfy the condition

$$|x_0 t_k - \sigma_j| \geq d.$$

This technique corresponds to truncated generalized hyperinterpolation and reduces the evaluation cost essentially.

Stability

We assume $s = 1$, again. The reconstruction formulae of Lorentz–Radon are problematic with respect to the necessary, but often unallowed partial differentiation.

Actually, $\mathcal{R}_0^{-1}$ is an unbounded operator, and so every reconstruction method must show some instability. This holds also for the reconstruction method suggested by Theorem 8.4.

We investigate the corresponding stability problem by introducing for $\mu \in \mathbb{N}_0$ the linear operators M_μ^r, defined by

$$\left(M_\mu^r G\right)(x) := \Omega_{r-1} \int\limits_{Z^r} G(\sigma, t)\, K_\mu^A(\sigma, tx)\, (1 - \sigma^2)^{\frac{r-1}{2}}\, d(\sigma, t) \qquad (8.32)$$

for arbitrary bounded integrable or square-integrable functions $G : Z^r \to \mathbb{R}$, which means that

$$\|G\|_\infty = \sup\{|G(\sigma, t) : (\sigma, t) \in Z^r\} < \infty$$

or

$$\|G\|_2^2 = \gamma_r \int\limits_{Z^r} [G(\sigma, t)]^2 (1 - \sigma^2)^{\frac{r-1}{2}} d(\sigma, t) < \infty$$

holds, respectively. The constant γ_r is chosen such that $\|1\|_2 = 1$ is valid. Because of (1.27) this yields

$$\gamma_r^{-1} = \omega_{r-1} \int\limits_{-1}^{1} (1 - \sigma^2)^{\frac{r-1}{2}}\, d\sigma = \frac{\omega_{r-1}\omega_{r+1}}{\omega_r}. \qquad (8.33)$$

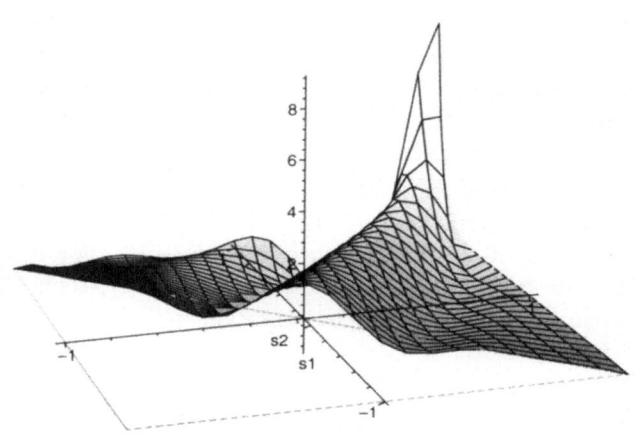

Figure 8.1. $K_\mu^A(s_1, s_2)(1 - s_1^2)^{\frac{r-1}{2}}$, the Kernel of M_μ^r.

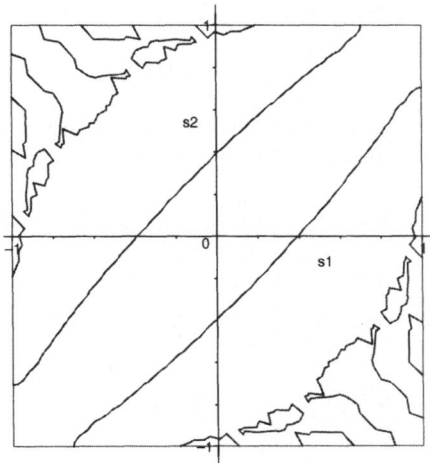

Figure 8.2. Zero Set of $K_\mu^A(s_1, s_2)$.

The definition of M_μ^r is such that (8.22) takes now the form

$$\left(L_\mu^{r,1} F\right)(x) = \left(M_\mu^r \mathcal{R} F\right)(x),$$

where we used (8.3), again. For the kernel see Figure 8.1 and Figure 8.2.

In what follows we assume that $\mathcal{R} F$ is not known exactly (for various reasons, of numerical or physical nature), but that $\mathcal{R} F$ is replaced by some approximant G. Then the reconstruction error can be estimated in the form

$$\|L_\mu^{r,1} F - M_\mu^r G\|_\infty \leq \|M_\mu^r\|_{p,\infty} \|\mathcal{R} F - G\|_\infty \tag{8.34}$$

for $p \in \{2, \infty\}$, where

$$\|M_\mu^r\|_{p,\infty} := \sup\{\|M_\mu^r G\|_\infty : \|G\|_p \leq 1\},$$

and the stability of our method can be measured by these operator norms. They are subject to the following theorem.

Theorem 8.6 (Reconstruction Norm). *Let $r \in \mathbb{N} \setminus \{1, 2\}$, and assume the summation matrix A satisfies the assumptions of Theorem 6.29. Then*

$$\|M_\mu^r\|_{\infty,\infty} \leq \|M_\mu^r\|_{2,\infty} \leq \sqrt{\frac{2}{r(r+1)}} \cdot \frac{(\mu + r + 1)^r}{(r-2)!} \tag{8.35}$$

holds for $\mu \in \mathbb{N}_0$.

Remark. The assumptions on A in Theorem 6.29 are stronger than in Theorem 7.2.

Proof. $\|G\|_\infty \le 1$ implies $\|G\|_2^2 \le \|1\|_2^2 = 1$, and we get

$$\|M_\mu^r\|_{\infty,\infty} \le \|M_\mu^r\|_{2,\infty}.$$

Therefore it suffices to estimate the operator norm on the right side. To this end let $\|G\|_2 \le 1$ and $x \in B^r$. From (8.32) we get

$$\left|(M_\mu^r G)(x)\right|^2 \le \Omega_{r-1}^2 \gamma_r^{-1} \int\limits_{Z^r} \left|K_\mu^A(\sigma, tx)\right|^2 (1 - \sigma^2)^{\frac{r-1}{2}} d(\sigma, t).$$

Using the kernel representation (8.23) and orthogonality we obtain

$$\left|(M_\mu^r G)(x)\right|^2 \le \Omega_{r-1}^2 \gamma_r^{-1} \sum_{\nu=0}^\mu \lambda_\nu^2 a_{\mu,\nu}^2 \int\limits_{-1}^1 \int\limits_{S^{r-1}} \left[C_\nu^{\frac{r}{2}}(\sigma)\right]^2 \left[C_\nu^{\frac{r}{2}}(tx)\right]^2 (1 - \sigma^2)^{\frac{r-1}{2}} d\omega(t) d\sigma.$$

From Theorem 6.29 we get $|a_{\mu,\nu}| \le 1$. Moreover, the integral can be factorized. The first factor is

$$\int\limits_{-1}^1 \left[C_\nu^{\frac{r}{2}}(\sigma)\right]^2 (1 - \sigma^2)^{\frac{r-1}{2}} d\sigma = C_\nu^{\frac{r}{2}}(1) \cdot c_\nu^{r,1} = \frac{r}{2\nu + r} \cdot \frac{\omega_{r+1}}{\omega_r} \cdot C_\nu^{\frac{r}{2}}(1).$$

Here we used (8.18) and (8.20). The second factor is given and estimated by

$$\int\limits_{S^{r-1}} \left[C_\nu^{\frac{r}{2}}(tx)\right]^2 d\omega(t) \le (r-1)^2 \omega_{r-1} C_\nu^{\frac{r}{2}}(1).$$

For the proof of this inequality we refer to Appendix (D), Lemma D.2. By inserting these results above and replacing λ_ν with the help of (8.21), γ_r^{-1} by (8.33), we get

$$\left|(M_\mu^r G)(x)\right|^2 \le (r-1)^2 \Omega_{r-1}^2 \cdot \left[\frac{\omega_{r+1}}{\omega_{r-1}\omega_r}\right]^2 \cdot r \cdot \sum_{\nu=0}^\mu (2\nu + r) \left[C_\nu^{\frac{r}{2}}(1)\right]^2.$$

From (1.8) we get $(r-1)\Omega_{r-1} = \omega_{r-2}$, and hence we obtain, in view of $C_0^{\frac{r}{2}}(1) < C_1^{\frac{r}{2}}(1) < \cdots$, see (2.13),

$$\left|(M_\mu^r G)(x)\right|^2 \le \left[\frac{\omega_{r-2}}{\omega_r} \cdot \frac{\omega_{r+1}}{\omega_{r-1}}\right]^2 \cdot r \cdot C_\mu^{\frac{r}{2}}(1) \sum_{\nu=0}^\mu (2\nu + r) C_\nu^{\frac{r}{2}}(1).$$

Again from (1.8) we get

$$\frac{\omega_{r+1}}{\omega_{r-1}} = \frac{2\pi}{r},$$

and using (4.28), (4.30), and (4.13) we obtain

$$\sum_{\nu=0}^{\mu}(2\nu+r)C_{\nu}^{\frac{r}{2}}(1) = r\cdot\left[C_{\mu}^{\frac{r+2}{2}}(1)+C_{\mu-1}^{\frac{r+2}{2}}(1)\right] \leq 2rC_{\mu}^{\frac{r+2}{2}}(1).$$

Inserting these results above we get finally

$$\begin{aligned}
\left|(M_{\mu}^{r}G)(x)\right|^2 &\leq 2r^2\left[\frac{r-1}{r}\right]^2\cdot C_{\mu}^{\frac{r}{2}}(1)C_{\mu}^{\frac{r+2}{2}}(1) \\
&= 2(r-1)^2\binom{\mu+r-1}{r-1}\binom{\mu+r+1}{r+1} \\
&\leq \frac{2}{r(r+1)}\cdot\frac{(\mu+r+1)^{2r}}{[(r-2)!]^2},
\end{aligned}$$

which yields (8.35), as claimed.   □

Remark 1. The estimate (8.35) might be too pessimistic, but says that $\|M_{\mu}^{r}\|_{\infty}$ increases at most at the order of dim $\mathbb{P}_{\mu}^{r}$. In the important case $r=3$, the estimate takes the form

$$\|M_{\mu}^{3}\|_{2,\infty} \leq \frac{(\mu+4)^3}{\sqrt{6}}.$$

In view of (8.34) this says that the error, caused by replacing $\mathcal{R}F$ by some G, is bounded for $p\in\{2,\infty\}$ by

$$\|L_{\mu}^{3,1}F-M_{\mu}^{3}G\|_{\infty} \leq \frac{(\mu+4)^3}{\sqrt{6}}\cdot\|\mathcal{R}F-G\|_{p}.$$

The estimate is valid, for instance, in case of the Newman–Shapiro operators.

Remark 2. In the case $r=3$, Theorem 8.4 establishes an approximation method for MR-tomography. In case of the Newman–Shapiro matrix $A=A^{(5)}$ the approximation order is $\mathcal{O}(\mu^{-2})$ for functions which are twice continuously differentiable in a neighbourhood of B^{r}. If (8.22) is evaluated by means of a product Gauß quadrature, then an operator $\hat{L}_{\mu}^{r,1}$ arises from $L_{\mu}^{r,1}$, as in the case of *generalized hyperinterpolation*. U. Maier [35] investigated $\hat{L}_{\mu}^{r,1}$-approximants up to the degree $\mu=160$ by numerical experiments, thus making visible the validity of our theory. For theoretical reasons she investigated the exact, i.e., *untruncated* operators. But actually, as in the case of generalized hyperinterpolation, truncation promises a significant reduction of the evaluation cost without loss of the accuracy order.

8.5   k-Plane Transform

The Radon transform $\mathcal{R}_0 = \mathcal{R}_0^{(r-1)}$ is defined by means of integrals stretched over the intersection of B^{r} and an $(r-1)$-dimensional affine hyperplane, which is orthogonal to $span\{t\}$. As a generalisation, the $(r-k)$-plane transform $\mathcal{R}_0^{(r-k)}$,

$k \in \{1, \ldots, r-1\}$, is defined by means of $(r-k)$-dimensional affine hyperplanes. For convenience we denote by k their co-dimension, but have to remark that in case of the k-plane transform $\mathcal{R}_0^{(k)}$ itself, the dimension of the hyperplanes is just k.

Let $T = (t_1, \ldots, t_k)$ consist of orthonormal columns $t_1, \ldots, t_k \in \mathbb{R}^r$. An element $v \in \mathbb{R}^r$ is orthogonal to $span\{t_1, \ldots, t_k\}$ exactly if $v'T = 0$ holds. In this case we write $v \perp T$. Every $a \in span\{t_1, \ldots, t_k\}$ has a uniquely determined representation $a = T\sigma$ with $\sigma = (\sigma_1, \ldots, \sigma_k)' \in \mathbb{R}^k$. Because of

$$|a|^2 = a'a = \sigma'T'T\sigma = \sigma'\sigma = |\sigma|^2,$$

$a \in B^r$ holds exactly for $\sigma \in B^k$. Therefore, every $(r-k)$-dimensional affine hyperplane which is orthogonal to $span\{t_1, \ldots, t_k\}$ has the form

$$\{x \in \mathbb{R}^r \,|\, x = T\sigma + v, \ v \perp T\},$$

and it intersects B^r exactly for $\sigma \in B^k$.

After these preliminaries, we generalize the definition (8.1) for $F \in C(B^r)$ by putting

$$(\mathcal{R}_0^{(r-k)}F)(\sigma, T) := \int\limits_{\substack{v \perp T \\ v^2 \leq 1-\sigma^2}} F(T\sigma + v)\, dv \tag{8.36}$$

for $\sigma \in B^r$ and T as above. Substituting $v = \sqrt{1-\sigma^2}\, u$ we obtain, again,

$$(\mathcal{R}_0^{(r-k)}F)(\sigma, T) = (1-\sigma^2)^{\frac{r-k}{2}} \int\limits_{u \perp T, \, u^2 \leq 1} F(T\sigma + \sqrt{1-\sigma^2}\, u)\, du. \tag{8.37}$$

In particular we get

$$(\mathcal{R}_0^{(r-k)}1)(\sigma, T) = \Omega_{r-k}\,(1-\sigma^2)^{\frac{r-k}{2}}. \tag{8.38}$$

And as above, we normalize $\mathcal{R}_0^{(r-k)}$ by the definition

$$\mathcal{R}^{(r-k)} := \mathcal{R}_0^{(r-k)} / \mathcal{R}_0^{(r-k)}1,$$

which implies

$$\mathcal{R}^{(r-k)}1 = 1.$$

We do not go into further details, except for showing how the spaces $\mathbf{V}_\mu^{r,s}$ are transformed, where it suffices in view of Theorem 7.6 to investigate the action of $\mathcal{R}^{(r-k)}$ to the basic functions $G_\mu^{r+s+1}(a' \cdot)$, $a \in S^{r-1}$.

Theorem 8.7 (k-Plane Transform). *Let* $r \in \mathbb{N} \setminus \{1\}$, $k \in \{1, \ldots, r\}$, $s \in \mathbb{N}_0$, $\mu \in \mathbb{N}_0$, *and* $a \in S^{r-1}$. *Then*

$$(\mathcal{R}^{(r-k)}G_\mu^{r+s+1}(a' \cdot))(\sigma, T) = \frac{1}{\omega_{r+s-k}} \cdot G_\mu^{k, r+s-k}(\sigma', a'T) \tag{8.39}$$

holds for $\sigma \in B^k$ and $T = (t_1, \ldots, t_k)$, $t_1, \ldots, t_k \in S^{r-1}$ pairwise orthogonal, where $G_\mu^{k, r+s-k}$ is the reproducing kernel function of $\mathbf{V}_\mu^{k, r+s-1-k}$.

Proof. Let $\kappa := s + 1$. From (8.37) and (8.38) we obtain for all rotations $A \in A_T^r$,

$$\left(\mathcal{R}^{(r-k)} G_\mu^{r+\kappa}(a' \cdot) \right)(\sigma, T)$$

$$= \frac{1}{\Omega_{r-k}} \cdot \int\limits_{u \perp T, \, u^2 \leq 1} G_\mu^{r+\kappa}\left(a'\left[T\sigma + \sqrt{1 - \sigma^2}\, u \right] \right) du$$

$$= \frac{1}{\Omega_{r-k}} \cdot \int\limits_{u \perp T, \, u^2 \leq 1} G_\mu^{r+\kappa}\left(a' A\left[T\sigma + \sqrt{1 - \sigma^2}\, u \right] \right) du.$$

Applying the average operator $\Pi_T^{r,\kappa}$ to both sides, and using Corollary 4.30 we obtain

$$\left(\mathcal{R}^{(r-k)} G_\mu^{r+\kappa}(a' \cdot) \right)(\sigma, T)$$

$$= \frac{1}{\Omega_{r-k}} \cdot \int\limits_{u \perp T, \, u^2 \leq 1} \frac{1}{\omega_{r+\kappa-1-k}} \cdot G_\mu^{k, r+\kappa-1-k}\left(a'T, \left[\sigma'T' + \sqrt{1 - \sigma^2}\, u' \right] T \right) du$$

$$= \frac{1}{\omega_{r+\kappa-1-k}} \cdot G_\mu^{k, r+\kappa-1-k}\left(a'T, \sigma' \right),$$

where we used $u'T = 0$ and $T'T = I$. In view of the symmetry of the kernel this is exactly (8.39), as claimed. $\qquad\square$

Remark The right side of (8.39) is given explicitly by

$$\frac{1}{\omega_{r+s-k}} \sum_{|n|=\mu} \hat{U}_n^{k, r+s-k}(\sigma_1, \ldots, \sigma_k)\, V_n^{k, r+s-k}(at_1, \ldots, at_k)$$

for $\sigma_1^2 + \cdots + \sigma_k^2 \leq 1$ and pairwise orthogonal $t_1, \ldots, t_k \in S^{r-1}$, see (4.100). For $k = 1$ this takes the form

$$\frac{1}{\omega_{r+s-1}} \cdot \hat{U}_\mu^{1, r+s-1}(\sigma_1)\, V_\mu^{1, r+s-1}(at_1) = const \cdot C_\mu^{\frac{r+s-1}{2}}(\sigma_1)\, C_\mu^{\frac{r+s-1}{2}}(at_1),$$

see (4.107). In view of $G_\mu^{r+s+1} = const \cdot C_\mu^{\frac{r+s-1}{2}}$ this result corresponds exactly to the result of Davison and Grünbaum, see Theorem 8.1.

8.6 Problems

Problem 8.1. Let $r \geq 2$, and assume the ball $A := \{x \in \mathbb{R}^r : |x - a| \leq \rho\}$, $a \in B^r$, $\rho > 0$, is a subset of B^r. For $\alpha > 0$ let $F \in C(B^r)$ be defined by

$$F(x) := (\rho^2 - |x - a|^2)_+^\alpha$$

for $x \in B^r$, where $(\cdot)_+^\alpha$ is the usual basic spline-function. Calculate $\mathcal{R}_0 F$.

Appendices

Appendix A

Legendre Basis

Let $r \in \mathbb{N} \setminus \{1\}$. For a fixed $\nu \in \{0, 1, \ldots, \mu\}$ let $\overset{*}{H}$ be an element of $\mathbb{H}_\nu^{r-1}$, and hence also of $\mathbb{H}_\nu^r$. It is not dependent on x_r, so we get

$$e_r \, grad \, \overset{*}{H} = 0. \tag{A.1}$$

Next let P be an arbitrary element of $\mathbb{Q}_{\mu-\nu}^1$, such that

$$\overset{*}{P}(x) = |x|^{\mu-\nu} P\left(\frac{x_r}{|x|}\right)$$

is a homogeneous polynomial of degree $\mu - \nu$, whose restriction onto S^{r-1} depends on x_r, only. It follows that

$$\overset{*}{F} := \overset{*}{H} \overset{*}{P}$$

is a homogeneous polynomial of degree μ with respect to all of the variables $x_1, \ldots, x_r$, and we ask for conditions on P which let this polynomial become harmonic. Note that $\Delta \overset{*}{F}$ is also homogeneous (of degree $\mu - 2$), such that it suffices to let this polynomial vanish on S^{r-1}.

Obviously, because of $\Delta \overset{*}{H} = 0$ we get

$$\Delta \overset{*}{F} = 2 \, grad \, \overset{*}{H} \cdot grad \, \overset{*}{P} + \overset{*}{H} \Delta \overset{*}{P}. \tag{A.2}$$

By a tedious, though elementary, calculation it is possible to evaluate the right side for the arguments $x \in S^{r-1}$. With the abbreviations $\overset{*}{H} = \overset{*}{H}(x)$, $P = P(x_r)$, and so on, we obtain the following equations,

$$grad \overset{*}{P} = (\mu - \nu)P \cdot x + P' \cdot (e_r - x_r \cdot x),$$

$$\Delta \overset{*}{P} = (1 - x_r^2)P'' - (r - 1)x_r P' + (\mu - \nu)(\mu - \nu + r - 2)P.$$

Using the first one, together with Euler's partial differential equation

$$x \, grad \overset{*}{H} = \nu \, \overset{*}{H},$$

see Theorem 3.3, and taking into account (A.1) we get

$$grad \, \overset{*}{H} \cdot grad \overset{*}{P} = [\nu(\mu - \nu)P - \nu x_r P'] \overset{*}{H},$$

again for $x \in S^{r-1}$. Inserting these results in (A.2) we obtain

$$\Delta \overset{*}{F} = [(1 - x_r^2)P'' - (2\nu + r - 1)x_r P' + (\mu - \nu)(\mu + \nu + r - 2)P] \cdot \overset{*}{H},$$

still for $x \in S^{r-1}$.

Now let P take the form

$$P = const \cdot \tilde{C}_{\mu-\nu}^{\frac{r-2}{2}+\nu}.$$

Then we obtain with the help of Theorem 2.1,

$$\Delta \overset{*}{F} = 0,$$

as yet for $x \in S^{r-1}$, only. But since $\Delta \overset{*}{F}$ is homogeneous, this equality is valid for arbitrary arguments. Moreover, in view of (4.12) and of (2.10) and (2.11) the constant can be chosen such that P takes the form

$$P = G_\mu^{(\nu)},$$

where $G_\mu^{(\nu)}$ is the ν-th derivative of the kernel function $G_\mu = G_\mu^r$. Our results can be summerized by the statement that

$$\overset{*}{F}(x) = \overset{*}{H}(x_1, \ldots, x_{r-1}) \cdot |x|^{\mu-\nu} G_\mu^{(\nu)}\left(\frac{x_r}{|x|}\right) \tag{A.3}$$

is a harmonic homogeneous polynomial of degree μ for arbitrary $\overset{*}{H}$ in $\overset{*}{\mathbb{H}}_\nu^{r-1}$.

Next let $\bar{x} = (x_1, \ldots, x_{r-1})'$. For $x \in S^{r-1}$ we get $|\bar{x}| = \sqrt{1 - x_r^2}$, and (A.3) is equivalent to

$$\overset{*}{F}(x) = \overset{*}{H}\left(\frac{\bar{x}}{|\bar{x}|}\right) \cdot (1 - x_r^2)^{\frac{\nu}{2}} G_\mu^{(\nu)}(x_r). \tag{A.4}$$

With respect to the first $r-1$ variables, the first factor is here a spherical harmonic of degree ν, it is combined with a univariate function of the last variable. Apart from a constant factor, the function

$$(1 - \xi^2)^{\frac{\nu}{2}} G_\mu^{(\nu)}(\xi),$$

$\nu = 0, 1, \ldots, \mu$, is a so-called *Legendre function*, see (5.101).

The presented construction of a harmonic homogeneous polynomial can be used to generate a complete basis.

Theorem A.1 (Legendre Bases). *Let* $r \in \mathbb{N} \setminus \{1\}$, $\mu \in \mathbb{N}$. *For* $\nu \in \{0, 1, \ldots, \mu\}$ *let* $\mathcal{B}_\nu^{r-1}$ *be a basis of* $\overset{*}{\mathbb{H}}{}_\nu^{r-1}$, *and define*

$$\mathcal{B}_{\mu,\nu}^r := \left\{ H(\bar{x}) \cdot |x|^{\mu-\nu} G_\mu^{(\nu)}\left(\frac{x_r}{|x|}\right) \right\}_{H \in \mathcal{B}_\nu^{r-1}}.$$

Then the family

$$\mathcal{B}_\mu^r := \mathcal{B}_{\mu,0}^r \cup \mathcal{B}_{\mu,1}^r \cup \cdots \cup \mathcal{B}_{\mu,\mu}^r$$

is a basis of $\overset{*}{\mathbb{H}}{}_\mu^r$.

Proof. The members of $\mathcal{B}_\mu^r$ have the form (A.3). So they are elements of $\overset{*}{\mathbb{H}}{}_\mu^r$. By the diagram (4.26) their number is

$$
\begin{aligned}
\sum_{\nu=0}^{\mu} \dim \overset{*}{\mathbb{H}}{}_\nu^{r-1} &= \sum_{\nu=0}^{\mu} \dim \overset{*}{\mathbb{H}}{}_\nu^{r-1}(S^{r-2}) \\
&= \dim \overset{*}{\mathbb{P}}{}_\mu^{r-1}(S^{r-2}) + \dim \overset{*}{\mathbb{P}}{}_{\mu-1}^{r-1}(S^{r-2}) \\
&= \binom{\mu+r-2}{r-2} + \binom{\mu+r-3}{r-2} \\
&= \dim \overset{*}{\mathbb{H}}{}_\mu^r,
\end{aligned}
$$

where we used (4.1) and Corollary 3.11. Therefore it suffices to prove that they are linearly independent.

So let us assume that a linear combination from $\mathcal{B}_\mu^r$ vanishes. This can be written in the form

$$\sum_{\nu=0}^{\mu} H_\nu(\bar{x})|x|^{\mu-\nu} G_\mu^{(\nu)}\left(\frac{x_r}{|x|}\right) = 0, \tag{A.5}$$

where every H_ν is a linear combination of basis elements from $\mathcal{B}_\nu^{r-1}$. Therefore it suffices to prove that $H_0, H_1, \ldots, H_\mu$ vanish, and it suffices even to prove that this holds on S^{r-2}. Actually, for $x \in S^{r-1}$ we can write (A.5) in the form

$$\sum_{\nu=0}^{\mu} H_\nu\left(\frac{\bar{x}}{|\bar{x}|}\right) \cdot (1 - x_r^2)^{\frac{\nu}{2}} G_\mu^{(\nu)}(x_r) = 0.$$

Now choose $x_r \in (-1, +1)$ such that none of the polynomials $G_\mu^{(\nu)}$ vanishes at x_r. The values of $\frac{\bar{x}}{|\bar{x}|}$ cover S^{r-2} while $\bar{x}$ varies in B^{r-1} under the side condition $|\bar{x}|^2 = 1 - x_r^2$. Moreover, by Theorem 4.10 the spherical harmonics

$$H_\nu\big|_{S^{r-2}}, \quad \nu = 0, 1, \ldots, \mu,$$

are pairwise orthogonal, see Theorem 4.10. Together this yields $H_\nu = 0$ for $\nu = 0, 1, \ldots, \mu$, and the theorem is proved. $\square$

Appendix B

Zeros of the Kernel Function

Lemma B.1. *For $\mu \in \mathbb{N}$ let $\xi_\mu < \xi_{\mu-1} < \ldots < \xi_1$ denote the zeros of the Jacobi polynomial $y = P_\mu^{(\frac{r-1}{2}, \frac{r-3}{2})}$. Then $\xi_1 < \xi_\mu^2$ holds in the particular case $r = 3$.*

Proof. Let $r = 3$ and $\mu \in \mathbb{N}$. For $\mu = 1$ the unique zero of y is given by $-\frac{1}{r}$. By the interlacing property of the zeros of orthogonal polynomials it follows that $\xi_\mu < 0$ holds for arbitrary $\mu \in \mathbb{N}$. Moreover, the polynomial

$$Y(x) := (-1)^\mu y\left(-\frac{1+x}{2}\right) \tag{B.1}$$

has a positive leading coefficient and the roots

$$x_\nu = -(1 + 2\xi_\nu), \quad \nu = 1, \ldots, \mu,$$

which are in the order

$$-3 < x_1 < \ldots < x_\mu < 1.$$

In particular, it follows from $\xi_\mu < 0$ that even

$$-1 < x_\mu < 1$$

holds, and it is easy to see that $x_\mu < \xi_\mu^2$ is also valid.

Now assume that the statement of the lemma is false. Then we get

$$x_\mu < \xi_1. \tag{B.2}$$

Moreover, it is well known that y satisfies the differential equation

$$(1 - x^2)y'' - (1 + rx)y' + \mu(\mu + r - 1)y = 0, \tag{B.3}$$

see, e.g., [13] or [75]. It follows that Y satisfies the differential equation

$$[2(1 - x) + (1 - x^2)]Y'' - [r - 2 + rx]Y' + \mu(\mu + r - 1)Y = 0.$$

Note that $Y''(x) > 0$ holds for $x_\mu < x < \infty$. So we get, in the interval $x_\mu < x \le 1$,

$$(1 - x^2)Y'' - (1 - rx)Y' + \mu(\mu + r - 1)Y < 0, \tag{B.4}$$

where we replaced $r-2$ by its actual value 1. Next we use the well-known transform

$$
\begin{aligned}
p(\phi) &:= (\mu + \tfrac{r-1}{2})^2 + \tfrac{1-(r-1)^2}{16(\sin\frac{\phi}{2})^2} + \tfrac{1-(r-3)^2}{16(\cos\frac{\phi}{2})^2}, \\
u(\phi) &:= (\sin\phi)^{\frac{r}{2}}(\cos\phi)^{\frac{r-2}{2}}\, y(\cos\phi), \\
U(\phi) &:= (\sin\phi)^{\frac{r}{2}}(\cos\phi)^{\frac{r-2}{2}}\, Y(\cos\phi),
\end{aligned}
$$

and define $\phi_\mu \in (0,\pi)$ by $x_\mu = \cos\phi_\mu$. Then (B.3) and (B.4) take the form

$$u'' + p(\phi)u = 0, \quad U'' + p(\phi)U < 0$$

for $0 < \phi < \phi_\mu$, where ϕ_μ is the lowest positive zero of U. Moreover, since $U(\phi)$ is positive in this interval, there exists a continuous function $P > p$ such that

$$U'' + P(\phi)U = 0$$

holds, again in this interval. Finally we get, using $r = 3$ again,

$$\lim_{\phi \to 0} \left[u'(\phi)U(\phi) - u(\phi)U'(\phi) \right] = 0,$$

and it follows by *Sturm's Theorem*, in the form presented by Tricomi [75], p.175, that U has a root ψ satisfying

$$0 \le \psi < \phi_\mu,$$

which is a contradiction. So the statement of the lemma is true.   □

Appendix C

Newman–Shapiro Operators

We repeat the lemma to be proved:

Lemma C.1. *Let $r \in \mathbb{N} \setminus \{1\}$. Given a constant $c_r > 0$, a constant d_r exists such that the following holds. If $A : [-1, 1] \to \mathbb{R}$ is an arbitrary monotonically nondecreasing function such that*

$$(i) \quad \int_0^\pi K_{2\nu}^{(r)}(\cos\phi)\, dA(\phi) = \frac{1}{\omega_{r-2}},$$

$$(ii) \quad \int_0^\pi \frac{G_{\nu+1}^2(\cos\phi)}{\cos\phi - \cos\chi_{\nu+1}}\, dA(\phi) = 0,$$

$$(iii) \quad \int_0^{\chi_{\nu+1}} dA(\phi) \leq \frac{c_r}{\nu+1},$$

holds for all $\nu \in \mathbb{N}$, then

$$\int_0^\pi K_{2\nu}^{(r)}(\cos\phi)(\mu\phi + 1)^j\, dA(\phi) \leq d_r$$

is valid for $j \in \{1, 2\}$, and again for arbitrary $\nu \in \mathbb{N}_0$. The integrals are defined in the sense of Riemann and Stieltjes.

Proof. Note that $K_\mu^{(r)}$ is defined by (6.88). We prove the statement of the lemma for a fixed $j \in \{1, 2\}$.

$\eta_{\nu+1} = \cos\chi_{\nu+1}$, $0 < \chi_{\nu+1} < \frac{\pi}{2}$, is the greatest zero of $G_{\nu+1}$, i.e., of $C_{\nu+1}^{\frac{r-2}{2}}$. In view of (6.93) a constant a_r exists such that

$$0 < \mu\chi_{\nu+1} + 1 \leq a_r \tag{C.1}$$

holds for all $\nu \in \mathbb{N}_0$. We split the integral in the form

$$I_{\nu,j} = I'_{\nu,j} + I''_{\nu,j},$$

with

$$I''_{\nu,j} := \int_0^{\chi_{\nu+1}} K^{(r)}_\mu(\cos\phi)(\mu\phi+1)^j dA(\phi).$$

Because of (C.1) and of the assumption (i), this term can be estimated as

$$I''_{\nu,j} \leq a_r^j \cdot \int_0^\pi K^{(r)}_\mu(\cos\phi)dA(\phi) = \frac{a_r^j}{\omega_{r-2}}.$$

In particular, $I''_{\nu,j}$ is bounded.

Next we investigate the complement

$$I'_{\nu,j} = \int_{\chi_{\nu+1}}^\pi K^{(r)}_\mu(\cos\phi)\Big(\mu(\phi-\chi_{\nu+1}) + \mu\chi_{\nu+1} + 1\Big)^j dA(\phi).$$

With the help of the integrals

$$B_{\nu,k} := \mu^k \int_{\chi_{\nu+1}}^\pi K^{(r)}_\mu(\cos\phi)(\phi-\chi_{\nu+1})^k dA(\phi)$$

it can be written in the form

$$I'_{\nu,j} = \sum_{k=0}^j \binom{j}{k} B_{\nu,k},$$

such that it suffices to prove that the $B_{\nu,k}$ are bounded for $\nu \in \mathbb{N}_0$, separately for $k \in \{0,1,2\}$.

For $k = 0$ we get, by the assumption (i),

$$B_{\nu,0} \leq \int_0^\pi K^{(r)}_\mu(\cos\phi)dA(\phi) = \frac{1}{\omega_{r-2}},$$

as wanted.

For $k = 1$ we have to investigate

$$B_{\nu,1} = \mu g_{\nu+1} \int_{\chi_{\nu+1}}^\pi \frac{G^2_{\nu+1}(\cos\phi)}{\cos\chi_{\nu+1} - \cos\phi} \cdot \frac{\phi-\chi_{\nu+1}}{2\sin\frac{\phi+\chi_{\nu+1}}{2}\sin\frac{\phi-\chi_{\nu+1}}{2}} \cdot dA(\phi),$$

where we used (6.88) to get the explicit form. Note that

$$\frac{\phi-\chi_{\nu+1}}{2} \leq \frac{\pi}{2}\sin\frac{\phi-\chi_{\nu+1}}{2}$$

and

$$\sin \frac{\phi + \chi_{\nu+1}}{2} \geq \min \left\{ \sin \chi_{\nu+1}, \ \sin \frac{\pi - \chi_{\nu+1}}{2} \right\}$$

hold for $\chi_{\nu+1} \leq \phi \leq \pi$. In view of

$$\sin \frac{\pi - \chi_{\nu+1}}{2} = \frac{\sin(\pi - \chi_{\nu+1})}{2 \cos \frac{\pi - \chi_{\nu+1}}{2}} \geq \frac{1}{2} \sin \chi_{\nu+1},$$

the last inequality implies

$$\sin \frac{\phi + \chi_{\nu+1}}{2} \geq \frac{1}{2} \sin \chi_{\nu+1}.$$

Inserting this above we obtain

$$
\begin{aligned}
B_{\nu,1} \ \leq \ & \mu \, g_{\nu+1} \frac{\pi}{\sin \chi_{\nu+1}} \int_{\chi_{\nu+1}}^{\pi} \frac{G_{\nu+1}^2(\cos \phi)}{\cos \chi_{\nu+1} - \cos \phi} \, dA(\phi) \\
= \ & \mu \, g_{\nu+1} \frac{\pi}{\sin \chi_{\nu+1}} \int_{0}^{\chi_{\nu+1}} \frac{G_{\nu+1}^2(\cos \phi)}{\cos \phi - \cos \chi_{\nu+1}} \, dA(\phi).
\end{aligned}
$$

Here we used the assumption (ii) in order to reflect the influence of the large interval $[\chi_{\nu+1}, \pi]$ to $[0, \chi_{\nu+1}]$, the small one, which happens to be possible because of the special choice of the kernel.

In view of (4.12), (2.10), (2.11) and of (2.16) it follows that

$$B_{\nu,1} \leq \mu \, g_{\nu+1} \frac{\pi}{\sin \chi_{\nu+1}} \cdot G_{\nu+1}(1) \cdot G'_{\nu+1}(1) \int_{0}^{\chi_{\nu+1}} dA(\phi),$$

and together with (iii) this yields

$$B_{\nu,1} \leq \mu \, g_{\nu+1} \frac{\pi}{\sin \chi_{\nu+1}} \cdot G_{\nu+1}(1) \cdot G'_{\nu+1}(1) \cdot \frac{c_r}{\nu+1}.$$

Now we use (6.93), (4.12), Corollary 3.11, Markoff's inequality, and Theorem 6.33, (iii), to prove that the right side is bounded, again.

For $k = 2$ we get likewise

$$
\begin{aligned}
B_{\nu,2} \ = \ & \mu^2 g_{\nu+1} \int_{\chi_{\nu+1}}^{\pi} \frac{G_{\nu+1}^2(\cos \phi)}{\cos \chi_{\nu+1} - \cos \phi} \cdot \frac{(\phi - \chi_{\nu+1})^2}{2 \sin \frac{\phi + \chi_{\nu+1}}{2} \sin \frac{\phi - \chi_{\nu+1}}{2}} \cdot dA(\phi) \\
\leq \ & \mu^2 g_{\nu+1} \cdot \frac{\pi^2}{2} \int_{\chi_{\nu+1}}^{\pi} \frac{G_{\nu+1}^2(\cos \phi)}{\cos \chi_{\nu+1} - \cos \phi} \cdot \frac{\sin \frac{\phi - \chi_{\nu+1}}{2}}{\sin \frac{\phi + \chi_{\nu+1}}{2}} \cdot dA(\phi),
\end{aligned}
$$

where the function

$$\frac{\sin \frac{\phi - \chi_{\nu+1}}{2}}{\sin \frac{\phi + \chi_{\nu+1}}{2}}$$

is monotonically increasing for $\chi_{\nu+1} \leq \phi \leq \pi$, such that

$$\frac{\sin \frac{\phi - \chi_{\nu+1}}{2}}{\sin \frac{\phi + \chi_{\nu+1}}{2}} \leq \frac{\sin \frac{\pi - \chi_{\nu+1}}{2}}{\sin \frac{\pi + \chi_{\nu+1}}{2}} = 1$$

holds. Inserting this above and using again the assumptions (ii) and (iii) we get

$$B_{\nu,2} \leq \mu^2 g_{\nu+1} \cdot \frac{\pi^2}{2} \int\limits_0^{\chi_{\nu+1}} \frac{G_{\nu+1}^2(\cos \phi)}{\cos \phi - \cos \chi_{\nu+1}} \, dA(\phi)$$

$$\leq \mu^2 g_{\nu+1} \cdot \frac{\pi^2}{2} \cdot G_{\nu+1}(1) \cdot G_{\nu+1}'(1) \cdot \frac{c_r}{\nu + 1},$$

and the $B_{\nu,2}$ prove to be bounded, again.

Summarizing our results we see that

$$I_{\nu,j} = I_{\nu,j}'' + \sum_{k=0}^{j} \binom{j}{k} B_{\nu,k}$$

is bounded above for $\nu \in \mathbb{N}_0$ by some constant d_r, which depends on c_r, only. We choose it so large that the statement of the lemma holds both for $j = 1$ and for $j = 2$. $\qquad \square$

Appendix D

Reconstruction

Theorem D.1 (A Gegenbauer Polynomial Identity). *Let $r \in \mathbb{N}$ be odd and $\mu \in \mathbb{N}_0$, and let $L := \frac{d}{d\sigma}$. Then the following identity holds,*

$$\Omega_{r-1} D^{r-1}\left[(1-\sigma^2)^{\frac{r-1}{2}} \tilde{C}_\mu^{\frac{r}{2}}(\sigma)\right] = 2(2\pi i)^{r-1} \cdot \frac{1}{\omega_{r-1}} \cdot C_\mu^{\frac{r}{2}}(\sigma).$$

Proof. By defining

$$f(\sigma) := \Omega_{r-1} D^{r-1}[(1-\sigma^2)^{\frac{r-1}{2}} \tilde{C}_\mu^{\frac{r}{2}}(\sigma)] \tag{D.1}$$

we obtain a univariate polynomial f of degree μ. It can be expanded in the form

$$f(\sigma) = \sum_{\nu=0}^{\mu} a_\nu \tilde{C}_\nu^{\frac{r}{2}}(\sigma). \tag{D.2}$$

We want to show that $a_\nu = 0$ holds for $\nu = 0, \ldots, \mu - 1$, where the formula of Rodrigues, see Problem 2.4, will help us. In the present case we bring it to the form

$$\tilde{C}_\nu^{\frac{r}{2}}(\sigma) = \gamma_\nu (1-\sigma^2)^{-\frac{r-1}{2}} D^\nu X_\nu(\sigma), \tag{D.3}$$

where $\nu \in \mathbb{N}_0$ is arbitrary, and where

$$X_\nu(\sigma) = (1-\sigma^2)^{\nu + \frac{r-1}{2}}, \tag{D.4}$$

$$\gamma_\nu = \frac{(-1)^\nu}{2^\nu \left(\frac{r+1}{2}\right)_\nu}. \tag{D.5}$$

Obviously, inserting (D.3) in (D.1) we obtain

$$f(\sigma) = \Omega_{r-1} \cdot \gamma_\mu \cdot D^{\mu+r-1} X_\mu. \tag{D.6}$$

The $C_\nu^{\frac{r}{2}}$ are orthogonal, see Theorem 2.3, so $a_0 = \cdots = a_{\mu-1} = 0$ holds if and only if the inner products

$$[f, \tilde{C}_\nu^{\frac{r}{2}}]_{\frac{r}{2}} = \int\limits_{-1}^{1} f(\sigma)\, \tilde{C}_\nu^{\frac{r}{2}}(\sigma)(1-\sigma^2)^{\frac{r-1}{2}} d\sigma$$

vanish for $\nu = 0, \ldots, \mu - 1$. Actually, for fixed $\nu \in \{0, \ldots, \mu - 1\}$ we get, using (D.6) and (D.3),

$$\frac{1}{\gamma_\mu \gamma_\nu} \cdot [f, \tilde{C}_\nu^{\frac{r}{2}}]_{\frac{r}{2}} = \int\limits_{-1}^{1} \left[D^{\mu+r-1} X_\mu \right] \cdot \left[D^\nu X_\nu \right] \cdot d\sigma,$$

and integration by parts yields

$$\frac{1}{\gamma_\mu \gamma_\nu} \cdot [f, \tilde{C}_\nu^{\frac{r}{2}}]_{\frac{r}{2}}$$

$$= \left[D^{\mu+r-2} X_\mu \cdot D^\nu X_\nu \right]_{-1}^{1} - \int\limits_{-1}^{1} D^{\mu+r-2} X_\mu \cdot D^{\nu+1} X_\nu \cdot d\sigma$$

$$= \cdots$$

$$= \pm \left[D^{\mu+\frac{r-1}{2}} X_\mu \cdot D^{\nu+\frac{r-3}{2}} X_\nu \right]_{-1}^{1} \mp \int\limits_{-1}^{1} D^{\mu+\frac{r-1}{2}} X_\mu \cdot D^{\nu+\frac{r-1}{2}} X_\nu \cdot d\sigma,$$

where we used that the expressions $[\,\cdot\,]_{-1}^{1}$ vanish since the second factor vanishes for $\sigma = \pm 1$. By a change in the argumentation, only, we can continue this process, and get

$$\frac{1}{\gamma_\mu \gamma_\nu} \cdot [f, \tilde{C}_\nu^{\frac{r}{2}}]_{\frac{r}{2}}$$

$$= \mp \left[D^{\mu+\frac{r-3}{2}} X_\mu \cdot D^{\nu+\frac{r-1}{2}} X_\nu \right]_{-1}^{1} \pm \int\limits_{-1}^{1} D^{\mu+\frac{r-3}{2}} X_\mu \cdot D^{\nu+\frac{r+1}{2}} X_\nu \cdot d\sigma$$

$$= \cdots$$

$$= (-1)^{\nu+r-1} \left\{ \left[D^{\mu-\nu} X_\mu \cdot D^{2\nu+r-2} X_\nu \right]_{-1}^{1} - \int\limits_{-1}^{1} D^{\mu-\nu} X_\mu \cdot D^{2\nu+r-1} X_\nu \cdot d\sigma \right\},$$

where we used now that the first factor in $[\,\cdot\,]_{-1}^{1}$ vanishes for $\sigma = \pm 1$. X_ν is a polynomial of degree $2\nu + r - 1$, such that $D^{2\nu+r-1} X_\nu$ is, finally, some constant. It follows that

$$\frac{1}{\gamma_\mu \gamma_\nu} \cdot [f, \tilde{C}_\nu^{\frac{r}{2}}]_{\frac{r}{2}} = const \cdot \int\limits_{-1}^{1} D^{\mu-\nu} X_\mu \cdot d\sigma = const \cdot \left[D^{\mu-\nu-1} X_\mu \right]_{-1}^{1} = 0,$$

and $a_0 = \cdots = a_{\mu-1}$ holds, as claimed. Now we get from (D.1) and (D.6)

$$a_\mu \tilde{C}_\mu^{\frac{r}{2}}(\sigma) = \Omega_{r-1} \cdot \gamma_\mu \cdot D^{\mu+r-1}(1-\sigma^2)^{\mu+\frac{r-1}{2}},$$

and together with (D.5) we obtain

$$a_\mu C_\mu^{\frac{r}{2}}(\sigma) = (-1)^\mu \cdot \frac{\Omega_{r-1}}{2^\mu \left(\frac{r+1}{2}\right)_\mu} \cdot C_\mu^{\frac{r}{2}}(1) \cdot D^{\mu+r-1}(1-\sigma^2)^{\mu+\frac{r-1}{2}}.$$

The leading coefficient of $C_\mu^{\frac{r}{2}}(\sigma)$ is known from (2.5), and a comparison with the leading coefficient on the right side yields

$$a_\mu \cdot \frac{\left(\frac{r}{2}\right)_\mu}{(1)_\mu} \cdot 2^\mu = (-1)^{\frac{r-1}{2}} \cdot \frac{\Omega_{r-1}}{2^\mu \left(\frac{r+1}{2}\right)_\mu} \cdot C_\mu^{\frac{r}{2}}(1) \cdot (\mu+1)_{\mu+r-1},$$

or, in view of (1.8),

$$a_\mu = (-1)^{\frac{r-1}{2}} \cdot \frac{2\pi^{\frac{r-1}{2}}}{2^{2\mu}(r-1)\Gamma(\frac{r-1}{2})} \cdot \frac{(2\mu+r-1)!}{\left(\frac{r}{2}\right)_\mu\left(\frac{r+1}{2}\right)_\mu} \cdot C_\mu^{\frac{r}{2}}(1)$$

$$= (-1)^{\frac{r-1}{2}} \cdot \frac{\pi^{\frac{r-1}{2}}}{2^{2\mu}} \cdot \frac{\Gamma(2\mu+r)\Gamma(\frac{r}{2})}{\Gamma(\mu+\frac{r}{2})\Gamma(\mu+\frac{r+1}{2})} \cdot C_\mu^{\frac{r}{2}}(1).$$

Now we apply Legendre's formula

$$\Gamma(2\mu+r) = \frac{2^{2\mu+r-1}}{\sqrt{\pi}} \cdot \Gamma(\mu+\tfrac{r}{2})\Gamma(\mu+\tfrac{r+1}{2}),$$

and obtain

$$a_\mu = (-1)^{\frac{r-1}{2}} \cdot 2^{r-1}\pi^{\frac{r-2}{2}}\Gamma(\tfrac{r}{2}) \cdot C_\mu^{\frac{r}{2}}(1) = (-1)^{\frac{r-1}{2}} \cdot 2^r \pi^{r-1} \cdot \frac{1}{\omega_{r-1}} C_\mu^{\frac{r}{2}}(1).$$

Inserting this above, together with $a_0 = \ldots = a_{\mu-1} = 0$, we get

$$f(\sigma) = 2(2\pi i)^{r-1} \cdot \frac{1}{\omega_{r-1}} C_\mu^{\frac{r}{2}}(\sigma),$$

which finishes the proof. $\qquad\square$

Lemma D.2. *For* $r \in \mathbb{N} \setminus \{1,2\}$, $\mu \in \mathbb{N}_0$, *and* $x \in B^r$ *the following inequality holds,*

$$\frac{1}{\omega_{r-1}} \cdot \int_{S^{r-1}} \left[C_\mu^{\frac{r}{2}}(tx)\right]^2 d\omega(t) \le (r-1)^2 \cdot C_\mu^{\frac{r}{2}}(1).$$

Proof. For $\mu = 0$ the statement is evident. So let $\mu \in \mathbb{N}$ in what follows.

The integral does not change its value if we replace t by At, A an arbitrary rotation. So it suffices to confirm the statement for $x = \xi e_1$, $0 \le \xi \le 1$. Moreover, if it holds for $0 < \xi \le 1$, then it is also valid for $\xi = 0$, by continuity. Therefore it suffices to prove it under the assumption $0 < \xi \le 1$.

In view of (1.27) the integral now takes the form

$$F(\xi) \;=\; \frac{\omega_{r-2}}{\omega_{r-1}} \int_{-1}^{1} \left[C_\mu^{\frac{r}{2}}(\xi\tau) \right]^2 (1-\tau^2)^{\frac{r-3}{2}} \, d\tau$$

$$=\; \frac{\omega_{r-2}}{\omega_{r-1}} \cdot \frac{1}{\xi} \cdot \int_{-\xi}^{\xi} \left[C_\mu^{\frac{r}{2}}(\sigma) \right]^2 \left(1 - \left(\frac{\sigma}{\xi}\right)^2\right)^{\frac{r-3}{2}} d\sigma.$$

Because of $r \ge 3$ it follows that

$$F(\xi) \;\le\; \frac{\omega_{r-2}}{\omega_{r-1}} \cdot \frac{1}{\xi} \cdot \int_{-\xi}^{\xi} \left[C_\mu^{\frac{r}{2}}(\sigma) \right]^2 (1-\sigma^2)^{\frac{r-3}{2}} \, d\sigma. \tag{D.7}$$

For $\xi = 1$, this means for $x = e_1$, we get in particular

$$F(1) \;=\; \frac{1}{\omega_{r-1}} \cdot \int_{S^{r-1}} \left[C_\mu^{\frac{r}{2}}(tx) \right]^2 d\omega(t) \;=\; C_\mu^{\frac{r}{2}}(1), \tag{D.8}$$

where we used the reproducing property of $\frac{1}{\omega_{r-1}} C_\mu^{\frac{r}{2}}(\cdot x)$.

For $\frac{1}{r} \le \xi \le 1$ we obtain from (D.7) and (D.8)

$$F(\xi) \;\le\; \tfrac{1}{\xi} \cdot F(1) \;\le\; r \cdot C_\mu^{\frac{r}{2}}(1) \;<\; (r-1)^2 \cdot C_\mu^{\frac{r}{2}}(1),$$

as claimed.

Next let $0 < \xi < \frac{1}{r}$. We want to apply the inequality (2.33) with $\lambda = \frac{r}{2}$. It is valid in the interval $[0, x_\mu]$, defined by (2.31), i.e., by

$$x_\mu^2 \;=\; 1 - \left(\frac{r}{2\mu + r} \right)^2 \;\ge\; 1 - \left(\frac{r}{r+2} \right)^2 \;>\; \frac{1}{r^2},$$

where the last inequalities hold in view of $\mu \ge 1$, and of $r \ge 3$, again. Because of $\frac{1}{r} < x_\mu$, the inequality (2.33) holds in particular for $0 < x < \frac{1}{r}$, which is just the interval of interest.

So we may use (2.33). In view of (2.13) and $\mu \ge 1$ it implies that

$$(1-\sigma^2)^{\frac{r}{2}} \left| C_\mu^{\frac{r}{2}}(\sigma) \right|^2 \;\le\; \tfrac{r}{4} \cdot C_\mu^{\frac{r}{2}}(1),$$

which is valid in particular for $0 \leq \sigma \leq \xi$. Inserting this in (D.7) we obtain, in view of $\xi^2 < \left(\frac{1}{r}\right)^2$,

$$
F(\xi) \; \leq \; \frac{\omega_{r-2}}{\omega_{r-1}} \cdot \frac{r}{4} \cdot C_{\mu}^{\frac{r}{2}}(1) \cdot \frac{1}{\xi} \int_{-\xi}^{\xi} (1 - \xi^2)^{-\frac{3}{2}} \, d\xi
$$

$$
\leq \; \frac{\omega_{r-2}}{\omega_{r-1}} \cdot \frac{r}{2} \cdot C_{\mu}^{\frac{r}{2}}(1) \cdot \left(\frac{r^2}{r^2 - 1}\right)^{\frac{3}{2}}.
$$

Finally we use (1.8) and the Beta function, see (1.4), in order to get

$$
\frac{\omega_{r-2}}{\omega_{r-1}} = \frac{\Gamma(\frac{r}{2})}{\sqrt{\pi}\,\Gamma(\frac{r-1}{2})} = \frac{r-2}{2\pi} \cdot B\left(\frac{1}{2}, \frac{r-2}{2}\right) \leq \frac{r-2}{2\pi} \cdot B\left(\frac{1}{2}, \frac{1}{2}\right) = \frac{r-2}{2},
$$

where we used $r \geq 3$, again. It follows that

$$
F(\xi) \leq \frac{r(r-2)}{4} \cdot \left(\frac{9}{8}\right)^{\frac{3}{2}} \cdot C_{\mu}^{\frac{r}{2}}(1) < (r-1)^2 \cdot C_{\mu}^{\frac{r}{2}}(1),
$$

again as claimed. $\qquad\qquad\qquad\qquad\qquad\qquad\qquad\qquad\qquad\qquad\qquad\qquad\qquad\square$

Appendix E

Solutions

Chapter 1

Problem 1.1

a) Let $r \in \mathbb{N} \setminus \{1\}$. By Fubini's Theorem we get

$$\Omega_r = \int\limits_{x_1^2 + \ldots + x_r^2 \leq 1} dx_1 \ldots dx_r = \int_{-1}^{+1} \left(\int\limits_{x_1^2 + \ldots + x_{r-1}^2 \leq 1 - x_r^2} dx_1 \ldots dx_{r-1} \right) dx_r.$$

Substituting

$$x_1 = \sqrt{1 - x_r^2} \cdot \xi_1, \ldots, x_{r-1} = \sqrt{1 - x_r^2} \cdot \xi_{r-1},$$

we obtain

$$\Omega_r = \int_{-1}^{+1} (1 - x_r^2)^{\frac{r-1}{2}} \Omega_{r-1} \, dx_r = \int_0^1 (1 - x_r^2)^{\frac{r-1}{2}} dx_r \cdot 2\Omega_{r-1}$$

$$= \int_0^1 \xi^{-\frac{1}{2}} (1 - \xi)^{\frac{r-1}{2}} d\xi \cdot \Omega_{r-1} = B(\tfrac{1}{2}, \tfrac{r+1}{2}) \cdot \Omega_{r-1},$$

where we used (1.4). It follows that

$$\Omega_r = B(\tfrac{1}{2}, \tfrac{r+1}{2}) B(\tfrac{1}{2}, \tfrac{r}{2}) \cdots B(\tfrac{1}{2}, \tfrac{3}{2}) \cdot \Omega_1$$

with $\Omega_1 = 2$. Using (1.4) again we obtain

$$\Omega_r = \frac{\pi^{\frac{r}{2}}}{\Gamma(\frac{r+2}{2})}.$$

b) Next we generate B^r from spheres of radius ρ, and obtain

$$\Omega_r = \int_0^1 (\rho^{r-1} \omega_{r-1}) d\rho = \frac{\omega_{r-1}}{r}.$$

Together with the result from above this yields

$$\omega_{r-1} = \frac{2\pi^{\frac{r}{2}}}{\Gamma(\frac{r}{2})}.$$

Problem 1.2

The hemispheres S_+^{r-1} and S_-^{r-1} are defined by $x_r \geq 0$ and $x_r \leq 0$, respectively, and have the parameter representation

$$x(\bar{x}) = (\bar{x}', \pm\sqrt{1 - |\bar{x}|^2})', \quad \bar{x} \in B^{r-1}.$$

The normal at the point $x = x(\bar{x})$ is given by x itself. So the surface element takes the form $d\omega(x) = d\bar{x}/|e_r x| = d\bar{x}/\sqrt{1 - |\bar{x}|^2}$, and we get

$$\int_{S_\pm^{r-1}} F(x)\, d\omega(x) = \int_{B^{r-1}} \bar{F}(\bar{x})\, \frac{d\bar{x}}{\sqrt{1 - |\bar{x}|^2}}.$$

This yields (1.24).

Problem 1.3

In view of (1.24) and of $\omega_0 = 2$, formula (1.25) is valid for $s = 0$. Next we assume that (1.25) holds for $s - 1 \in \{0, 1, \ldots, r - 3\}$, and that $F(x) = \bar{F}(x_1, \ldots, x_{r-s-1})$ does not depend on $x_{r-s}, \ldots, x_r$. Putting $\bar{x} = (x_1, \ldots, x_{r-s-1})'$, and using (1.25) and Fubini's theorem, we get

$$\int_{S^{r-1}} F(x)\, d\omega(x) = \omega_{s-1} \int_{B^{r-s}} \bar{F}(\bar{x})(1 - |\bar{x}|^2 - x_{r-s}^2)^{\frac{s-2}{2}}\, d(\bar{x}, x_{r-s})$$

$$= \omega_{s-1} \int_{B^{r-s-1}} \bar{F}(\bar{x}) \left(\int_{-\sqrt{1-|\bar{x}|^2}}^{+\sqrt{1-|\bar{x}|^2}} (1 - |\bar{x}|^2 - \xi^2)^{\frac{s-2}{2}}\, d\xi \right) d\bar{x}.$$

Substituting $\xi = \sqrt{1 - |\bar{x}|^2}\, \eta$ we obtain

$$\int_{S^{r-1}} F(x)\, d\omega(x) = \omega_{s-1} \int_{-1}^{1} (1 - \eta^2)^{\frac{r-2}{2}}\, d\eta \cdot \int_{B^{r-s-1}} \bar{F}(\bar{x})(1 - |\bar{x}|^2)^{\frac{s-1}{2}}\, d\bar{x}.$$

Finally we use (1.4) and (1.8) to get

$$\omega_{s-1} \int_{-1}^{1} (1 - \eta^2)^{\frac{r-2}{2}}\, d\eta = \omega_{s-1} \int_{0}^{1} \tau^{-\frac{1}{2}} (1 - \tau)^{\frac{s-2}{2}}\, d\tau$$

$$= \omega_{s-1} \cdot B(\tfrac{1}{2}, \tfrac{s}{2}) = \frac{2\pi^{\frac{s}{2}}}{\Gamma(\frac{s}{2})} \cdot \frac{\Gamma(\frac{1}{2})\Gamma(\frac{s}{2})}{\Gamma(\frac{s+1}{2})} = \omega_s.$$

Therefore, (1.25) holds for $s \in \{1, \ldots, r-2\}$ instead of $s-1$, and recursively we obtain this formula for $s = 0$ and $s = 1, \ldots, r-2$.

Problem 1.4

With $f(\xi) = \xi^\mu$ formula (1.27) yields, in view of (1.4) and (1.8),

$$
\int\limits_{S^{r-1}} (x)^\mu d\omega(x) = \omega_{r-2} \cdot \int\limits_{-1}^{1} \xi^\mu (1 - \xi^2)^{\frac{r-3}{2}} d\xi
$$

$$
= \omega_{r-2} \cdot \int\limits_{0}^{1} \tau^{\frac{\mu-1}{2}} (1-\tau)^{\frac{r-3}{2}} d\tau = \omega_{r-2} \cdot B(\tfrac{\mu+1}{2}, \tfrac{r-1}{2})
$$

$$
= \frac{2\pi^{\frac{r-1}{2}}}{\Gamma(\frac{r-1}{2})} \cdot \frac{\Gamma(\frac{\mu+1}{2}) \Gamma(\frac{r-1}{2})}{\Gamma(\frac{\mu+r}{2})} = 2\pi^{\frac{r-1}{2}} \frac{\Gamma(\frac{\mu+1}{2})}{\Gamma(\frac{\mu+r}{2})}.
$$

Chapter 2

Problem 2.1

First assume $\lambda \neq 0$. With the help of the operator $D = z\frac{\partial}{\partial z}$ we obtain, with $C_{-1}^\lambda = 0$ and similar to the proof of Theorem 2.1,

$$
\sum_{\mu=0}^{\infty} z^\mu \left\{ (\mu+1)C_{\mu+1}^\lambda - 2(\mu+\lambda)xC_\mu^\lambda + (\mu+2\lambda-1)C_{\mu-1}^\lambda \right\}
$$

$$
= \sum_{\mu=1}^{\infty} \mu\, z^{\mu-1} C_\mu^\lambda - 2x \sum_{\mu=0}^{\infty} (\mu+\lambda)\, z^\mu C_\mu^\lambda + z \sum_{\mu=0}^{\infty} (\mu+2\lambda)\, z^\mu C_\mu^\lambda
$$

$$
= G_z^\lambda - 2x(D+\lambda)G^\lambda + z(D+2\lambda)G^\lambda
$$

$$
= 2\lambda(z-x)G^\lambda + (1 - 2xz + z^2)G_z^\lambda = 0,
$$

and the first recurrence relation is proved.

Next let $\lambda > -\frac{1}{2}$, such that $\tilde{C}_\mu^\lambda$ is defined, see (2.13). Inserting

$$
C_\mu^\lambda = \frac{(2\lambda)_\mu}{(1)_\mu} \tilde{C}_\mu^\lambda
$$

in the equation just proved, we obtain the second recurrence relation after a multiplication by $(1)_\mu/(2\lambda)_\mu$. The initial values are easily obtained from (2.5).

Problem 2.2

First let $\lambda \neq 0$, and recall (2.3) and (2.4). Using the operator $D := z\frac{\partial}{\partial z}$ we get

$$\sum_{\mu=0}^{\infty}\{-\mu x C_{\mu}^{\lambda} + (\mu + 2\lambda - 1)C_{\mu-1}^{\lambda}\} \cdot z^{\mu}$$

$$= -x \cdot D \frac{1}{(1 - 2xz + z^2)^{\lambda}} + z\sum_{\mu=0}^{\infty}(\mu + 2\lambda)C_{\mu}^{\lambda} \cdot z^{\mu}$$

$$= -x \cdot D \frac{1}{(1 - 2xz + z^2)^{\lambda}} + z\left(D + 2\lambda\right)\frac{1}{(1 - 2xz + z^2)^{\lambda}}$$

$$= \frac{2\lambda(1 - x^2)z}{(1 - 2xz + z^2)^{\lambda+1}} \quad = \quad \sum_{\mu=0}^{\infty}2\lambda(1 - x^2)C_{\mu-1}^{\lambda+1} \cdot z^{\mu}.$$

Now the statement follows in view of (2.10) by a comparison of the coefficients.
Next let $\lambda = 0$. For $\mu = 0$ the equation is valid, obviously. For $\mu \in \mathbb{N}$ it is equivalent to

$$\tfrac{1}{\mu}(1 - x^2)T_{\mu}' = -xT_{\mu} + T_{\mu-1},$$

see (2.8). By the substitution $x = \cos\phi$ this equation takes the form

$$-\sin\phi(\cos\mu\phi)' = -\cos\phi\cos\mu\phi + \cos(\mu - 1)\phi,$$

and so it is valid because of the addition theorem.

Problem 2.3

For $\mu \in \mathbb{N}$ it follows from the recurrence relation that $xC_{\mu}^{\lambda}(x)$ is a positive linear combination of $C_{\mu+1}^{\lambda}(x)$ and of $C_{\mu-1}^{\lambda}(x)$. In the Chebyshev case $\lambda = 0$ this is well known, for $\lambda > 0$ we refer to the result of Problem 2.1. Using this argument twice we get for $\mu \in \mathbb{N} \setminus \{1\}$ that $x^2 C_{\mu}^{\lambda}(x)$ is a positive linear combination of

$$C_{\mu+2}^{\lambda}(x), \ C_{\mu}^{\lambda}(x), \ and \ C_{\mu-2}^{\lambda}(x).$$

Now we turn to the statement of the problem. Obviously, because of $1 = C_0^{\lambda}(x)$, $x = \frac{1}{2}C_1^0(x)$, and $x = \frac{1}{2\lambda}C_1^{\lambda}(x)$ for $\lambda > 0$, it is valid for $\mu \in \{0, 1\}$.
Next we assume that it holds for some $\mu \in \mathbb{N}_0$, i.e., x^{μ} is a positive linear combination of the polynomials

$$C_{\mu}^{\lambda}(x), \ C_{\mu-2}^{\lambda}(x), \ \ldots,$$

for this value of μ. Then $x^{\mu+2}$ is a positive linear combination of

$$x^2 C_{\mu}^{\lambda}(x), \ x^2 C_{\mu-2}^{\lambda}(x), \ \ldots,$$

and by our preceding result also of

$$C_{\mu+2}^{\lambda}(x), \ C_{\mu}^{\lambda}(x), \ \ldots.$$

So the statement holds for $\mu + 2$ instead of μ, and mathematical induction finishes the proof.

Problem 2.4

Let $\lambda > -\frac{1}{2}$, $\mu \in \mathbb{N}$, and define the functions

$$Y(x) := (1 - x^2)^{\mu + \lambda - \frac{1}{2}} \quad \text{and} \quad F(x) := (1 - x^2)^{-\lambda + \frac{1}{2}} Y^{(\mu)}(x).$$

Applying the Leibniz rule in order to get $Y^{(\mu)}$ from

$$Y(x) = (1 - x)^{\mu + \lambda - \frac{1}{2}} \cdot (1 + x)^{\mu + \lambda - \frac{1}{2}},$$

we obtain

$$F(x) = \sum_{\nu=0}^{\mu} (-1)^{\nu} \binom{\mu}{\nu} (\mu + \lambda - \nu + \tfrac{1}{2})_{\nu} (\lambda + \nu + \tfrac{1}{2})_{\mu - \nu} (1 - x)^{\mu - \nu} (1 + x)^{\nu},$$

and hence $F \in \mathbb{P}^1_\mu$. We want to show that F is orthogonal to $\mathbb{P}^1_{\mu - 1}$. Actually, for arbitrary $G \in \mathbb{P}^1_{\mu - 1}$ we get by integration by parts

$$
\begin{aligned}
[G, F]_\lambda &= \int_{-1}^{1} GF \cdot (1 - x^2)^{\lambda - \frac{1}{2}} dx = \int_{-1}^{+1} G Y^{(\mu)} dx \\
&= +\left[G Y^{(\mu-1)} \right]_{-1}^{+1} - \int_{-1}^{+1} G' Y^{(\mu-1)} dx = -\int_{-1}^{+1} G' Y^{(\mu-1)} dx \\
&= -\left[G' Y^{(\mu-2)} \right]_{-1}^{+1} + \int_{-1}^{+1} G'' Y^{(\mu-2)} dx = +\int_{-1}^{+1} G'' Y^{(\mu-2)} dx \\
&\quad \cdots \\
&= \pm \left[G^{(\mu-1)} Y \right]_{-1}^{+1} \mp \int_{-1}^{+1} G^{(\mu)} Y \, dx = 0,
\end{aligned}
$$

where we used that the derivatives $Y^{(\mu-1)}, \ldots, Y^{(0)}$ vanish for $x \in \{+1, -1\}$. So F is contained in the orthogonal complement of $\mathbb{P}^1_{\mu-1}$ in $\mathbb{P}^1_\mu$, which has the dimension 1. In view of Theorem 2.3 it follows that

$$F = const \cdot \tilde{C}^\lambda_\mu.$$

We get the constant from the equation

$$const = F(1) = (-2)^\mu (\lambda + \tfrac{1}{2})_\mu.$$

Together this yields

$$\tilde{C}_\mu^\lambda(x) = \frac{(-1)^\mu}{2^\mu(\lambda + \frac{1}{2})_\mu}\,(1 - x^2)^{-\lambda+\frac{1}{2}}\,\left(\tfrac{d}{dx}\right)^\mu(1 - x^2)^{\mu+\lambda-\frac{1}{2}}\;,$$

as claimed.

Problem 2.5

We have to prove that the expression

$$A(x, z) := (1 - x^2)G_{xx}^\lambda + z^2 G_{zz}^\lambda + (2\lambda + 1)[zG_z^\lambda - xG_x^\lambda]$$

vanishes identically, where G^λ is defined by

$$G^\lambda(x, z) := \frac{1}{(1 - 2xz + z^2)^\lambda}.$$

Actually we get

$$G_x^\lambda = \frac{2\lambda z}{(1 - 2xz + z^2)^{\lambda+1}}, \qquad G_z^\lambda = \frac{2\lambda(x - z)}{(1 - 2xz + z^2)^{\lambda+1}},$$

$$G_{xx}^\lambda = \frac{4\lambda(\lambda + 1)z^2}{(1 - 2xz + z^2)^{\lambda+2}}, \qquad G_{zz}^\lambda = \frac{4\lambda(\lambda + 1)(x - z)^2}{(1 - 2xz + z^2)^{\lambda+2}} - \frac{2\lambda}{(1 - 2xz + z^2)^{\lambda+1}}.$$

It follows that

$$zG_z^\lambda - xG_x^\lambda = \frac{-2\lambda z^2}{(1 - 2xz + z^2)^{\lambda+1}},$$

$$(1 - x^2)G_{xx}^\lambda + z^2 G_{zz}^\lambda = \frac{2\lambda(2\lambda + 1)z^2}{(1 - 2xz + z^2)^{\lambda+1}},$$

and together this yields $A(x, z) = 0$, as claimed.

Problem 2.6

By assumption we have $\alpha > \frac{1}{2}$. The $j_{\alpha,k}$, $k \in \mathbb{N}_0$, are the nonnegative zeros of u. Moreover,

$$v(j_{\alpha,k}) = 0 = v(2j_{\alpha,k} - j_{\alpha,k-1})$$

holds for $k \in \mathbb{N}$. We want to show that u has a zero in the interval $j_{\alpha,k} < x < 2j_{\alpha,k} - j_{\alpha,k-1}$ for $k \in \mathbb{N}$.

By the transform $z = x^{-\alpha-\frac{1}{2}}u$ we get from (2.24) the differential equation

$$u'' + f(x)u = 0, \quad \text{where } f(x) := [1 + (\tfrac{1}{4} - \alpha^2)\tfrac{1}{x^2}].$$

v satisfies, correspondingly,

$$v'' + g(x)v = 0, \quad \text{where } g(x) := f(x + j_{\alpha,k-1} - j_{\alpha,k}).$$

In the interval $j_{\alpha,k} < x < 2j_{\alpha,k} - j_{\alpha,k-1}$ we get $g(x) < f(x)$. So it follows by Sturm's theorem that u has a zero in this interval. The lowest zero of this kind is $j_{\alpha,k+1}$, and we get

$$j_{\alpha,k+1} < 2j_{\alpha,k} - j_{\alpha,k-1},$$

equivalently to what is claimed.

Chapter 3

Problem 3.1

The multinomial coefficients are the coeffients in the expansion

$$(x_1 + x_2 + \ldots + x_r)^\mu = \sum_{|m|=\mu} \binom{\mu}{m} x^m.$$

Using (3.26) we get

$$\binom{\mu}{m} = \frac{1}{m!} D^m (x_1 + x_2 + \ldots + x_r)^\mu = \frac{\mu!}{m!}.$$

Problem 3.2

From

$$(x_1 + x_2 + \ldots + x_r)^\nu = \sum_{|n|=\nu} \binom{\nu}{n} x^n,$$

$$(x_1 + x_2 + \ldots + x_r)^{\mu-\nu} = \sum_{|k|=\mu-\nu} \binom{\mu-\nu}{k} x^k,$$

$$(x_1 + x_2 + \ldots + x_r)^\mu = \sum_{|m|=\mu} \binom{\mu}{m} x^m,$$

we obtain

$$\sum_{|n|=\nu} \sum_{|k|=\mu-\nu} \binom{\nu}{n} \binom{\mu-\nu}{k} x^{n+k} = \sum_{|m|=\mu} \binom{\mu}{m} x^m,$$

and the statement follows by a comparison of the coefficients occurring with x^m.

Problem 3.3

In the case $m \neq 2n$, $n \in \mathbb{N}_0^r$, at least one component m_ν is odd, and x^m is an odd function with respect to x_ν. Hence the integral vanishes. Next assume $m = 2n$, $n \in \mathbb{N}_0^r$, and let $\nu := |n|$. We integrate the expansion

$$(tx)^{2\nu} = \sum_{|m|=2\nu} \binom{2\nu}{m} t^m x^m,$$

$t, x \in S^{r-1}$, with respect to x. The integral from the left side is given by (1.26).
In view of $(t_1^2 + \cdots + t_r^2)^\nu = 1$ we may write it in the form

$$\int_{S^{r-1}} (tx)^{2\nu} \, dx = 2\pi^{\frac{r-1}{2}} \frac{\Gamma(\frac{\mu+1}{2})}{\Gamma(\frac{\mu+r}{2})} \cdot \sum_{|k|=\nu} \binom{\nu}{k} t^{2k}.$$

In view of the result from above, the integral from the right side takes the form

$$\sum_{|k|=\nu} \binom{2\nu}{2k} t^{2k} \int_{S^{r-1}} x^{2k} \, d\omega(x).$$

By a comparison of the coefficients occurring with t^{2n} we get now the result
wanted.

Problem 3.4

It suffices to prove the formula for pairwise different x_j. In this case the divided
difference has the representation

$$\left[\xi^{\mu+r-1}; x_1, \ldots, x_r\right] = \sum_{j=1}^{r} x_j^{\mu+r-1} / \prod_{\substack{i=1 \\ i \neq j}}^{r} (x_j - x_i),$$

with empty products to be put to unity. The assertion is trivial for $r = 1$, and
evident for $r = 2$ because of

$$\sum_{|m|=\mu} x^m = \frac{x_1^{\mu+1} - x_2^{\mu+1}}{x_1 - x_2}.$$

Next assume that the statement is true for $r - 1 \in \mathbb{N}$. Then we get

$$\sum_{|m|=\mu} x^m = \sum_{\nu=0}^{\mu} x_r^{\mu-\nu} \sum_{|\bar{m}|=\nu} \bar{x}^{\bar{m}},$$

where $\bar{x} = (x_1, \ldots, x_{r-1})'$, $\bar{m} = (m_1, \ldots, m_{r-1})'$. We use our knowledge with
respect to the inner sum, and obtain by some calculation

$$\sum_{|m|=\mu} x^m = \sum_{j=1}^{r-1} x_j^{r-2} \left(x_j^{\mu+1} - x_r^{\mu+1} \right) / \prod_{\substack{i=1 \\ i \neq j}}^{r} (x_j - x_i).$$

From

$$\sum_{j=1}^{r} x_j^{r-2} / \prod_{\substack{i=1 \\ i \neq j}}^{r} (x_j - x_i) = [\xi^{r-2}; x_1, \ldots, x_r] = 0$$

we obtain

$$-\sum_{j=1}^{r-1} x_j^{r-2} / \prod_{\substack{i=1 \\ i \neq j}}^{r} (x_j - x_i) = x_r^{r-2} / \prod_{i=1}^{r-1} (x_r - x_i).$$

Inserting this result above we obtain the statement for r instead of $r-1$. Mathematical induction finishes the proof.

Problem 3.5

To begin with, it is easy to see that if a nontrivial bivariate polynomial $P(x,y)$ vanishes for $y = 0$ identically in x, then it can be written in the form $P(x,y) = y\,Q(x,y)$ with a bivariate polynomial $Q(x,y)$ of lower degree. Using affine linear transforms, we can generalize this result as follows. If $P(x,y)$ vanishes on the affine line $p(x,y) = 0$, defined by the non-constant polynomial $p \in \mathbb{P}_1^2$, then $P(x,y)$ can be written in the form $P(x,y) = p(x,y)\,Q(x.y)$, where $Q(x,y)$ is a polynomial of lower degree.

Now let A, B, C be the vertices of the non-degenerating triangle D, and define the map $L : \mathbb{P}_2^2 \to \mathbb{P}_2^2(D)$ by $LP := P|_D$. Moreover, recall the identity

$$\dim \mathbb{P}_2^2 = \dim \mathbb{P}_2^2(D) + \dim \ker(L).$$

Now let $P \in \ker(L)$, and let $p_1(x,y) = 0$ be the affine line defined by the vertices A and B. The restriction of P to this line is a univariate polynomial, which vanishes on an open interval, and hence identically, i.e., on the whole line. Therefore, P contains p_1 as a factor. Likewise P contains the non-constant polynomial factors p_2 and p_3 defined by the vertices B, C and C, A, respectivly, and there exists a factorisation

$$P = p_1\,p_2\,p_3\,Q,$$

where Q is again a polynomial. Now we see by a comparison of the degrees that $Q = 0$ must hold, which implies $P = 0$ and hence $\ker(L) = [0]$. It follows that

$$\dim \mathbb{P}_2^2 = \dim \mathbb{P}_2^2(D),$$

where the common dimension equals 6, see (3.8). In particular, there is no reduction of the dimension, though D does not contain an interior point.

Problem 3.6

It is obvious that $\|M_0\| = 1$ holds. Next assume $m \neq 0$, and put $\mu := |m|.$ $|x^m|$ attains its maximum value on B^r for $|x| = 1$, only, i.e., under the side-condition $|x|^2 - 1 = 0$. Lagrange's maximality conditions take the form

$$m_\nu x^{m-e_\nu} - \lambda x_\nu = 0, \quad \nu = 1, \ldots, r.$$

Together with the side-condition, multiplication by x_ν and addition of the resulting equalities yields $\lambda = \mu x^m$. So

$$x_\nu^2 = \frac{m_\nu}{\mu}$$

must hold for $\nu = 1, \dots, r$ at every extreme point. This implies

$$\|M_m\| = m^{\frac{m}{2}}/\mu^{\frac{\mu}{2}}.$$

Chapter 4

Problem 4.1

Note that

$$F(x) = |x|^\mu f\left(\frac{tx}{|x|}\right)$$

is homogeneous of degree μ. For $t \in S^{r-1}$, $0 \neq x \in \mathbb{R}^r$, put $\xi := \frac{tx}{|x|}$. For $\nu = 1, \dots, r$ we obtain

$$\frac{\partial F}{\partial x_\nu} = |x|^\mu \left\{ f' \cdot \left(\frac{t_\nu}{|x|} - \xi \frac{x_\nu}{|x|^2}\right) + \mu f \cdot \frac{x_\nu}{|x|^2}\right\},$$

$$\frac{\partial^2 F}{\partial x_\nu^2} = |x|^\mu \left\{ f'' \cdot \left(\frac{t_\nu}{|x|} - \xi \frac{x_\nu}{|x|^2}\right)^2 \right.$$

$$+ f' \cdot \left[-\frac{t_\nu x_\nu}{|x|^3} - \xi\left(\frac{1}{|x|^2} - 2\frac{x_\nu^2}{|x|^4}\right) - \left(\frac{t_\nu}{|x|} - \xi\frac{x_\nu}{|x|^2}\right)\frac{x_\nu}{|x|^2}\right]$$

$$+ \mu f' \cdot \left(\frac{t_\nu}{|x|} - \xi\frac{x_\nu}{|x|^2}\right)\frac{x_\nu}{|x|^2} + \mu f \cdot \left(\frac{1}{|x|^2} - 2\frac{x_\nu^2}{|x|^4}\right)\right\}$$

$$+ \mu|x|^{\mu-2}x_\nu\left\{f'\cdot\left(\frac{t_\nu}{|x|} - \xi\frac{x_\nu}{|x|^2}\right) + \mu f\cdot\frac{x_\nu}{|x|^2}\right\}.$$

For $|x| = 1$, and hence for $\xi \in [-1, 1]$, we get

$$\Delta F = (1 - \xi^2)\, f'' - (r - 1)\,\xi\, f' + \mu(\mu + r - 2)\, f,$$

and (4.10) implies (4.11). Vice versa, (4.11) implies $\Delta F = 0$ for $x \in S^{r-1}$. But ΔF is a homogeneous polynomial, so it must vanish identically, and (4.10) is valid.

Problem 4.2

We use the abbreviation $K := G_\mu^r$, and write $\xi \in [-1, 1]$ in the form $\xi = xy$, $x, y \in S^{r-1}$. By the reproducing property of K we get, using Cauchy's inequality,

$$K^2(\xi) = K^2(xy) = \langle K(x\,\cdot\,), K(\,\cdot\,y)\rangle^2 \leq \langle K(x\,\cdot\,), K(x\,\cdot\,)\rangle\langle K(\,\cdot\,y), K(\,\cdot\,y)\rangle = K^2(1).$$

Problem 4.3

Let $F \in \mathbf{V}$. For $x \in S^{r-1}$ we represent $F(x)$ in the form

$$F(x) = \int\limits_{S^{r-1}} F(y)K(xy)d\omega(y),$$

where $K(xy)$ is the reproducing kernel of **V**. By Cauchy's inequality we obtain

$$F^2(x) \leq \int_{S^{r-1}} F^2(y)d\omega(y) \cdot \int_{S^{r-1}} K^2(xy)d\omega(y) \leq \|F\|_\infty \cdot \omega_{r-1} \cdot K(1) = N \cdot \|F\|_\infty^2.$$

Problem 4.4

In (3.56) we replace x by $\frac{x}{|x|}$ and t by $|x|t$. Then we get

$$\frac{1}{(1 - 2xt + x^2t^2)^{\frac{r+s-1}{2}}} = \sum_{\mu=0}^{\infty} \sum_{|m|=\mu} \overset{*}{V}_m^{r,s}(x)\, t^m$$

for $x, t \in B^r$, $|t| < 1$. Note that $\lambda := \frac{r+s-1}{2}$ is positive by assumption. With the abbreviation $N := 1 - 2xt + x^2t^2$ we get for $\nu = 1, \ldots, r$

$$\frac{\partial}{\partial x_\nu} \frac{1}{N^\lambda} = 2\lambda \frac{t_\nu - x_\nu t^2}{N^{\lambda+1}},$$

$$\frac{\partial^2}{(\partial x_\nu)^2} \frac{1}{N^\lambda} = \frac{2\lambda}{N^{\lambda+2}} \left\{ 2(\lambda+1)\left(t_\nu^2 - 2x_\nu t_\nu t^2 + x_\nu^2 |t|^4\right) - t^2 N \right\},$$

$$\Delta \frac{1}{N^\lambda} = \frac{2\lambda t^2}{N^{\lambda+1}} \left\{ 2(\lambda+1) - r \right\}.$$

Therefore, $\Delta \frac{1}{N^\lambda} = 0$ holds if and only if $\lambda = \frac{r-2}{2}$, i.e., exactly for $s = -1$. The family $\overset{*}{V}_m^{r,-1}$ is the unique family of its kind whose members are all harmonic.

Problem 4.5

In (3.63) we replace x by $\dfrac{x}{[x_1^2 + \cdots + x_{r+s+1}^2]^{\frac{1}{2}}}$ and t by $[x_1^2 + \cdots + x_{r+s+1}^2]^{\frac{1}{2}}t$. Then we get

$$\frac{1}{\left[(1 - xt)^2 + (x_{r+1} + \cdots + x_{r+s+1}^2)t^2\right]^{\frac{s}{2}}} = \sum_{\mu=0}^{\infty} \sum_{|m|=\mu} H_m^{r,s}(x_1, \ldots, x_{r+s+1})\, t^m,$$

where

$$H_m^{r,s} \in \overset{*}{\mathbb{P}}_\mu^{r+s+1}$$

holds by definition. It is left to prove that $H_m^{r,s}$ is harmonic with respect to the variables $x_1, \ldots, x_{r+s+1}$, and it suffices to prove this for the generating function. Using the abbreviations

$$N := (1 - xt)^2 + (x_{r+1}^2 + \cdots + x_{r+s+1}^2)\, t^2$$

and $\lambda := \frac{s}{2}$, we get

$$\frac{\partial}{\partial x_\nu} \frac{1}{N^\lambda} = 2\lambda \frac{(1 - xt)t_\nu}{N^{\lambda+1}} \text{ for } \nu = 1, \dots, r,$$

$$\frac{\partial}{\partial x_\nu} \frac{1}{N^\lambda} = -2\lambda \frac{x_\nu t^2}{N^{\lambda+1}} \text{ for } \nu = r+1, \dots, r+s+1,$$

$$\frac{\partial^2}{(\partial x_\nu)^2} \frac{1}{N^\lambda} = \frac{2\lambda}{N^{\lambda+2}} \left\{ 2(\lambda+1)(1-xt)^2 t_\nu^2 - t_\nu^2 N \right\} \text{ for } \nu = 1, \dots, r,$$

$$\frac{\partial^2}{(\partial x_\nu)^2} \frac{1}{N^\lambda} = \frac{2\lambda}{N^{\lambda+2}} \left\{ 2(\lambda+1)x_\nu^2 |t|^4 - t^2 N \right\} \text{ for } \nu = r+1, \dots, r+s+1.$$

Together this yields

$$\Delta \frac{1}{N^\lambda} = \sum_{\nu=1}^{r+s+1} \frac{\partial^2}{(\partial x_\nu)^2} \frac{1}{N^\lambda} = \frac{2\lambda t^2}{N^{\lambda+1}} \left\{ 2(\lambda+1) - (s+2) \right\} = 0.$$

The generating function is harmonic with respect to $x_1, \dots, x_{r+s+1}$, and so are the $H_m^{r,s}$, and we obtain $H_m^{r,s} \in \overset{*}{\mathbb{H}}_\mu^{r+s+1}$.

Chapter 5

Problem 5.1

In the case $\|A\| = 0$ the statement is evident. So let us assume $\|A\| > 0$ in what follows. Then we obtain for $x \in X$, $\|x\|_X \leq 1$,

$$\|(B \circ A)x\|_Z = \|B(Ax)\|_Z = \|A\| \cdot \left\| B\left(\tfrac{1}{\|A\|} \cdot Ax \right) \right\|_Z,$$

where

$$\left\| \tfrac{1}{\|A\|} \cdot Ax \right\|_Y \leq 1.$$

It follows that

$$\|(B \circ A)x\|_Z \leq \|A\| \cdot \|B\|.$$

Since this holds for arbitrary $x \in X$ with $\|x\|_X \leq 1$, the statement is true.

Problem 5.2

The B_j form a basis of $\mathbf{V}$. For fixed $x \in D$, $G(x, \cdot)$ has a representation

$$G(x, \cdot) = \sum_{j=0}^N a_j(x) B_j(\cdot)$$

with real coefficients $a_j(x)$. By the reproducing property of $G(x, \cdot)$ we get for $k = 1, \dots, N$,

$$A_k(x) = \langle G(x, \cdot), A_k \rangle = \sum_{j=0}^N a_j(x) \langle B_j, A_k \rangle = a_k(x),$$

and hence

$$G(x, \cdot) = \sum_{j=1}^{N} A_j(x) \, B_j(\cdot),$$

as claimed.

Problem 5.3

The Lagrange elements and the kernel functions are biorthogonal, and the statement follows from the result of Problem 5.2.

Problem 5.4

$[\cdot, \cdot]$ is a positive semidefinite bilinear form, where

$$[F, F] = \sum_{j=1}^{N} F^2(t_j)$$

vanishes for $F \in \mathbf{V}$ if and only if $F(t_j) = 0$ holds for $j = 1, \ldots, N$. Since the t_j form a fundamental system, this is equivalent to $F = 0$. So, $[\cdot, \cdot]$ is an inner product. Moreover we get $[L_j, L_k] = \delta_{j,k}$ for $j, k = 1, \ldots, N$, obviously.

Problem 5.5

From

$$(xy)^\mu = \sum_{|m|=\mu} \binom{\mu}{m} x^m y^m$$

we get for arbitrary $a_1, \ldots, a_N \in \mathbb{R}$

$$\sum_{j=1}^{N} \sum_{k=1}^{N} a_j (t_j t_k)^\mu a_k = \sum_{|m|=\mu} \binom{\mu}{m} \sum_{j=1}^{N} \sum_{k=1}^{N} a_j t_j^m t_k^m a_k = \sum_{|m|=\mu} \binom{\mu}{m} \left[\sum_{j=1}^{N} a_j t_j^m \right]^2 \geq 0,$$

and the matrix is positive semidefinite. Equality is valid if and only if

$$\sum_{j=1}^{N} a_j t_j^m = 0$$

holds for $|m| = \mu$. This system has a nontrivial solution if and only if the nodes form a fundamental system, see Theorem 5.14.

Problem 5.6

We identify $F = L_j$ and $V^* = 0$. Then t_j is an extreme point of $F - V^*$, and we get

$$\left(F(t_j) - V^*(t_j) \right) G(t_j) = 0$$

for arbitrary $G \in X_j$. In view of (5.80) the Kolmogroff criterion is satisfied, and we could decide this with the help of a single extreme point.

Problem 5.7

Every monomial M_m, $|m| = \mu$, $m \neq \mu e_1$, vanishes at e_1, which is an extreme point of $M_{\mu e_1}$. So the statement follows by the Criterion of Kolmogoroff.

Problem 5.8

Because of the assumptions on D, we may define $\bar{F} \in C(D)$ for $F \in C(D)$ by

$$\bar{F}(\ldots, x_j, \ldots) := F(\ldots, -x_j, \ldots).$$

The condition on $\mathbf{V}$ can be written in the form

$$\mathbf{V} \ni V \text{ implies } \bar{V} \in \mathbf{V}.$$

Now assume that $F \in C(D)$ satisfies $\bar{F} = (-1)^\epsilon F$ for some $\epsilon \in \{0, 1\}$, such that

$$F = \tfrac{1}{2}(F + (-1)^\epsilon \bar{F}),$$

and let V be a best approximation to F in $\mathbf{V}$. Then the following inequalities are valid,

$$\|F - V\| \;\leq\; \|F - \tfrac{1}{2}(V + (-1)^\epsilon \bar{V})\| \;=\; \|\tfrac{1}{2}(F - V) + \tfrac{1}{2}(-1)^\epsilon(\bar{F} - \bar{V})\|$$

$$\leq\; \tfrac{1}{2}\|F - V\| + \tfrac{1}{2}\|\bar{F} - \bar{V}\| \;=\; \|F - V\|.$$

So, equality must hold everywhere. In particular the element $W \in \mathbf{V}$ defined by $W := \tfrac{1}{2}(V + (-1)^\epsilon \bar{V})$ is a best approximation to F. It satisfies

$$\bar{W} = \tfrac{1}{2}(\bar{V} + (-1)^\epsilon V) = (-1)^\epsilon \tfrac{1}{2}(V + (-1)^\epsilon \bar{V}) = (-1)^\epsilon W,$$

and the problem is solved.

Problem 5.9

By the result of Problem 5.8 a best approximation to $M_{2,1,1}$ in $\mathbb{P}_3^3$ exists which is even in x_1, and odd in x_2 and in x_3. So it must have the form cx_2x_3, $c \in \mathbb{R}$, and we have to determine c such that the norm on B^3 of the polynomial $F(x) = (x_1^2 - c)x_2x_3$ attains its minimum value with respect to c.

First let us determine this norm for fixed c. It is easy to see that if F has a relative extremum at an interior point $x \in B^3$, then $F(x) = 0$ must hold. Therefore $|F|$ attains its absolute maximum value on the surface S^2.

Under the restriction $x_1^2 + x_2^2 + x_3^2 = 1$ we get

$$F(x) = f(x_2, x_3) := (1 - c - x_2^2 - x_3^2)x_2x_3,$$

and we have to determine the extreme points of f on the disk $x_2^2 + x_3^2 \leq 1$. An inner point furnishes a relative extremum $f(x_2, x_3) \neq 0$ only under the conditions $-1 < c < +1$ and $x_2^2 = x_3^2 = \frac{1-c}{4}$, where the value of $|f|$ is

$$|f(x_2, x_3)| = \tfrac{1}{8}(1 - c)^2.$$

On the border $x_2^2 + x_3^2 = 1$ we have $|f(x_2, x_3)| = |c||x_3|\sqrt{1 - x_3^2}$, and it is easy to see that the maximum of this function is given by $\frac{1}{2}|c|$. Together we obtain

$$\|F\| = \begin{cases} \max\left\{\frac{|c|}{2}, \frac{(1-c)^2}{8}\right\} & \text{for } -1 < c < 1, \\ \frac{|c|}{2} & \text{for } |c| \geq 1. \end{cases}$$

The right side attains its minimum value for $c = 3 - \sqrt{8}$. The remaining follows immediately.

Problem 5.10

For $\nu \in \{-, \ldots, \mu\}$ we get

$$\|S_{\mu\nu}\|_2^2 = \int\limits_0^\pi \int\limits_0^{2\pi} (\sin \nu\phi)^2 \cdot (\sin \psi)^{2\nu+1} \left[P_\mu^{(\nu)}(\cos \psi)\right]^2 d\phi\, d\psi$$

$$= \pi \int\limits_0^\pi (\sin \psi)^{2\nu+1} \left[P_\mu^{(\nu)}(\cos \psi)\right]^2 d\psi,$$

where

$$P_\mu^{(\nu)} = (2\nu - 1)!! \cdot C_{\mu-\nu}^{\frac{2\nu+1}{2}},$$

see (2.10). Putting $r := 2\nu + 3$, we obtain

$$\|S_{\mu\nu}\|_2^2 = \pi \cdot [(2\nu - 1)!!]^2 \cdot \int\limits_0^\pi \left[C_{\mu-\nu}^{\frac{r-2}{2}}(\cos \psi)\right]^2 (\sin \psi)^{r-2}\, d\psi$$

$$= \frac{\pi}{\omega_{r-2}} \cdot [(2\nu - 1)!!]^2 \cdot \int\limits_{S^{r-1}} \left[C_{\mu-\nu}^{\frac{r-2}{2}}(tx)\right]^2 d\omega(x)$$

with an arbitrary $t \in S^{r-1}$, see (1.27). Because of (4.13) it follows that

$$\|S_{\mu\nu}\|_2^2$$

$$= \pi \cdot \frac{\omega_{r-1}}{\omega_{r-2}} \cdot [(2\nu - 1)!!]^2 \cdot \frac{r-2}{2(\mu - \nu) + r - 2} \cdot \int\limits_{S^{r-1}} C_{\mu-\nu}^{\frac{r-2}{2}}(tx)\, G_{\mu-\nu}^r(tx)\, d\omega(x)$$

$$= \pi^{\frac{3}{2}} \cdot \frac{\Gamma(\frac{r-1}{2})}{\Gamma(\frac{r}{2})} \cdot [(2\nu - 1)!!]^2 \cdot \frac{2\nu + 1}{2\mu + 1} \cdot C_{\mu-\nu}^{\frac{r-2}{2}}(1),$$

where we used (1.8) and the reproducing property of $G_{\mu-\nu}^r$. Finally we get with the help of (2.13)

$$
\begin{aligned}
\|S_{\mu\nu}\|_2^2 &= \pi^{\frac{3}{2}} \cdot \frac{\nu!}{(\nu + \frac{1}{2})(\nu - \frac{1}{2}) \cdots \frac{1}{2}\Gamma(\frac{1}{2})} \cdot [(2\nu - 1)!!]^2 \cdot \frac{2\nu + 1}{2\mu + 1} \cdot \frac{(2\nu + 1)_{\mu-\nu}}{(1)_{\mu-\nu}} \\
&= \frac{2\pi}{2\mu + 1} \cdot [(2\nu - 1)!!] \cdot [(2\nu)!!] \cdot \frac{(2\nu + 1)_{\mu-\nu}}{(1)_{\mu-\nu}} \\
&= \frac{2\pi}{2\mu + 1} \cdot \frac{(\mu + \nu)!}{(\mu - \nu)!}.
\end{aligned}
$$

Moreover, we obtain

$$
\begin{aligned}
\|C_{\mu\nu}\|_2^2 &= \int_0^\pi \int_0^{2\pi} (\cos \nu\phi)^2 \cdot (\sin \psi)^{2\nu+1} \Big[P_\mu^{(\nu)}(\cos \psi) \Big]^2 d\phi d\psi \\
&= \pi \int_0^\pi (\sin \psi)^{2\nu+1} \Big[P_\mu^{(\nu)}(\cos \psi) \Big]^2 d\psi = \|S_{\mu\nu}\|_2^2.
\end{aligned}
$$

For $\nu = 0$ we get by similar arguments

$$
\|C_{\mu,0}\|_2^2 = \frac{4\pi}{2\mu + 1}.
$$

Chapter 6

Problem 6.1

Without restriction of generality we may assume $t_0 = e_1$. Then the nodes are given by

$$
t_j = \begin{pmatrix} \cos \phi_j \\ \sin \phi_j \end{pmatrix}, \quad \phi_j = \frac{2\pi j}{2\mu + 1},
$$

for $j = 0, 1, \ldots 2\mu$. Now let $F \in \mathbb{P}_{2\mu}^2$, and put

$$
f(\phi) := F(\cos \phi, \sin \phi).
$$

Then f is a trigonometric polynomial of degree 2μ, and in view of

$$
\int_{S^1} F(x)d\omega(x) = \int_0^{2\pi} f(\phi)d\phi
$$

it suffices to prove that

$$
\int_0^{2\pi} f(\phi)d\phi = \frac{2\pi}{2\mu + 1} \sum_{j=0}^{2\mu} f(\phi_j)
$$

holds for all $f \in \{1, \sin\phi, \cos\phi, \ldots, \sin 2\mu\phi, \cos 2\mu\phi\}$. For $f = 1$ this is evident. For the remaining functions, the integral on the left side vanishes, and it suffices to prove that the right side vanishes for the functions

$$e_k(\phi) := \cos k\phi + i \sin k\phi = e^{k\phi i}, \quad k \in \{1, \ldots, 2\mu\}.$$

But actually we get

$$\sum_{j=0}^{2\mu} e_k(\phi_j) = \sum_{j=0}^{2\mu} e^{j \frac{2k\pi i}{2\mu+1}} = \left(e^{2k\pi i} - 1\right) \cdot \left(e^{\frac{2k\pi i}{2\mu+1}} - 1\right)^{-1} = 0,$$

and the quadrature formula holds as claimed.

Problem 6.2

It suffices to prove the formula for the Chebyshev polynomials T_k, $k = 0, 1, \ldots, 2\mu + 1$.

For T_0 it is obvious. For T_k, $k \in \{1, \ldots, 2\mu+1\}$, the integral vanishes, so it suffices to prove the same for the right side. But, actually, we get

$$
\begin{aligned}
\sum_{\nu=0}^{\mu} T_k\left(\cos\frac{2\nu+1}{2\mu+2}\pi\right) &= \sum_{\nu=0}^{\mu} \cos\frac{2\nu+1}{2\mu+2}k\pi = \Re\left\{\sum_{\nu=0}^{\mu} e^{\frac{2\nu+1}{2\mu+2}k\pi i}\right\} \\
&= \Re\left\{e^{\frac{k\pi i}{2\mu+2}} \sum_{\nu=0}^{\mu} e^{\nu \frac{k\pi i}{\mu+1}}\right\} = \Re\left\{e^{\frac{k\pi i}{2\mu+2}} \cdot \frac{e^{k\pi i} - 1}{e^{\frac{k\pi i}{\mu+1}} - 1}\right\} \\
&= ((-1)^k - 1) \cdot \Re \frac{1}{e^{\frac{k\pi i}{2\mu+2}} - e^{-\frac{k\pi i}{2\mu+2}}} = 0.
\end{aligned}
$$

Problem 6.3

The unit points $e_1, \ldots, e_{r+1}$ in $\mathbb{R}^{r+1}$ are the vertices of the regular simplex Σ^r, which is contained in the hyperplane $H := \{x \in \mathbb{R}^{r+1} | ex = 1\}$, $e := e_1 + \cdots + e_{r+1}$. The point $\frac{1}{r+1} \cdot e$ is the center of this simplex, and the vertices have the distance $a := |e_j - \frac{1}{r+1}| = \sqrt{\frac{r}{r+1}}$ from it. So the points

$$t_j := \frac{1}{a}\left(e_j - \frac{1}{r+1}e\right),$$

$j = 1, \ldots, r+1$, are the vertices of a regular simplex contained in $H \cap S^r$, which is a unit sphere S^{r-1} by a proper identification. It is easy to see that

$$t_j t_k = \begin{cases} 1 & , \text{ if } k = j, \\ -\frac{1}{r} & , \text{ if } k \neq j \end{cases}$$

holds for $j, k = 1, \ldots, r+1$.

The reproducing kernel of $\mathbb{P}_1^r(S^{r-1})$ is given by

$$\Gamma_1^r(xy) = \frac{1}{\omega_{r-1}}\left\{C_1^{\frac{r}{2}}(xy) + C_0^{\frac{r}{2}}(xy)\right\} = \frac{1}{\omega_{r-1}}\left\{r(xy) + 1\right\}.$$

This implies

$$\Gamma_1^r(t_j, t_k) = \frac{r+1}{\omega_{r-1}}\delta_{j.k}$$

for $j, k = 1, \ldots, r + 1$. In view of $\dim \mathbb{P}_1^r(S^{r-1}) = r + 1$, it follows now from Corollary 5.34 that $t_1, \ldots, t_{r+1}$ support a Gauß quadrature. It is exact of the degree 2.

Problem 6.4

For $r \geq 2$, the content of a spherical cap $C(x, \phi)$, $x \in S^{r-1}$, $0 < \phi \leq \pi$, is given by

$$|C(x, \phi)| = \int\limits_{tx \geq \cos \phi} 1 \cdot d\omega(t) = \omega_{r-2}\int\limits_0^\phi (\sin \psi)^{r-2}d\psi.$$

This implies

$$|C(x, \phi)| < \omega_{r-2}\int\limits_0^\phi \psi^{r-2}d\psi = \frac{\omega_{r-2}}{r-1}\cdot\phi^{r-1}.$$

From below we get the estimate

$$|C(x, \phi)| > \omega_{r-2}\int\limits_0^\phi (\sin \psi)^{r-2}\cos \psi \, d\psi = \frac{\omega_{r-2}}{r-1}\cdot(\sin \phi)^{r-1}.$$

Both together yield

$$|C(x, \phi)| \sim \frac{\omega_{r-2}}{r-1}\cdot\phi^{r-1} \text{ as } \phi \to 0+.$$

Problem 6.5

Using well-known trigonometric formulas, we obtain

$$\frac{1}{\sin\phi}\sum_{\nu=0}^{\mu}\big(\sin(\nu+1)\phi+\sin\nu\phi\big)$$

$$= \frac{2}{\sin\phi}\sum_{\nu=0}^{\mu}\sin(2\nu+1)\frac{\phi}{2}\cdot\cos\frac{\phi}{2}$$

$$= \frac{1}{\sin^2\frac{\phi}{2}}\sum_{\nu=0}^{\mu}\sin(\nu+\frac{1}{2})\phi\cdot\sin\frac{\phi}{2}$$

$$= \frac{1}{2\sin^2\frac{\phi}{2}}\sum_{\nu=0}^{\mu}\big(\cos\nu\phi-\cos(\nu+1)\phi\big)$$

$$= \frac{1}{2\sin^2\frac{\phi}{2}}\big(1-\cos(\mu+1)\phi\big) = \left[\frac{\sin(\mu+1)\frac{\phi}{2}}{\sin\frac{\phi}{2}}\right]^2.$$

Chapter 7

Problem 7.1

Let $r \in \mathbb{N}$, and let $Z = \text{circ}(0,1,0,\ldots,0)$ be the circulant $(r+1)\times(r+1)$ basic matrix. The eigenvalues of the $(r+1)\times(r+1)$-matrix

$$E := \begin{pmatrix} 1 & \cdots & 1 \\ \vdots & & \vdots \\ 1 & \cdots & 1 \end{pmatrix} = I + Z + Z^2 + \cdots + Z^r =: f(Z)$$

are given by $f(\omega)$, where $\omega^{r+1} = 1$. This yields

$$spec\,(E) = \{0, r+1\}.$$

Now we use that the fundamental matrix can be written in the form

$$\left(\Gamma_1^{r,0}(t_j,t_k)\right) = \frac{1}{r\,\omega_r}\big\{(r+1)^2 I - E\big\},$$

see Section 7.3, Example (D). It follows that

$$spec\left(\Gamma_1^{r,0}(t_j,t_k)\right) = \frac{r+1}{r\,\omega_r}\big\{r+1,\,r\big\} = \frac{1}{\Omega_{r+1}}\big\{1, 1+\tfrac{1}{r}\big\},$$

see (1.8).

Problem 7.2

In view of the result of Problem 7.1, the minimum eigenvalue of the fundamental matrix is

$$\lambda_{\min} = \frac{r+1}{\omega_r}.$$

Besides we get from (7.20) for $x \in B^r$,

$$\Gamma_1^{r,0}(x,x) = \tfrac{1}{2}\left\{\Gamma_1^{r+1}(1) + \Gamma_1^{r+1}(2x^2 - 1)\right\} \leq \Gamma_1^{r+1}(1),$$

where

$$\Gamma_1^{r+1}(1) == \frac{1}{\omega_r}\,\dim \mathbb{P}_1^{r+1} = \frac{r+2}{\omega_r}.$$

Now we use Theorem 5.17, and obtain

$$\sum_{j=1}^{r+1} L_j^2(x) \leq \frac{r+2}{\omega_r} \cdot \frac{\omega_r}{r+1} = \frac{r+2}{r+1}$$

for arbitrary $x \in B^r$.

Problem 7.3

Let $t_1, \ldots, t_{r+1} \in S^{r-1}$ be the vertices of a regular simplex, and assume $L_1, \ldots, L_{r+1} \in \mathbb{P}_1^r(\mathbb{R}^r)$ are the corresponding Lagrange elements. Since $t_j t_k = -\frac{1}{r}$ holds for $k \neq j$, see Problem 6.3, we get

$$L_j(x) = \frac{1}{r+1}\left\{r(t_j x) + 1\right\}.$$

Together with $t_1 + \cdots + t_{r+1} = 0$, this yields

$$\sum_{j=1}^{r+1} L_j^2(x) = \frac{1}{(r+1)^2}\left\{r^2 \sum_{j=1}^{r+1}(t_j x)^2 + r + 1\right\}.$$

The maximum value is attained under the side condition $x^2 - 1 = 0$. The corresponding Lagrange condition is given by

$$\sum_{j=1}^{r+1}(t_j x)t_j = \lambda \cdot x,$$

and implies

$$\sum_{j=1}^{r+1}(t_j x)(t_j t_k) = \lambda (t_k x)$$

for $k = 1, \ldots, r+1$. Therefore, λ is an eigenvalue of the matrix

$$\left(t_j t_k\right) = (1 + \tfrac{1}{r})\,I - \tfrac{1}{r}\,E.$$

The spectrum of E is known from Problem 7.1. It follows that

$$spec\left(t_j t_k\right) = \{0, 1 + \tfrac{1}{r}\}.$$

On the other side we obtain

$$\sum_{j=1}^{r+1}(t_j x)^2 = \lambda,$$

where $\lambda = 0$ is definitely does not furnish the maximum value. Therefore,

$$\sum_{j=1}^{r+1}(t_j x)^2 = 1 + \tfrac{1}{r}$$

must be valid, and since x is a maximum point, this yields

$$\max\left\{\sum_{j=1}^{r+1}L_j^2(x)\,\Big|\,x \in B^r\right\} = 1,$$

as claimed.

Problem 7.4

See characterisation (ii) of Gauß quadratures in Theorem 6.16.

Chapter 8

Problem 8.1

Let $G(\sigma, t) := (\mathcal{R}_0 F)(\sigma, t)$ for $(\sigma, t) \in Z^r$. Using formula (8.1) we get

$$G(\sigma, t) = \int_{v \perp t} \left(\rho^2 - |v - (a - \sigma t)|^2\right)_+^\alpha dv.$$

By the substitution $v = u + a - (at)t$ it follows that

$$G(\sigma, t) = \int_{u \perp t} \left(\rho^2 - [u^2 + (\sigma - at)^2]\right)_+^\alpha du.$$

Apparently we obtain

$$G(\sigma, t) = 0 \quad for \quad |\sigma - at| \geq \rho.$$

So we may assume $|\sigma - at| < \rho$ in what follows. Then we get by the substitution $u = \sqrt{\rho^2 - (\sigma - at)^2} \cdot x$,

$$G(\sigma, t) = \int_{x \perp t} (1 - x^2)_+^\alpha \, dx \cdot \left(\rho^2 - (\sigma - at)^2\right)_+^{\alpha + \frac{r-1}{2}},$$

with the constant

$$\int\limits_{x\perp t} (1-x^2)_+^\alpha\, dx \;\; = \;\; \int\limits_{B^{r-1}} (1-\bar{x}^2)^\alpha\, d\bar{x}$$

$$= \;\; \omega_{r-2} \int\limits_0^1 (1-\xi^2)^\alpha\, \xi^{r-2}\, d\xi$$

$$= \;\; \tfrac{1}{2}\omega_{r-2} \int\limits_0^1 (1-\eta)^\alpha \eta^{\frac{r-3}{2}}\, d\eta \;\; = \;\; \tfrac{1}{2}\omega_{r-2} B(\alpha+1, \tfrac{r-1}{2}).$$

Using (1.4) and (1.8), again, we get the final result

$$G(\sigma, t) = \pi^{\frac{r-1}{2}} \frac{\Gamma(\alpha+1)}{\Gamma(\alpha+\frac{r+1}{2})} \cdot \left(\rho^2 - (\sigma - at)^2\right)_+^{\alpha+\frac{r-1}{2}}.$$

Bibliography

1] A. Angelescu: *Sur les polynomes généralisant le polynomes de Legendre et d'Hermite et sur le calcul approché des intégrales multiples.* Diss. Paris 1916.

2] P. Appell et J. Kampé de Feriét: *Fonctions Hypergéometriques et Hypersphériques. Polynomes d'Hermite,* Gauthier-Villars: Paris 1926.

3] E. Bannai, E. M. Damerell: *Tight spherical designs, I.* J. Math. Soc. Japan 31 (1979), 199–207.

4] E. Bannai, E. M. Damerell: *Tight spherical designs, II.* J. London Math. Soc. (2) 21 (1980), 13–39.

5] W. Beekmann, K. Zeller: *Theorie der Limitierungsverfahren.* 2nd ed., Springer: Berlin, Heidelberg, New York 1970.

6] H. Berens, P. L. Butzer, P. Pawelke: *Limitierungsverfahren von Reihen mehrdimensionaler Kugelfunktionen und deren Saturationsverhalten.* Publ. RIMS, Kyoto Univ. Ser. A 4 (1968), 201–268.

7] D. L. Berman: *On a class of linear operators.* Dokl. Akad. Nauk SSSR 85 (1952), 13–16 (Russian).

8] L. Bos: *On Kergin-interpolation in the disk.* J. Approx. Th. 37 (1983), 251–261.

9] L. Bos: *Some remarks on the Fejér problem for Lagrange interpolation in several variables.* J. Approx. Th. 60 (1990), 133–140.

[10] E. W. Cheney: *Projection operators in approximation theory.* In: Studies in functional analysis (R. G. Bartle, ed.), Studies in Mathematics 21, Math. Ass. Amer. 1980, 50–80.

[11] E. W. Cheney: *Multivariate Approximation Theory: Selected Topics.* Philadelphia Soc. Ind. Appl. Math. 1986.

[12] I. K. Daugavet: *Some applications to the Marcinkiewicz-Berman identity.* Vestnik Leningrad Univ., Math. 1 (1974), 321–327.

[13] Ph. J. Davis: *Interpolation and Approximation*. Blaisdell: New York, Toronto, London 1963.

[14] M. E. Davison and F. A. Grünbaum: *Tomographic reconstruction with arbitrary directions*. Comm. Pure Appl. Math. 34 (1983) 428–448.

[15] H. Ehlich, K. Zeller: *Čebyčev-Polynome in mehreren Veränderlichen*. Math. Zeitschr. 93 (1966), 142–143.

[16] H. Ehlich, W. Haußmann: *Čebyěv-Approximation stetiger Funktionen in zwei Veränderlichen*. Math. Zeitschr. 117 (1970), 21–34.

[17] G. Faber: *Über die interpolatorische Darstellung stetiger Funktionen*. Jahresbericht DMV 23 (1914), 192–210.

[18] L. Fejér: *Lagrange Interpolation und die zugehörigen konjugierten Punkte*. Math. Ann. 106 (1932), 1–55.

[19] J. Fromm: *Tschebyscheff-Polynome in mehreren Veränderlichen*. Diss. Dortmund 1975.

[20] G. Gasper: *Rogers' linearization formula for the continuous q-ultraspherical polynomials and quadratic transformation formulas*. SIAM J. Math. Anal. 16 (1985), 1061–1071.

[21] L. Gegenbauer: The work of G. is scattered. Most of it can be found in Wiener Berichte. Cf. Jahrbuch über die Fortschritte der Mathematik 4 (1872), 6 (1874), 9 (1877), 16 (1884), 25 (1893–94).

[22] W. G. Gearhart: *Some Chebyshev approximations by polynomials in two variables*. J. Approx. Th. 8 (1973), 195–209.

[23] A. Grundmann, H. M. Möller: *Invariant integration formulas for the n-simplex by combinational methods*. SIAM J. Numer. Anal. 15 (1978), 282–290.

[24] H. Hakopian: *On a theorem on bivariate homogeneous polynomials*. Bull. Polish Acad. Sci. 42 (1994), 129–132.

[25] C. Hermite: Œuvre t. II, 309–346.

[26] J. R. Isbell and Z. Semadeni: *Projection constants and spaces of continuous functions*. Trans. Amer. Math Soc. 107 (1963), 38–48.

[27] O. D. Kellogg: *On bounded polynomials in several variables*. Math. Z. 27 (1927), 55–64.

[28] P. Kergin: *A natural interpolation in C^k-functions*. J. Approx. Th. 19 (1980), 278–283.

[29] E. Kogbetliantz: *Recherches sur la sommabilité de séries ultrasphériques par le méthode de moyennes arithmétiques.* J. Math. Pure Appl. Ser. 9, V. 3 (1924), 107–187.

[30] P. P. Korovkin: *Linear Operators and Approximation Theory.* Fitzmatgiz: Moscow 1959.

[31] L. Koschmieder: *Über die C–Summierbarkeit gewisser Reihen von Didon und Appell.* Math. Ann. 104 (1931), 387–402.

[32] V. I. Krylov: *Approximate Calculation of Integrals.* Macmillan: New York, London 1962 (Translation).

[33] U. Linde, M. Reimer and B. Sündermann: *Numerische Berechnung extremaler Fundamentalsysteme für Polynomräume über der Vollkugel.* Computing 43 (1939), 37–45.

[34] G. G. Lorentz: *Approximation of Functions.* Holt, Rinehart and Winston: New York etc. 1966.

[35] U. Maier: *Tomographic reconstruction using Cesàro means and Newman–Shapiro operators.* In: Algorithms for Approximation IV (J. Levesley, I. J. Anderson, J. C. Mason, eds.), University of Huddersfield 2002, 478-485.

[36] H. M. Möller: *Polynomideale und Kubaturformeln.* Diss. Dortmund 1973.

[37] H. M. Möller: *Kubaturformeln mit minimaler Knotenzahl.* Numer. Math. 25 (1976), 185–200.

[38] C. Müller: *Spherical Harmonics.* Springer: Berlin etc. 1966.

[39] F. Natterer: *The Mathematicas of Computerized Tomography.* Teubner: Stuttgart etc. 1986.

[40] D. J. Newman and H. S. Shapiro: *Jackson's theorem in higher dimensions.* In: Über Approximationstheorie (P. L. Butzer, J. Korevaar, eds.), Birkhäuser Verlag: Basel 1964, 208–219.

[41] S. Pawelke: *Über die Approximationsordnung bei Kugelfunktionen und algebraischen Polynomen.* Tôhoku Math. Journ. 24 (1972), 473–486.

[42] J. Radon: *Über die Bestimmung von Funktionen durch ihre Integralwerte längs gewisser Mannigfaltigkeiten.* Ber. Sächs. Akad. Wiss. Leipzig 69 (1917), 262–277.

[43] D. L. Ragozin: *Uniform convergence of spherical harmonic expansions.* Math. Ann. 195 (1972), 87–94.

[44] M. Reimer: *Auswertungsverfahren für Polynome in mehreren Variablen.* Numer. Math. 23 (1975), 321–336.

[45] M. Reimer: *On multivariate polynomials of least deviation from zero on the unit ball.* Math. Zeitschr. 153 (1977), 51–58.

[46] M. Reimer: *On multivariate polynomials of least deviation from zero on the unit cube.* J. Approx. Th. 23 (1978), 65–69.

[47] M. Reimer: *Grundlagen der Numerischen Mathematik I.* Akademische Verlagsgesellschaft: Wiesbaden 1980.

[48] M. Reimer: *Constructive Theory of Multivariate Functions with an Application to Tomography.* B.I. Wissenschaftsverlag: Mannheim 1990.

[49] M. Reimer: *On the existence–problem for Gauss–quadrature on the sphere.* In: Approximation by Solutions of Partial Differential Equations (B. Fuglede, M. Goldstein, W. Haußmann, W. K. Hayman, and L. Rogge, eds.), Kluwer: Dordrecht 1992, 169 – 184.

[50] M. Reimer: *Quadrature rules for the surface integral of the unit sphere on extremal fundamental systems.* Math. Nachr. 169 (1994), 235–241.

[51] M. Reimer: *A short proof of a result of Kogbetliantz on the positivity of certain Cesàro means.* Math. Zeitschr. 221, 189–192 (1996).

[52] M. Reimer: *Leading coefficients and extreme points of homogeneous polynomials.* Constr. Approx. 13 (1997), 357 – 362.

[53] M. Reimer: *Discretized Newman–Shapiro operators and Jackson's inequality on the sphere.* Result. Math. 36 (1999), 331–341.

[54] M. Reimer: *Hyperinterpolation on the sphere at the minimal projection order.* J. Approx. Th. 104 (2000), 272–286.

[55] M. Reimer: *Generalized hyperinterpolation on the sphere and the Newman–Shapiro operators.* Constr. Approx. 18 (2002), 183 – 204.

[56] M. Reimer and B. Sündermann: *A Remez–type algorithm for the calculation of extremal fundamental systems for polynomial spaces on the sphere.* Computing 37 (1986), 43–58.

[57] M. Reimer and B. Sündermann: *Günstige Knoten für die Interpolation mit homogenen harmonischen Polynomen.* Result. Math. 11 (1987), 254 – 266.

[58] M. Riesz: *Eine trigonometrische Interpolationsformel und einige Ungleichungen für Polynome.* Jahresbericht DMV 23 (1914), 354–368.

[59] Th. Rivlin and H. S. Shapiro: *A unified approach to certain problems of approximation and minimization.* J. Soc. Ind. Appl. Math. 9 (1961),670–699.

[60] M. Rosier: *Biothogonale Zerlegungen der Radon-Transformation mit Appell–Polynomen.* Diss. Dortmund 1992.

[61] A. F Ruston: *Auerbach's theorem and tensor products in Banach spaces.* Proc. Cambridge Phil. Soc. 58 (1962), 476–480.

[62] H. Schubert: *Topologie.* Teubner: Stuttgart 1964.

[63] I. H. Sloan: *Polynomial interpolation and hyperinterpolation over general regions.* J. Approx. Th. 83 (1995), 238–254.

[64] I. H. Sloan: *Interpolation and hyperinterpolation on the sphere.* In: Multivariate Approximation (W. Haußmann, K. Jetter, M. Reimer, eds.), Akademie Verlag: Berlin 1997, 255–268.

[65] I. H. Sloan and R. S. Womersley: *Constructive polynomial approximation on the sphere.* J. Approx. Th. 103 (2000), 91–98.

[66] I. H. Sloan and R. S. Womersley: *How good can polynomial interpolation on the sphere be?* Adv. Comp. Math. 14 (2001), 195–226.

[67] I. H. Sloan and R. S. Womersley: *Extremal systems of points and numerical integration on the sphere.* Electronic Publication (2001).

[68] J. M. Sloss: *Chebyshev approximation to zero.* Pacific J. Math. 15 (1965), 305–313.

[69] E. M. Stein and G. Weiss: *Introduction to Fourier Analysis on Euclidean Spaces.* Princeton University Press: Princeton 1971.

[70] A. E. Stroud: *Quadrature methods for functions of more than one variable.* Ann. New York Acad. Sci. 86 (1960), 776–791.

[71] A. H. Stroud: *Approximate Calculation of Multiple Integrals.* Prentice-Hall, Englewood Cliffs 1971.

[72] B. Sündermann: *Projektionen auf Polynomräume in mehreren Veränderlichen.* Diss. Dortmund 1983.

[73] G. Szegö: *Orthogonal Polynomials.* Amer. Math. Soc.: Providence 1991.

[74] A. E. Taylor: *A geometric theorem and its application to biorthogonal systems.* Bull. Amer. Math. Soc. 53 (1947), 614–616.

[75] F. G. Tricomi: *Vorlesungen über Orthogonalreihen.* Springer: Berlin 1955.

[76] V. A. Yudin: *Covering a sphere and extremal properties of orthogonal polyno mials (translated from the Russian)* Discrete Math. Appl. 5 (1995), 371–379

[77] G. Watson: *A Treatise on the Theory of Bessel Functions.* Cambridge Univ Press, Cambridge 1966.

Index